Lecture Notes in Networks and Systems

Volume 163

The series "Lecture Notes in Networks and Systems" publishes the latest developments in Networks and Systems—quickly, informally and with high quality. Original research reported in proceedings and post-proceedings represents the core of LNNS.

Volumes published in LNNS embrace all aspects and subfields of, as well as new challenges in, Networks and Systems.

The series contains proceedings and edited volumes in systems and networks, spanning the areas of Cyber-Physical Systems, Autonomous Systems, Sensor Networks, Control Systems, Energy Systems, Automotive Systems, Biological Systems, Vehicular Networking and Connected Vehicles, Aerospace Systems, Automation, Manufacturing, Smart Grids, Nonlinear Systems, Power Systems, Robotics, Social Systems, Economic Systems and other. Of particular value to both the contributors and the readership are the short publication timeframe and the world-wide distribution and exposure which enable both a wide and rapid dissemination of research output.

The series covers the theory, applications, and perspectives on the state of the art and future developments relevant to systems and networks, decision making, control, complex processes and related areas, as embedded in the fields of interdisciplinary and applied sciences, engineering, computer science, physics, economics, social, and life sciences, as well as the paradigms and methodologies behind them.

Indexed by SCOPUS, INSPEC, WTI Frankfurt eG, zbMATH, SCImago.

All books published in the series are submitted for consideration in Web of Science.

More information about this series at http://www.springer.com/series/15179

Mohuya Chakraborty · Moutushi Singh ·
Valentina E. Balas · Indraneel Mukhopadhyay
Editors

The "Essence" of Network Security: An End-to-End Panorama

Springer

Editors
Mohuya Chakraborty
Department of Information Technology
Institute of Engineering & Management
Kolkata, West Bengal, India

Moutushi Singh
Department of Information Technology
Institute of Engineering & Management
Kolkata, West Bengal, India

Valentina E. Balas
Department of Automatics & Applied Software
Aurel Vlaicu University of Arad
Arad, Arad, Romania

Indraneel Mukhopadhyay
Department of Information Technology
Institute of Engineering & Management
Kolkata, West Bengal, India

ISSN 2367-3370 ISSN 2367-3389 (electronic)
Lecture Notes in Networks and Systems
ISBN 978-981-15-9316-1 ISBN 978-981-15-9317-8 (eBook)
https://doi.org/10.1007/978-981-15-9317-8

This Springer imprint is published by the registered company Springer Nature Singapore Pte Ltd.
The registered company address is: 152 Beach Road, #21-01/04 Gateway East, Singapore 189721, Singapore

"To my daughter Rheya, the love of my life Jayanta for his encouragement and my parents for their inspiration and support"

Mohuya Chakraborty

"To my sweet & loving Father & Mother whose affection love, encouragement & prayers of day & night make me able to get success"

Moutushi Singh

"To my husband Marius and to our twin daughters Sanda and Dana"

Valentina E. Balas

"To my family for constant support during the preparation of the book and to my father"

Indraneel Mukhopadhyay

Preface

It all started in 2018–2019 after we organized two International Ethical Hacking Conferences—eHaCON 2018 and eHaCON 2019. The conferences helped us to meet lot of experts in the field of network security across the globe and gather knowledge, both theoretical and practical. Along with this, our expertise in teaching this subject for past 15 years prompted us to write the book—*The "Essence" of Network Security: An End-To-End Panorama*. Essentially this book combines cutting-edge technologies of network security as outcome of research and development as well as from industry perspective.

The reader will be able to grasp advanced technologies used in network security like Blockchain, Cryptography, Digital Forensics, Artificial Intelligence, Machine Learning, and Deep Learning. The book gives thrust upon security aspects of profound areas like Internet of Things (IoT), Cloud Computing, Cyberspace, Software Defined Networking, Anonymous Traffic Network, and Named Data Networking. Let us briefly explain how these advanced technologies help in protecting our devices connected to the network.

Part I Introduction

Chapter "Introduction to Network Security Technologies" provides an overview of network security attacks and highlights the newer technologies in the related area. The objective of this chapter is to guide the reader to the overall content of the chapters to follow.

Part II Review of Recent Trends in Forensics

Chapter "A Systematic Review of Digital, Cloud and IoT Forensics" deals with the review of digital forensics. Digital forensics is a branch of forensic science that involves the recuperation and examination of valuable information found in digital devices related to the computer as well as cybercrimes, as a part of the investigation. From a technical standpoint, the main goal of digital forensics is to identify, collect, preserve, and analyze data in a way that preserves the integrity of the evidence collected so it can be used effectively in a legal case. Just as central and state authorities look for used digital evidence to convict lawbreakers, IT managers, security, and legal teams can use digital forensics to collect and preserve evidence to analyze and defend against a cyberattack, stop an insider threat, or complete an internal investigation.

Part III Blockchain and Cryptography

Chapter "Blockchain-Based Framework for Managing Customer Consent in Open Banking" discusses about blockchain framework and chapter "A Comprehensive Study of Pros and Cons on Implementation of Blockchain for IoT Device Security" touches upon blockchain implementation for IoT device security. Blockchain technology can be used to guard our systems, like computers, laptops, and devices like routers, switches, etc., connected to the network from attacks. Everything that occurs on the blockchain is passed through rigorous encryption process making it possible to ensure that data has not been altered during course of transaction or transmission. It uses distributed ledger technology and due to its distributed nature, one may check file signatures across all the ledgers on all the nodes in the network and verify that they have not been altered. Furthermore, blockchain security means that there is no longer a centralized authority controlling the network and verifying the data going through it. Every transaction that happens on blockchain happens in a decentralized manner and goes through consensus mechanism ensuring integrity and transparency. This means we do not trust a single entity but rather trust the entire system as a whole. Such is the power of blockchain technology.

Chapter "Role of Cryptography in Network Security" talks about use of cryptography in network security. Cryptography comes from the Greek word "çryptos" meaning hidden and "graphy" means study. So in short it is the study of hidden messages. Cryptography protects information by transforming it into a format that is not recognizable by anyone who is not intended to. A basic cryptography may be an encrypted message called plaintext in which letters are replaced with other characters. This is called encryption. To decode the encrypted contents, one would need a grid or table that defines how the letters are converted. This is the reverse process of encryption and is called decryption. Contemporary cryptography uses complicated mathematical equations (algorithms) and secret keys (public and private) to encrypt and decrypt data. In today's world, cryptography is used to provide secrecy, authenticity, and integrity to our data during the course of our communications.

Part IV Machine Learning and Artificial Intelligence in Network Security

Chapter "Cyber Security with AI—Part I" and "Cyber Security with AI—Part II" are devoted to the use of artificial intelligence in cybersecurity. Chapter "Detection of Malicious URLs Using Deep Learning Approach" highlights the use of deep learning methodology in detecting malicious URLs. Artificial Intelligence (AI) is used in building machines that can mimic human cognitive functions and perform "smart" or "intelligent" things on their own without human guidance. In effect, AI security involves leveraging AI to identify and stop cyberthreats with less human intervention than is typically expected or needed with traditional security approaches. Machine learning is a subset of AI, and it comprises the methods that enable computers to figure things out from the data and deliver AI-based applications. In machine learning, we feed lot of data to an algorithm to analyze things out on its own just as we started learning grammar before picking up our first book in childhood. Deep learning, a subset of machine learning, uses artificial neural networks (mathematical expressions) with additional neurons, layers, and interconnectivity just like our brain, which enables computers to solve more complex problems.

These enumerated technologies can help to solve a lot of security-related problems of the following areas that are covered in this book. One of the areas is IoT. It may be defined as a system of interconnected computing devices, mechanical and digital machines, embedded with software, sensors, and network connectivity to collect and exchange data. These are provided with exclusive identifiers and the ability to transfer data over a network without requiring manual intervention. Hardware, software, and network connectivity need to be protected against attacks for IoT objects to work efficiently. If they are vulnerable then hackers may acquire control of these devices and may disrupt the object's functionality and steal the user's digital data. So IoT security is an important aspect.

Cloud computing may be defined as the on-demand availability of computer system resources, especially data storage and computing power, without direct active management by the user. Importance of cloud security arises from the fact that both personal and business users want to ensure that their information is protected and secured. Moreover, keeping client data secure is first and foremost legal obligations of businesses.

Cyberspace refers to the virtual computer world, and more specifically, it is an electronic medium used to form a global computer network to facilitate online communication. Hence, cyberspace security or simply cybersecurity is important as it encompasses everything that pertains to protecting our sensitive data, such as Personally Identifiable Information (PII), Protected Health Information (PHI), Personal

Information, Intellectual Property Data, and Governmental and Industry Information Systems from theft and damage attempted. The main objective of cybersecurity is to achieve three elements (confidentiality, integrity, and availability) also known as CIA Triad. For any organization, it is necessary to defend its data and information using security tools.

Part V Security Networking

Chapter "Software-Defined Network Vulnerabilities" provides the concept of Software Defined Networking (SDN) and its vulnerabilities. SDN is a network architecture approach that enables the network to be intelligently and centrally controlled or "programmed" using software applications. This helps operators to manage the entire network constantly regardless of the underlying network technology. The SDN layer essentially acts as a virtual software switch or router in place of (or in conjunction with) the physical network devices. So instead of software embedded in the routers and switches managing the traffic, software from outside the devices takes over the job. SDN security needs to be built into the architecture, as well as delivered as a service to protect the availability, integrity, and privacy of all connected resources and information.

Chapter "Demystifying Security on NDN: A Survey of Existing Attacks and Open Research Challenges" shows how Named Data Networking (NDN) may be used as a fully secured Internet architecture. NDN is a proposed future Internet architecture enthused by many years of pragmatic investigation into network usage and a rising awareness of unsolved problems in current Internet architectures like IP. NDN names the data instead of data locations for network packet forwarding, so communications in NDN are consumer-driven. Each piece of sensitive data (packet) is cryptographically signed by its creator and hence NDN communications are secured in a data-centric manner. NDN adopts intelligent stateful forwarding strategies where forwarders maintain a state for each data request and wipe away the state when a corresponding data packet comes back eliminating loop.

Chapter "Anonymous Traffic Networks" introduces Anonymous Traffic Network (ATN). ATN also called onion router or TOR, which forwards Internet traffic through a cost-free overlay network having several thousand relays that conceal a user's location and traffic pattern usage from hackers who monitor network surveillance or do traffic analysis for illegal activities. In TOR, the data, including the next node destination address, is encrypted multiple times at the application layer of a communication protocol stack, nesting like the layers of an onion and sent through a virtual circuit comprising of successive, randomly selected TOR relays. Each relay decrypts a layer of encryption to reveal the next relay in the circuit to pass the remaining encrypted data onto it. Decryption of the innermost layer of encryption and transmission of the original data to its destination is done by the final relay by keeping the source IP address hidden. As the routing of the communication was partially hidden at every hop in the TOR network, this method helps to eradicate the communicating peers from getting determined through network surveillance that relies hugely upon knowledge of source and destination addresses.

Hope the readers would find it extremely useful in understanding the newer technologies and application areas from network security perspective.

Kolkata, India — Mohuya Chakraborty
Kolkata, India — Moutushi Singh
Arad, Romania — Valentina E. Balas
Kolkata, India — Indraneel Mukhopadhyay

Acknowledgements

First and foremost, praises and thanks to the God, the Almighty, for His showers of blessings throughout our endeavor to complete the book successfully. We would like to express our deep and sincere gratitude to Springer team, for giving us the opportunity to edit this technical book—*The "Essence" of Network Security: An End-to-End Panorama*—and providing invaluable guidance throughout this process. Their dynamism, vision, sincerity, and motivation have deeply inspired us. They taught us the methodology to carry out the process and to present the research works as clearly as possible. We are extending our heartfelt thanks to them for their acceptance and patience during the discussion we had on research work and blueprint of the book preparation.

It was a great privilege and honor to work and study with them. We are extremely grateful for what they have offered us. We would also like to thank our colleagues for their friendship, empathy, and continuous encouragement.

We are extremely grateful to our family members for their love, prayers, care, and sacrifices for educating and preparing us in this path. We are very much thankful to all our relatives for their love, understanding, prayers, and continuing support to complete this process. Also, we express our thanks to them for their support and valuable prayers.

We express our special thanks to all the coworkers for their genuine support throughout this research work.

We are extending our thanks to all the contributors of the book for their support and cooperation during the process of preparation of the book. We also thank all the teachers and mentors of the authors for their kindness.

We thank the management of Institute of Engineering & Management, Kolkata for their support to do this work. Finally, our thanks go to all the people who have supported us to complete the book directly or indirectly.

Mohuya Chakraborty
Moutushi Singh
Valentina E. Balas
Indraneel Mukhopadhyay

Description

Network security is the method of taking physical and software preventative measures to protect the underlying networking infrastructure from unauthorized access, misuse, malfunction, modification, destruction, or improper disclosure, thereby creating a secure platform for computers, users, and programs to perform their permitted critical functions within a secure environment. *The "Essence" of Network Security—An End-to-End Panorama* focuses on advanced security technologies to combat security threats of networks. The book is split into five parts. Part I of the book is an end-to-end overview of network security introducing the new technologies to the reader. The goal of computer forensics is to examine digital media, in a forensically sound manner with the aim of identifying, preserving, recovering, analyzing, and presenting facts and opinions about the digital information. Part II reviews the recent trends in forensics. Part III concentrates on newer concepts of cryptography and blockchain technologies. Artificial intelligence is a topic of interest in today's world which basically mimics human cognitive functions by analyzing data and surroundings to solve a variety of problems. Part IV is fully dedicated to this topic. Part V introduces the reader to security networking.

- One of the first books to cover cutting-edge technologies of network security in the same book.
- Introduces the primary concepts of network security and also includes discussion of recent advancements in the field.
- Written for researchers, teachers, engineers, and professionals answering practical questions from industry and academia.

Contents

Editors and Contributors

About the Editors

Mohuya Chakraborty has specialized in computer networking and security aspects throughout her 25 year career in teaching and research. She has gained practical experience in management and leadership in industry and academia. In recent years she has worked with several Indian Universities and Institutes in various administrative and academic positions. Presently she holds the position of Dean of Faculty and Professor at Institute of Engineering & Management, Kolkata. Prior to this she worked as Professor and Head of the department of Information Technology for twelve years in the same institute. She is the recipient of Paresh Lal Dhar Bhowmik Award for ranking first in post graduate programme from Calcutta University and Rashtriya Gaurav Award from India International Friendship Society for her contribution in the field of Information Technology in March 2019. She is one of the top 20 out of 200 participants of AICTE UKIERI Technical Leadership Programme organized during 2018–2019 by AICTE, British Council and Dudley College of Technology, Dudley, UK, selected for a 1 week study tour to UK in October 2019. She holds B.Tech and M.Tech degrees in Electronics and Communication Engineering from Calcutta University and Ph.D. in Engineering in the field of Mobile Computing from Jadavpur University. She holds CMI Level 5 certificate in Leadership and Management from UK. She is a senior member of IEEE and faculty advisor of various IEEE student branches. Her research interests are in the areas of network security, cognitive radio, mobile communication, artificial neural network. She holds a number of published patents and has published more than 80 scientific and technical books, special journal issues and proceedings, as well as chairing many international conferences and workshops. She has bagged several funded projects from AICTE, DST, NRDC—Government of India worth Rs 35 lakhs. She has organized several International conferences, national seminars and faculty development programmes in varied areas. She heads the recruitment team of the Institute as well.

Moutushi Singh received her B.E.(Hons.) degree in Information Technology from The University of Burdwan and M.Tech degree from the Institute of Radio Physics and Electronics, University of Calcutta, India. She received her Ph.D (Engg) degree in the field of Network Security from the Department of Computer Science & Engineering, Maulana Abul Kalam Azad University of Technology. She has received Government Certificate for securing high marks in Madhyamik Examination. Currently she is working as an Assistant Professor in the Department of Information Technology in Institute of Engineering & Management, Kolkata. Before joining IEM, she has also worked in The University Institute of Technology, The University of Burdwan. She has attended and received many certificates for outstanding performances in various Upskilling and Development Programs from different Organizations of National and International repute. She has authored several research papers in Technical Journals and referred conference proceedings and many Book Chapters. She is a member of IEEE & ACM. Her research interest includes Intrusion Detection System and Network Security. She is presently serving as the editorial board member and reviewer of several international and national journals and conferences of repute.

Valentina E. Balas is currently Full Professor in the Department of Automatics and Applied Software at the Faculty of Engineering, "Aurel Vlaicu" University of Arad, Romania. She holds a Ph.D. in Applied Electronics and Telecommunications from Polytechnic University of Timisoara. Dr. Balas is author of more than 300 research papers in refereed journals and International Conferences. Her research interests are in Intelligent Systems, Fuzzy Control, Soft Computing, Smart Sensors, Information Fusion, Modeling and Simulation. She is the Editor-in Chief to International Journal of Advanced Intelligence Paradigms (IJAIP) and to International Journal of Computational Systems Engineering (IJCSysE), member in Editorial Board member of several national and international journals and is evaluator expert for national, international projects and Ph.D. Thesis. Dr. Balas is the director of Intelligent Systems Research Centre in Aurel Vlaicu University of Arad and Director of the Department of International Relations, Programs and Projects in the same university. She served as General Chair of the International Workshop Soft Computing and Applications (SOFA) in eight editions 2005–2020 held in Romania and Hungary. Dr. Balas participated in many international conferences as Organizer, Honorary Chair, Session Chair and member in Steering, Advisory or International Program Committees. She is a member of EUSFLAT, SIAM and a Senior Member IEEE, member in TC—Fuzzy Systems (IEEE CIS), chair of the TF 14 in TC—Emergent Technologies (IEEE CIS), member in TC—Soft Computing (IEEE SMCS). Dr. Balas was past Vice-president (Awards) of IFSA International Fuzzy Systems Association Council (2013–2015) and is a Joint Secretary of the Governing Council of Forum for Interdisciplinary Mathematics (FIM),—A Multidisciplinary Academic Body, India.

Indraneel Mukhopadhyay is Professor at Institute of Engineering and Management, Kolkata. He is associated with the education profession for more than 16 years and 8 months now, first in NIST (Berhampore) and then IEM, Kolkata. Prior to that, he has almost 4 years and 5 month of experience in the software industry. He has earned his Doctorate of Philosophy (Ph.D.), in Computer Science & Engineering on "Network Security in Business Application", from Maulana Abul Kalam Azad University of Technology, West Bengal (Formerly known as West Bengal University of Technology). He has done his B.E. in Computer Science & Engineering (Amravati University, Maharashtra) and M.S. in Information Technology (Clark University, Worcester, MA, USA). He is an Entrepreneurship Educator (STVP, Stanford University, USA). He has been the recipient of various awards during student and professional life.He has one on-going project, approved by AMRI, Mukundapur and in the past also completed various Government/DST/Research Projects. Apart from this, he has several international, and national journal publications, book chapters and international conference publications. He has published two patents. His professional membership includes Chartered Management Institute Level 5 Certified in Management and Leadership, UK, 2019, Senior Member ACM, Life Member ISTE. He has conducted multiple "Workshops on Raspberry Pi3" in Canada and United States of America between 2015 and 2017 at University of British Columbia, Columbia University. In February 2020 he has conducted a workshop on "Digital Marketing" at National University of Singapore. His research interests are Machine Learning and Cyber Security.

Contributors

Madhurima Buragohain Computer Science and Engineering Department, Indian Institute of Technology Guwahati, Guwahati, India

Swati Chakraborti Netaji Subhash Engineering College, Kolkata, India

Mohuya Chakraborty Institute of Engineering & Management, Kolkata, India

Bhanu Chander Computer Science and Engineering, Pondicherry University, Pondicherry, India

Swagata Roy Chatterjee Netaji Subhash Engineering College, Kolkata, India

Debashis De Department of Computer Science and Engineering, Maulana Abul Kalam Azad University of Technology, Kolkata, West Bengal, India

Abir Ghosh Larsen & Toubro Infotech Ltd., Pune, India

Atonu Ghosh Department of Computer Science and Engineering, Maulana Abul Kalam Azad University of Technology, Kolkata, West Bengal, India

Gopalakrishnan Kumaravelan Computer Science and Engineering, Pondicherry University, Pondicherry, India

Rajni Kusshwaha R Systems International Ltd., Greater Noida, Uttar Pradesh, India

Hrithik Lall Institute of Engineering & Management, Kolkata, India

Koushik Majumder Department of Computer Science and Engineering, Maulana Abul Kalam Azad University of Technology, Kolkata, West Bengal, India

Indraneel Mukhopadhyay Institute of Engineering & Management, Kolkata, India

Sukumar Nandi Computer Science and Engineering Department, Indian Institute of Technology Guwahati, Guwahati, India

Anand Raje India Internet Foundation (IIFON), Kolkata, India

Atrayee Majumdar Ray Netaji Subhash Engineering College, Kolkata, India

Kumar Sankar Ray Electronics and Communication Sciences Unit, Natural Computing Lab, Indian Statistical Institute, Kolkata, India

Anindita Sarkar Netaji Subhash Engineering College, Kolkata, India

Shaurya Institute of Engineering & Management, Kolkata, India

Moutushi Singh Institute of Engineering & Management, Kolkata, India

Sushanta Sinha India Internet Foundation (IIFON), Kolkata, India

Acronyms

ADFA	Australian Defense Force Academy
ADMM	Alternating Direction Method for Multipliers
AE	Authenticated Encryption
AES	Advance Encryption Standard
AI	Artificial Intelligence
ANN	Artificial Neural Network
AOL	America Online
ATN	Anonymous Traffic Networks
AUC	Area under the Curve
BEUC	European Consumer Organization
Bps	Bits per Second
CART	Classification and Regression Trees
CFS	Correlation Feature Selection
CFTT	Computer Forensic Tool Testing
CIA	Confidentiality, Integrity, Availability
CIDS	Collaborative Intrusion Detection Systems
CNN	Convolution Neural Networks
CPA	Content Poisoning Attack
CREnS	Cognitive Radio Encryption Standard
CS	Content Store
CTU-13	Czech Technical University
CVV	Card Verification Values
DARPA	Defense Advanced Research Projects Agency
DBN	Deep Belief Network
D-CPI	Data Controller Programming Interface
DDoS	Distributed Denial of Service
DES	Data Encryption Standard
DFI	Digital Forensic Investigation
DFRW	Digital Forensics Research Workshop
DL	Deep Learning
DLP	Data Loss Prevention
DNN	Deep Neural Network

DNS	Domain Name System
DO	Data Obfuscation
DoB	Date of Birth
DoS	Denial of Service
DPE	Disabling Pit Exhaustion
DPID	Data Path Identifier
DSM	Digital Single Market
DTE	Distribution Transforming Encoder
ECAES	Elliptic Curve Authenticated Encryption Scheme
ECC	Elliptic Curve Cryptography
ECDSA	Elliptic Curve Digital Signature Algorithm
FC	Fully Connected
FIB	Forwarding Interest Base
FN	False Negative
FNR	False Negative Rate
FP	False Positive
FPGA	Field Programmable Gate Arrays
FPR	False Positive Rate
GA	Genetic Algorithm
GBRBM	Gaussian-Bernoulli RBM
GDPR	General Data Protection Regulation
GEP	Gene Expression Programming
GISWS	Global Information Security Workforce Study
GP	Genetic Programming
GPG	Gnu Privacy Guard
GPU	Graphics Processing Unit
HAN	Hospital Area Network
HE	Honey Encryption
HECC	Hyperelliptic Curve Cryptography
HMAC	Hash-Based Message Authentication Code
HMM	Hidden Markov Model
HT	Hardware Trojan
HTS	Host Tracking Service
I2P	Invisible Internet Project
IC	Integrated Circuit
ICT	Information and Communication Technology
IDIP	Integrated Digital Investigation Process
IDS	Intrusion Detection System
IFA	Interest Flooding Attack
IoT	Internet of Things
IPS	Intrusion Prevention Program
IRCbot	IRC Trojans
ISC	International Security Consortium
ISOT	Information Security and Object Technology
ISP	Internet Service Provider

ISR	Interest Satisfaction Ratio
ISTR	Internet Security Threat Report
IT	Internet Technology
LAN	Local Area Network
LBNL	Lawrence Berkeley National Lab
LGP	Linear Genetic Programming
LLDP	Link Layer Discovery Protocol
LPT	Left Plain Text
LSTM	Long Short-Term Memory
LWE	Learning-With-Errors
MAC	Message Authentication Code
MAN	Metropolitan Area Network
MD	Message Digest
MD5	Message Digest 5
MEP	Multi-expression Programming
ML	Machine Learning
MVAr	Mega Volt-Amps
NAC	Network Access Control
NATO	North Atlantic Treaty Organization
NDN	Named Data Networking
NI	Network Interface
NIJ	National Institute of Justice
NIST	National Institute of Standards and Technology
NLP	Natural Language Processing
NoC	Network-on-Chips
NSA	National Bureau of Standards
NSW	University of New South Wales
OCF	Open Cloud Forensics
ONF	Open Networking Foundation
OSI	Open System Interconnection
OVF	Open Virtualization Format
P2P	Peer-to-Peer
PAN	Personal Area Network
PBE	Password-Based Encryption
PC	Personal Computers
PCI DSS	Payment Card Industry Data Security Standard
PGP	Pretty-Good-Privacy
PIT	Pending Interest Table
PKI	Public Key Infrastructure
PMU	Phasor Measurement Units
PN	Pseudonoise
PPS	Packets Per Second
QKD	Quantum Key Distribution
QoE	Quality of Experience
QR	Quick Response Code

QSDC	Quantum Secret Direct Communication
QSS	Quantum Secret Sharing
RAT	Remote Access Trojans
RBF	Radial Basis Function
RBN	Restricted Boltzmann Machine
RC5	Rivest Cipher 5
RDB	Relational Database
RFID	Radio Frequency Identification
RLWE	Ring-Learning-with-Errors
RNN	Recurrent Neural Networks
RNN-IDS	RNN-Based Intrusion Detection System
ROC	Receiver Operating Curve
RPS	Requests Per Second
RPT	Right Plain Text
RREP	Routing Replay
RREQ	Routing Request
RSA	Adi Shamir and Leonard Adleman
SCADA	Supervisory Control and Data Acquisition
SDN	Software Defined Network
SEA	Scalable Encryption Algorithm
SHA	Secure Hash Algorithm
SHE	Somewhat Homomorphic Encryption
SoC	System-on-Chips
SOM	Self-organizing Maps
SP	Service Provider
SSL	Secure Socket Layer
SVC	Support Vector Classification
SVM	Support Vector Machine
SVR	Support Vector Regression
TCAM	Ternary Content Addressable Memory
TCB	Trusted Computing Bases
TE	Transforming Encoder
TLS	Transport Layer Security
TLV	Type Length and Value
TN	True Negative
TNR	True Negative Rate
TOR	Onion Router
TP	True Positive
TPP	Third Party Providers
UDP	User Datagram Protocol
VANET	Vehicular Ad hoc Networks
VPN	Virtual Private Network
VUCA	Volatility, Uncertainty, Complexity, and Ambiguity
WAN	Wide Area Network
WSN	Wireless Sensor Network

List of Figures

Blockchain-Based Framework for Managing Customer Consent in Open Banking

A Comprehensive Study of Pros and Cons on Implementation of Blockchain for IoT Device Security

Role of Cryptography in Network Security

Cyber Security with AI–Part II

Detection of Malicious URLs Using Deep Learning Approach

List of Tables

Demystifying Security on NDN: A Survey of Existing Attacks and Open Research Challenges

List of Graphs

Part I
Introduction

Introduction to Network Security Technologies

Mohuya Chakraborty and Moutushi Singh

Abstract The proliferation of computer networks and increased dependence on the Internet has brought in vulnerabilities in the systems connected to the network. Opponents pitch various types of security-related attacks to the organizations, thereby stealing and modifying their valuable information. There is a great need of understanding the various attack types, taking precautionary measures to counter them and protect our systems from malicious parties. This chapter discusses the various types of attacks and their countermeasures. The chapter throws light upon the newer cutting edge technologies in the field of network security as well.

Keywords Network attacks · Malwares · Blockchain · IoT · Cloud computing · Cryptography · Artificial intelligence · Machine learning · Software-defined networking · Named data networking · Anonymous traffic networking

1 Introduction

All of us need the security of our belongings. So we keep our valuables in safety vaults or lockers. Today is the age of information. Businesses, educational systems, government organizations, military organizations thrive on information that is more digitally advanced than ever. Hence, as technology improves, information security aspects of various organizations be it public or private must be enhanced as well. For passing the information from one entity to another it is required that many devices communicate with each other over wired, wireless or cellular networks. It is where network security comes into play. Inherently network security is all about different methods of taking preventative measures to protect the underlying networking infrastructure from illegal access, misappropriation, fault, alteration, annihilation or indecorous revelation [1].

With the advent of Internet Technology (IT) we are too much dependent on it. The Internet has certainly become a part and parcel of our lives. The use of Information and

M. Chakraborty (✉) · M. Singh
Institute of Engineering & Management, Kolkata, India
e-mail: mohuyacb@iemcal.com

M. Chakraborty et al. (eds.), *The "Essence" of Network Security: An End-to-End Panorama*, Lecture Notes in Networks and Systems 163,
https://doi.org/10.1007/978-981-15-9317-8_1

Communication Technology (ICT) has changed our lives to a large extent. In today's world, we depend largely on the Internet for many of our personal, social and professional activities. We depend on Internet banking, E-mails, cab booking, online shopping and several other Governmental activities that use IT. As E-commerce and World Wide Web on the Internet are increasing we have a huge storehouse of several kinds of information such as flight tickets, stock markets, immigration databases, book stores, apparel collections, medical database, college database, student database, military, etc. The systems that are connected to the Internet are much more vulnerable and prone to attacks by interlopers as compared to the standalone machines of earlier days. Several populaces try to cause harm to our Internet-connected computers and laptops, infringe our confidentiality and disrupt the Internet services. Is it possible for us to verify that the party we are talking to over the network is really the one that we want to talk to? Can we ensure that no one else is listening and learning the data that we are sending over the network? Can we stay relaxed that no hacker has entered our network to play havoc? Whether the website we are downloading information from a legitimate one or a fake one? How do I guarantee that the person we just did a monetary transaction refutes from having done it at a later time? Hence, several security concerns crop in. We need security to safeguard the confidentiality, integrity, authenticity and availability of the data transmitted over insecure networks. However, besides the Internet, other insecure networks in this world are internal networks in organizations that are prone to insider attacks. Actually, insider attacks are higher both in terms of probability of occurrence and severity of damage being caused. These are mainly performed by disgruntled employees who leak data to the opponent or rival organizations. They in fact try to take confidential and highly valuable data for malevolent intent or financial gain [2].

As per data from a review of IT professionals by Ipswitch, a Lexington, Mass-based global provider of secure file transfer solutions about 40% of the human resources admit that they have sent sensitive information privately by using their personal email from the place of work. Above 25% confessed that sent patented files have been transferred to their personal email accounts, with the objective of placing that information at their next place of employment. Almost 50% of workers send confidential information via normal email weekly, thus jeopardizing payroll information, financial data and social security numbers due to lack of security. 41% of IT officers use personal external storage devices to back up work-related files monthly. This leads to complications when there are orphaned accounts that remain open and available even after the employees leave organizations. The dangers caused by disgruntled employees leaking information amplify when they get unnecessary access rights due to promotions or transfers within an organization or company [3].

In addition to this, there are other types of attacks that put our systems at risk. Network security has turned out to be a vital subject in the field of cyber security, cloud computing, the Internet of Things (IoT), etc., due to the increased occurrence and diversity of existing attacks as well as the threat of unknown and more critical expected future attacks. Implementing network security measures allows computers, users and programs to perform their permitted critical functions within a secure environment.

The organization of the chapter is as follows. After the introduction in Sect. 1, an overview of different types of network security attacks are explained in Sect. 2. In Sect. 3, we introduce the various cutting-edge technologies, mechanisms and services of network security that are of use in today's world. Basically, this section would introduce the readers to the rest of the chapters of the book. Section 4 concludes this chapter.

2 Network Security Attacks

Network Technology is the key technology for a wide variety of applications such as email, file transfer, web browsing, online transactions, form fill up for various governmental or private activities, cab booking, etc. However, there is a significant lack of easy implementation of security methods for these applications. So a huge communication gap exists between network developers and security developers. Network design is a truly developed process that depends on Open System Interconnection (OSI) model. OSI has certain advantages such as modularity, ease-of-use, flexibility, standardization of protocols, etc. Modular development is carried out by producing stacks upon combining protocols of different layers. In this process, the security of the complete network must be emphasized upon. It should be understood that the signal is weakest at the communication channel where hackers attack, get the data, modify and reinsert duplicate data as shown in Fig. 1.

Hence, not only the nodes but the communication channel should also be secured. While developing a secure network, Confidentiality, Integrity, Availability (CIA) needs to be considered [4].

Network security attacks are the communication network illegal activities operated by unauthorized agents or parties against private, corporate or governmental IT assets in order to destroy them, modify them or steal sensitive data. Public and private organizational networks become vulnerable to data-stealing or total annihilation of the data or network if employees have the right to access data from mobile devices of various enterprises. Attacks may be the cause of slow network performance, uncontrolled traffic, viruses, etc.[5].

There are mainly two types of network attacks – passive attack and active attack.

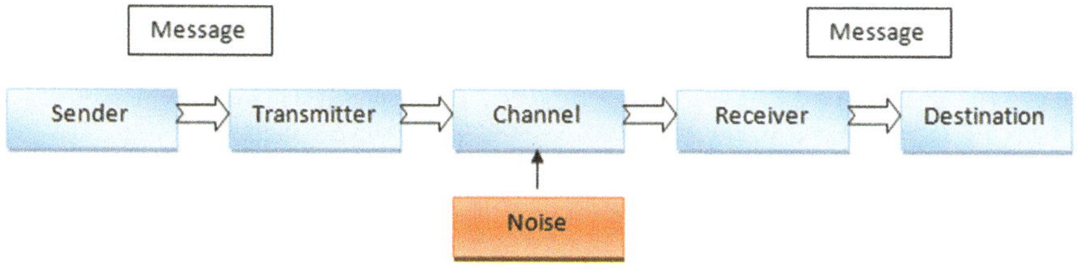

Fig. 1 Communication network model

- *Passive*: This type of attack happens when sensitive information is monitored and analyzed, possibly compromising the security of enterprises and their customers. In short, network intruder intercepts data traveling through the network.
- *Active*: This type of attack happens when information is modified, altered or demolished entirely by a hacker. Here the interloper starts instructions to disturb the network's regular process.

So the motives behind passive attackers and active attackers are totally different. Whereas the motive of passive attackers is simply to steal sensitive information and to analyze the traffic to steal future messages, the motive of active attackers is to stop normal communication between two legitimate entities.

There is another term associated with network security which is called a threat. The basic difference between threat and attack may be understood properly in order to protect and secure our networks. The threat is a possible security condition or violation to exploit the vulnerability of a system or asset. A threat may occur from any state viz., mishap, fire incident, environmental such as natural calamity, human error/carelessness, etc. The attack is a planned illegal deed on a system or organization. An attack always has a motivation to exploit a system and usually remain silent until a chance happens. Table 1 highlights the differences between threats and attacks vis-à-vis various key parameters [6].

Table 1 Threat versus attacks vis-à-vis key parameters

Sl. No	Key parameters	Threat	Attack
1	Planning	May be deliberate such as human carelessness/error or accidental such as a natural calamity	A deliberate action. An attacker has a motive and he plans the attack accordingly
2	Nature	May or may not be malevolent	Always malicious
3	Explanation	A state/situation that can cause harm to the system/organization	An intended action to cause damage to system/asset
4	Severity of damage	Varies from low to very high	Very high
5	Detection	Difficult to detect	Comparatively easy to detect
6	Prevention	May be prevented by managing the risky points	Cannot be prevented by merely managing the risky or weak points. Other actions such as backup, sense, act, etc., are needed to maneuver a cyber-attack

2.1 Passive Attacks

Passive attackers are mainly interested in stealing sensitive information. This happens without the knowledge of the victim. As such passive attacks are difficult to detect and thereby secure the network. The following are some of the passive attacks that are in existence [7].

- Traffic Analysis—Attacker senses the communication path between the sender and receiver.
- Monitoring—Attacker can read the confidential data, but he cannot edit or modify the data.
- Eavesdropping—This type of attack occurs in the mobile ad-hoc network where basically the attacker finds out some secret or confidential information from communication.

2.2 Active Attacks

The active attacks happen in such a manner so as to notify the victims that their systems have been compromised. As a result, the victim stops communication with the other party. Some of the active attacks are as follows [8].

- Modification—Some alterations in the routing route is performed by the malicious node. This results in causing the sender to send messages through the long route, which causes communication delay. This is an attack on integrity as shown in Fig. 2.
- Wormhole—This attack is also called a tunneling attack. A packet is received by an attacker at one point. He then tunnels it to another malicious node in the network. This causes a beginner to assume that he found the shortest path in the network as shown in Fig. 3.
- Fabrication—A malicious node generates a false routing message that causes the generation of incorrect information about the route between devices. This is an attack on authenticity as shown in Fig. 4.
- Spoofing—A malicious node miss-present his identity so that the sender changes his topology as shown in Fig. 5.
- Denial of services—A malicious node sends a message to the node and consumes the bandwidth of the network as given in Fig. 6.

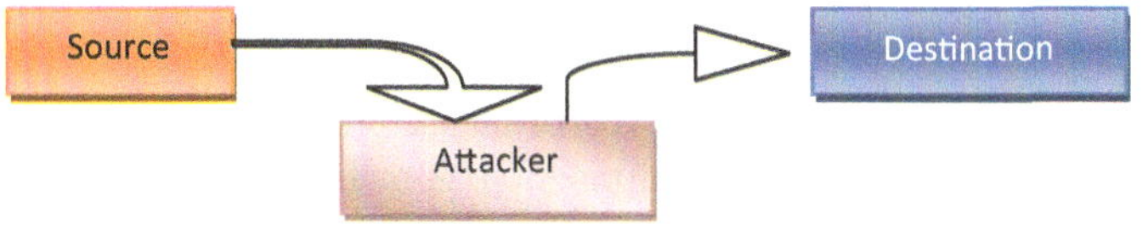

Fig. 2 Modification

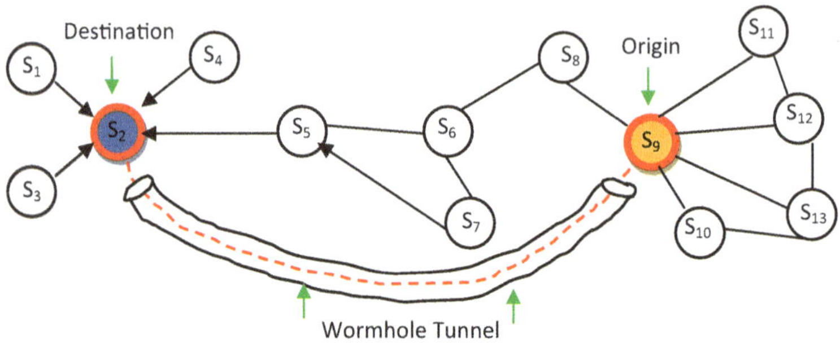

Fig. 3 Wormhole

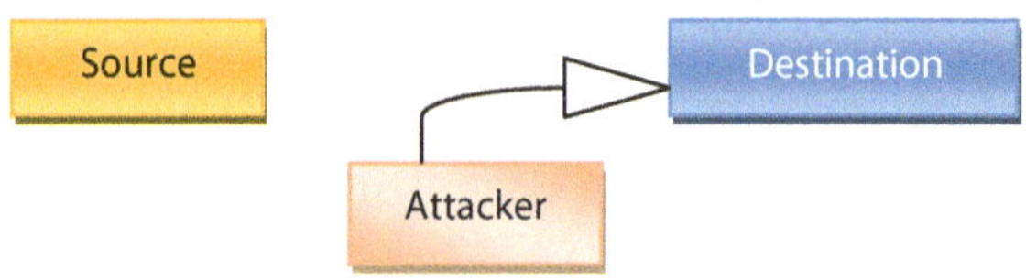

Fig. 4 Fabrication

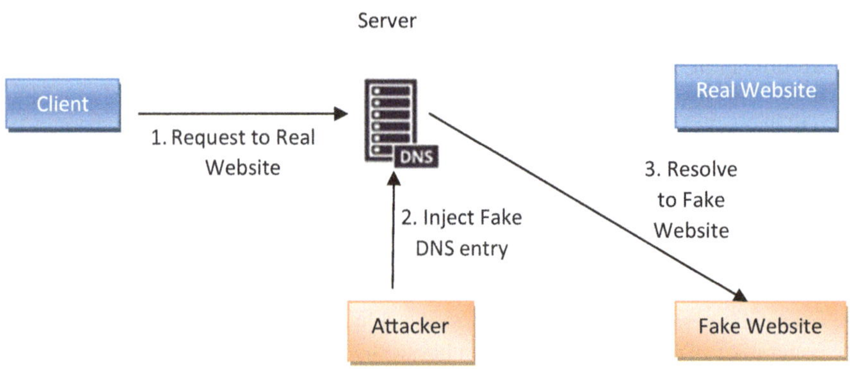

Fig. 5 Spoofing

- Sinkhole (service attack)—This type of attack prevents the base station from obtaining complete and correct information. A malicious node attempts to draw the data or packets from his all adjacent nodes. Packet dropping, selective alteration or data forwarding may be done on account of this attack as depicted in Fig. 7.
- Sinkhole (service attack)—This type of attack prevents the base station from obtaining complete and correct information. A malicious node attempts to draw the data or packets from his all adjacent nodes. Packet dropping, selective alteration or data forwarding may be done on account of this attack as depicted in Fig. 7.

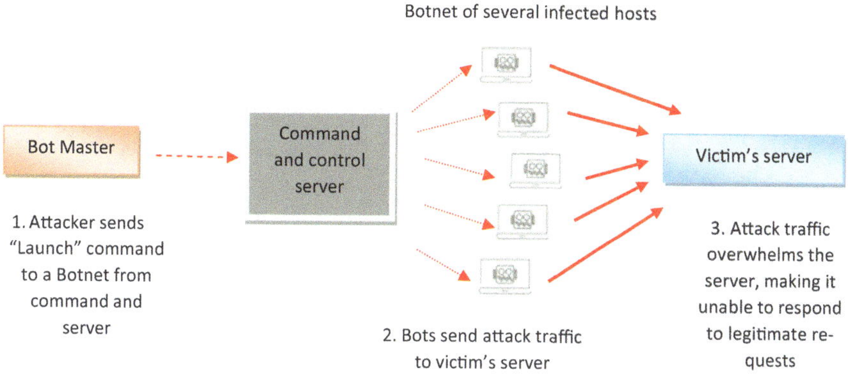

Fig. 6 Denial of service

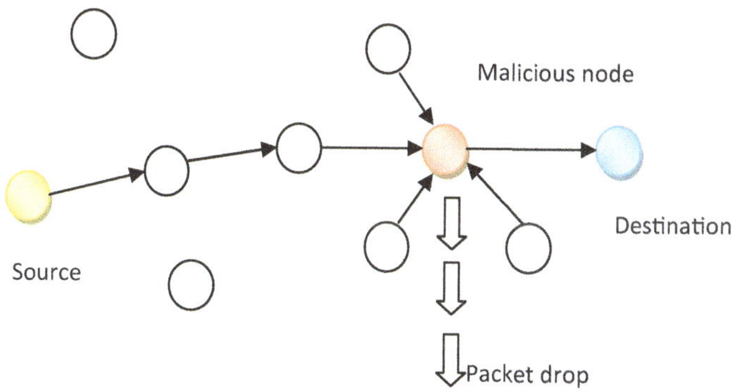

Fig. 7 Sinkhole

- Sybil—This attack is related to the multiple copies of the malicious node. A malicious node secretly sends its covert key with other malicious nodes that then use it to attack victims by using multiple routing. In this way, the network of malicious nodes increases thereby increasing the possibility of attacks. The likelihood of selecting a path by the malicious node will be increased in the network as depicted in Fig. 8.

2.3 *Advanced Attacks*

An advanced persistent threat is an attack in which an unauthorized user gains access to a system or network and remains there for an extended period of time without being detected. Advanced persistent threats are particularly dangerous for enterprises, as hackers have ongoing access to sensitive company data. Having awareness of how this

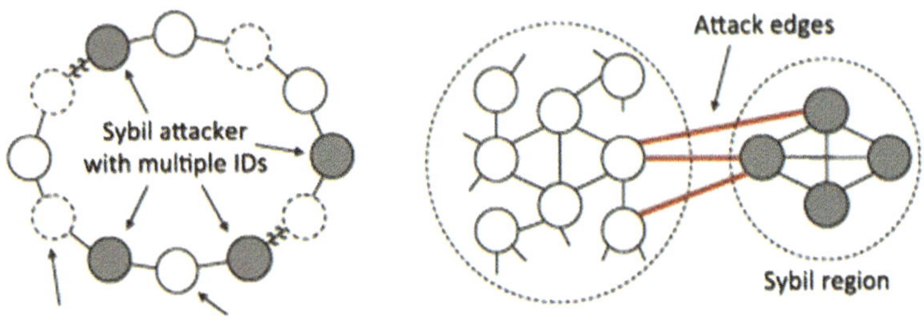

Fig. 8 Sybil attack

type of attack operates will give the reader a significant advantage when defending against it.

- Black Hole Attack—The best routing path to a node is advertised by an attacker by using the routing protocol. This node is actually the node whose packets it wants to intercept. The attacker uses the flooding-based protocol for listing the request for a route from the originator called Routing Request (RREQ). Hacker creates a reply message called Routing Replay (RREP) indicating that he has the shortest path to the receiver. As this message from the hacker reaches the initiator before the reply from the actual node, the initiator will consider that, it is the shortest path to the receiver. A malicious fake route is thus created as given in Fig. 9 [9].
- Rushing Attack—When the sender sends the packet to the receiver, then the attacker alters the packet and forwards it to the receiver. The attacker performs the duplicate. Sends the copy to the receiver over and over. The receiver assumes that packets are coming from the sender. The receiver becomes busy continuously [11].
- Replay Attack—A malicious node may repeat the data or delay the data. This can be done by the originator who intercepts the data and retransmits it. At that time an attacker can intercept the password [12]. The replay attack is shown in Fig. 10.
- Byzantine attack—In between the sender and the receiver there happens to be a set of transitional nodes. These nodes modify some of the tasks viz., they may create

Fig. 9 Blackhole attack

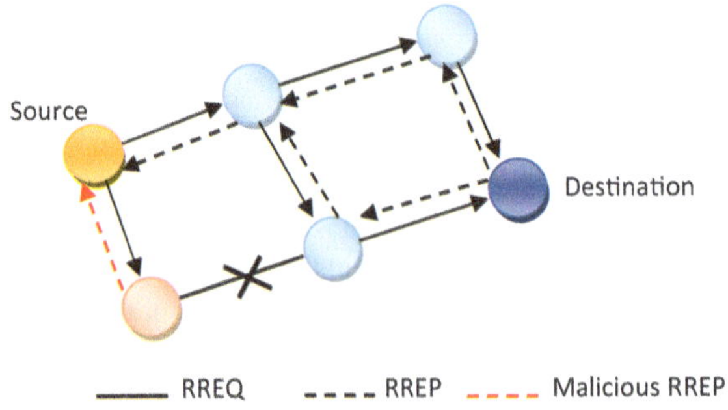

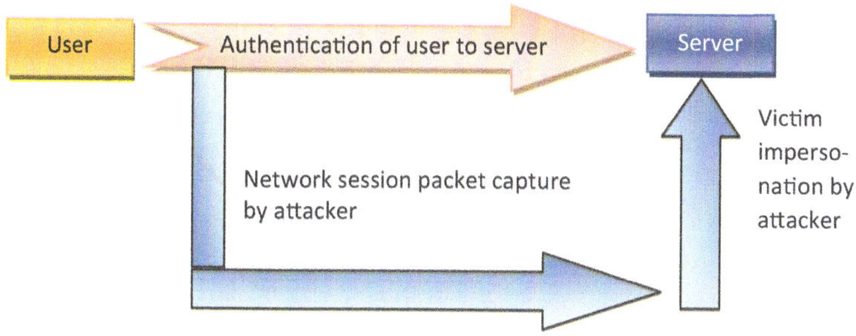

Fig. 10 Replay attack

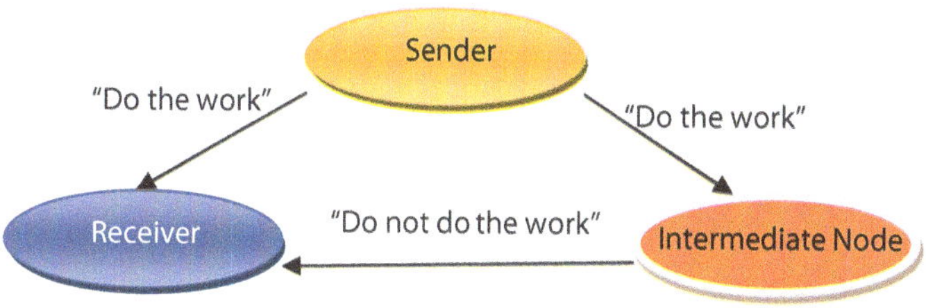

Fig. 11 Byzantine attack

additional routing loops or may drop legitimate packets or may even try to send packets through a non-optimal path. This produces disturbance or deprivation of routing services [12]. This attack is depicted in Fig. 11.

- Location Disclosure Attack—The location of nodes or the configuration of the network infrastructure is disclosed by an attacker. It gathers the node location information, such as a route map and then plans further attack scenarios. Hackers analyze the traffic between the communicating parties to understand network structure. The seepage of this information is disturbing in security-aware environments [5].
- Man-in-the-middle Attack—Also called a hijacking attack, it is an attack where the attacker secretly alters and relays the communications between two legitimate parties without their knowledge. These parties in turn are unaware of the secret hacker consider that they are doing direct communication with each other [13]. Figure 12 depicts this attack.

The various attacks that occur in a different layer of the OSI model are enumerated in Table 2 [14].

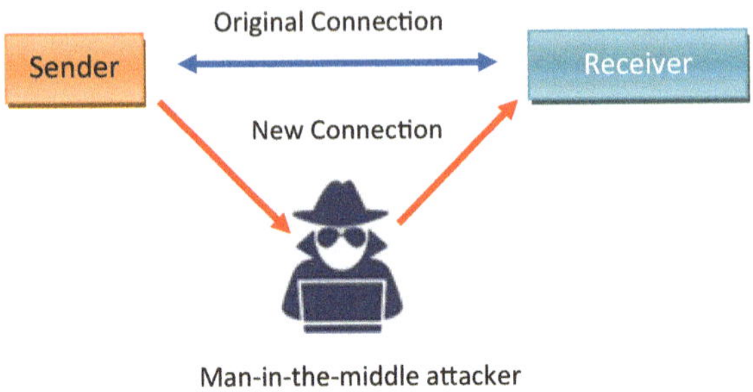

Fig. 12 Man-in-the-middle attack

Table 2 Attacks in OSI layers

Layer	Attacks
Application	Repudiation, Data Corruption
Transport	Session Hijacking, SYN Flooding
Network	Wormhole, Black Hole, Byzantine, Flooding, Resource Consumption, Location Disclosure
Data link	Traffic Analysis, Monitoring
Physical	Jamming, Eavesdropping
Multiple layers	Replay, Impersonation, DoS, Man-in-the-Middle

2.4 *Malwares*

Malware is a short form of malicious software. It is basically software used or created to disrupt computer operation, gather sensitive information and gain access to private computer systems. It can appear in the form of software scripts, codes, active web document and other-related software program. 'Malware' is a general term used to refer to a variety of forms of hostile, intrusive or annoying software [15].

The primary usage of malware may include many early infectious programs, including the first Internet Worm, which were written as experiments or pranks. Malware is first and foremost used to pinch sensitive personal, financial or business information for the benefit of others. Malware is sometimes used broadly against the government or corporate websites to gather guarded information or to disrupt their operation in general. However, malware is frequently used to get personal information such as social security numbers, bank or credit card numbers and so on of victims. There are various types of malware viz., Viruses, Trojan horses, Worms, Spyware, Zombie, Phishing, Spam, Adware and Ransomware to name a few. Let us now give an overview of these types.

Virus—It is a program or piece of code that is loaded onto our computer without the knowledge and runs against our wishes. Viruses can also replicate themselves. All computer viruses are man-made. Viruses copy themselves to other disks to spread to other computers. They can be merely annoying or they can be vastly destructive to our files. Examples of viruses are Macrovirus, Bootvirus, LogicBombvirus, Directoryvirus, and Residentvirus [16].

Trojan Horse—A Trojan Horse program has the appearance of having a useful and desired function. A Trojan Horse neither replicates nor copies itself, but causes damage or compromises the security of the computer. A Trojan Horse must be sent by someone or carried by another program and may arrive in the form of a joke program or software of some sort. These are often used to capture our logins and passwords. Key logging Trojans, Backdoor Trojans, IRC Trojans and Remote access Trojans are some examples of Trojan Horses [17].

Worms—A computer worm is a self-replicating computer program. It uses a network to send copies of itself to other nodes (computers on the network) and it may do so without any user intervention. It does not need to attach itself to an existing program [18].

Spyware—Spyware is a type of malware installed on computers that collects information about users without their knowledge. The presence of spyware is typically hidden from the user and can be difficult to detect. Spyware programs prowl onto our computer to lift secret information, such as our login ids, passwords and other private identification information and then secretly transfers them to others involved in criminal activities [19].

Zombie—Zombie programs take control of victim computers and use them to create a network to attack other computers connected to them without their knowledge to perform other malicious acts [20].

Phishing—These are attacks in the form of an innocent message that fools us into providing valuable and secret information such as our bank account details, social security number or user id and password for a website. The message may state that upon not logging onto a monetary website and providing the required information the account may become inactive and data or money may be confiscated [21].

Spam—Spam is an email that is not requested for and unwanted. If it is spam for me it may be an important message for someone else. It is widely used to multiply trojan horses, viruses and other malware [22].

Adware—Adware (short for advertising-supported software) is a type of malware that automatically delivers advertisements. Common examples of adware include pop-up ads on websites and advertisements that are displayed by software. Very often software and applications offer "free" versions that come bundled with adware [23].

Ransomware—Ransomware is a form of malware that essentially holds a computer system captive while demanding a ransom. The malware restricts user access to the computer either by encrypting files on the hard drive or locking down the system and displaying messages that are intended to force the user to pay the malware creator to remove the restrictions and regain access to their computer. In May 2017, a worldwide cyber-attack happened by The WannaCry cryptoworm, which targeted computers running the Microsoft Windows operating system by encrypting data and demanding ransom payments in the Bitcoin cryptocurrency [24, 25].

Malware spreads through a number of ways into our system to create havoc and steal sensitive information. Some of the ways are through websites, social networking sites, emails, removable disks, pirated software, etc. However, these programs cannot infect our systems or spread to others unless and until they are activated or executed. The damage caused by them is severe and may cause data loss, account theft, harm via Botnets, financial losses, etc.

- *Data Loss*—There are cases where files from the computer may be completely deleted by viruses and Trojans. They may even attempt to wipe hard drives. These events happen when the malware is activated.
- *Account Theft*—Many types of malware include key logger functions, designed to steal accounts and passwords from their targets. This eventually may give the hacker right of entry to any of the user's online accounts, including email servers from which the hacker can launch new attacks.
- *Botnets*—There are several types of malware that take control of the user's computers and turn them into a "bot" or "zombie". These fall prey to hackers who may create networks of these attacked computers, using their collective processing power for carrying out various malicious activities such as cracking password files or sending out bulk emails as shown in Fig. 13.
- *Financial Losses*—As online banking and bill payment services are becoming more and more popular, a hacker may gain access to the user's financial account,

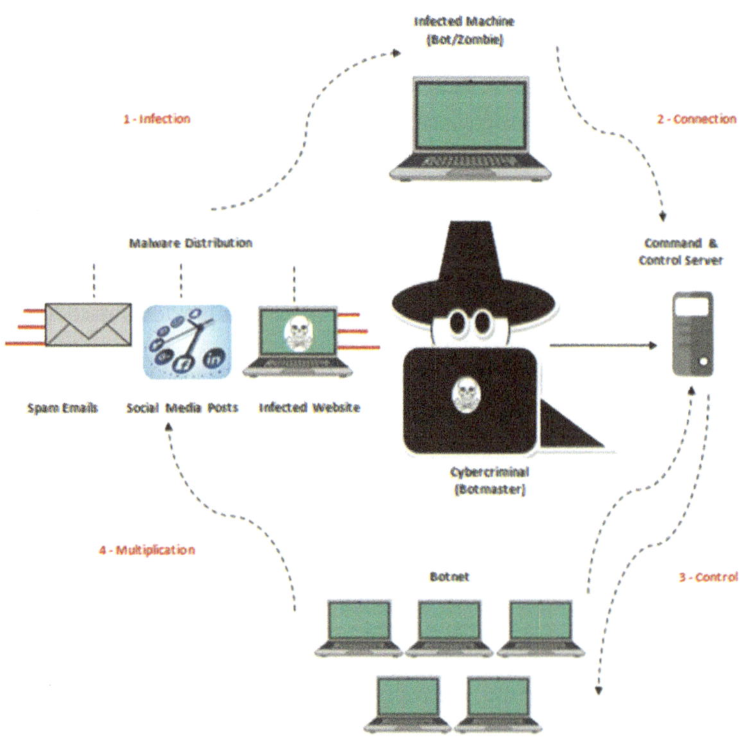

Fig. 13 Principle of operation of Botnets

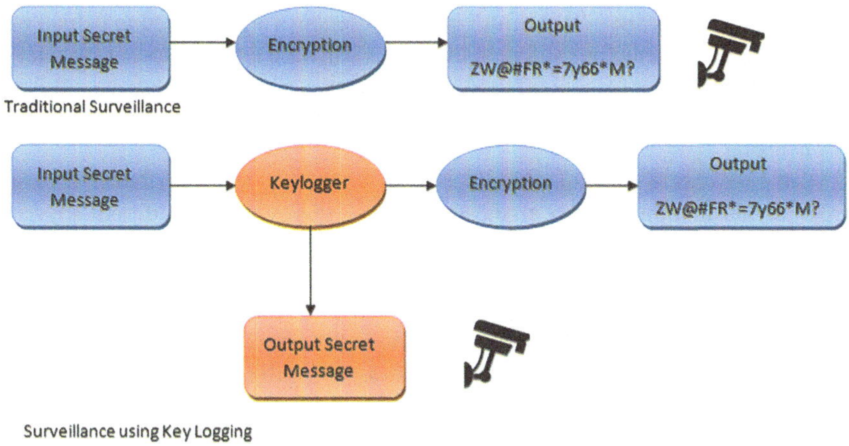

Fig. 14 Keylogger

or credit card or bank account with the help of a key logger. This would allow him to run up charges or drain the account as given in Fig. 14 in a single attack [26].

2.4.1 Protection of Our Systems from Attacks

In order to protect our system and computer from attacks, we need to install protection software, practice caution when working with files from unknown or questionable sources, do away with opening e-mail from an unrecognized sender, download files only from reputable Internet sites, install firewalls as well as can our hard drives for viruses on a monthly basis.

There are several symptoms by which it is possible for us to detect whether our system has been compromised or not. We may find increased CPU usage, slow computer or web browser speeds, problems connecting to networks, freezing or crashing of computers, modified or deleted files, the appearance of strange files, programs, or desktop icons, programs running, turning off, or reconfiguring themselves (malware may often reconfigure or turn off antivirus and firewall programs), strange computer behavior, emails/messages being sent automatically and without user's knowledge (a friend may receive a strange email from a person that was not sent), a lot of network activity when you are not using the network, the available memory on our computer is lower than it should be, programs or files appear or disappear without our knowledge, file names are changed, etc. On noticing these symptoms we should take proper measure to combat the damage. Some of the measures that may be taken are as follows.

Anti-Malware Program—In order to Anti Malware program is used to avoid, spot, and eliminate computer viruses, worms, trojan horses and any other type of malware, anti-malware programs are used. Examples of anti-malware programs are Antivirus program, Anti-spyware program, Anti-spam program, Firewall.

Antivirus Program—"Antivirus" is protective software designed to defend our computer against malicious software. Antivirus software must be updated regularly to make it suitable for recognizing and detecting new versions of malicious software. It must always run in the background in order that it becomes an effective defense mechanism.

Anti-Spyware Program—In order to avoid and sense superfluous spyware program installations and to eradicate those programs if installed, anti-spyware programs are used.

Anti-Spam Program—In order to recognize ineffective or hazardous messages, we may use Anti-spam software.

Firewall—A firewall block attempts to access our files over a network or internet connection. It helps to block incoming attacks as well. Infection may spread through shared disks or other computers connected to the network to which our computer is connected as well. So we need to monitor what our computer is putting out over the network or the Internet too.

Let us explore in depth these various cutting edge technologies that are used for network security in the following section.

3 Cutting Edge Network Security Technologies

In this section, we shall survey the hot technologies and concepts making headlines in network security. However, what is hot now will be lukewarm tomorrow, but the new era has already been started.

3.1 Topical Forensics Approaches

The contemporary scenario is incredible without various electronic gadgets. The exponential development in the global revenue generation in this area, evidences that human beings cannot continue their daily activity without different electronic devices. The old IT world has disappeared. Cyber security specialists are now dealing with coercions coming from the Cloud, Internet of Things, mobile/wireless and wearable technologies. Previously data were contained within the systems. But now they are traveling through various routers, data centers and hosts. Therefore, firewalls, anti-virus measures and tool-based security approaches are no longer enough. Nowadays, major service failures happen because of the inability of the IT security team to manage digital risk in new technology and use cases.

Hence, with the arrival of cloud computing and Internet of Things (IoT) technology, electronic device's capabilities increase and they start communicating with each other. The devices used by people are the nodes, which store different data, which are very much confidential and sensitive. As a result, data leakage becomes a common phenomenon and developed the main interest of network attackers. Attackers steal those data and use them for their own benefit. These sorts of criminal activities are rising rapidly. So, law enforcement agencies came into the picture to restrict these criminal activities.

Digital Forensics is a subdivision of forensic science that encompasses the recovery and enquiry of quantifiable found in digital devices. To retrieve and strengthen one's case, Digital Forensics Specialists are required. There are many subdivisions of this branch such as Computer Forensics, IoT Forensics, Network Forensics, Cloud Forensics, etc. [27]. Similar to the Central and State authorities who look for evidences to find out lawbreakers, Digital Forensic Specialists and legal teams apply Digital Forensics techniques to collect and reserve indications to analyze and defend against a cyber-attack, both from internal and outside threats [28]. Digital forensic specialists do their investigation in a way that the electronic evidences can be placed in court.

Now, there is another branch of Forensic sciences known as Cloud Forensics, which is even more challenging than Digital Forensics due to its distributed nature. Devices in cloud systems are geologically spread across the world. So, multijurisdictional issues came into the scenario and become impossible to recognize and grab the suitable digital devices that might have noteworthy evidences in the digital crime. There are huge numbers of pending cases involving the cloud computing environment. Evidence exclusion is a challenging task here as multiple clients share the same hardware resources. Nowadays, most of the company's serious data stirred to Cloud service providers. Hence, the major concern is to deal with security matters. So, Companies must swiftly respond and report events that may proceed to legal issues and, in extreme cases, even take in law execution. This is a difficult task as one has to trust the Cloud provider's capability to bring digital forensics data in the case of any legal argument all through the cyberattacks or if any data breach arises [29]. Due to the sharing of resources, it becomes hard to identify the devices that need to be seized for forensic investigation [30].

The IoT is a famous model that describes a self-motivated setting of interconnected computing devices with dissimilar machineries for continuous connectivity and data transmission [31]. With the complexity of cloud forensics, IoT has led to the detonation of data. Transmission of data over a network without any human intervention brings trustworthiness and suitability to customers. At the same time, new doors are also open for intruders. While IoT data becomes a huge source of evidence, forensics experts manage diverse problems, initiating from various IoT devices to the multi-resident cloud infrastructure.

The advantages of these computer forensics are its capability to the pursuit and scrutinize a huge data very quickly and efficiently. These fields are relatively new and criminal matters habitually deal with physical evidences. Hence, electronic evidence is very new. Fortunately, all these forensic tools and methods are very helpful, where important data for a case that has been lost, deleted or damaged may be recovered.

3.2 Blockchain and Cryptography

Blockchain is a distributed database, which has features such as decentralization, traceability, non-tamperability, security and reliability. It assimilates Peer-to-Peer (P2P) protocol, digital encryption, consensus mechanism, smart contract and supplementary technologies together. Blockchain mechanism is used for safeguarding systems such as computers, laptops and devices such as routers, switches, etc., which are connected to the network from different cyber-attacks. This blockchain platform is divided into three chains: (i) public chain, (ii) private chain and (iii) alliance chain. Nodes in the public chain join or withdraw themselves freely; whereas, the private chain stringently restricts the qualification of participating nodes; and the alliance chain mutually brings about by several participating institutions. Bitcoin, which is the utmost effective digital currency was first projected by Nakamoto in 2008 [32] and is the most typical application of blockchain [33]. The whole thing that occurs on the blockchain must be passed through a severe encryption procedure and ensures that the data has not been transformed during the course of the transaction. File signatures of all the ledgers in the network are verified to prove that they have not been altered. In the blockchain, centralized authority for controlling the network is absent. Table 3 enumerates different Blockchain architectures vis-à-vis protocols used in different layers.

Cryptography is the study of hidden writings. This technique protects information by transforming it into an arrangement that is unrecognizable to a person who is not intended to. In cryptography, letters of plaintext are replaced with other characters to prepare ciphertext. This methodology is called encryption. For decoding that encrypted message, one must know the reverse algorithm, called decryption that gets back the original message from the conversion. Modern cryptographic techniques use many types of intricate mathematical algorithms and undisclosed keys for encryption and decryption purposes. Encryption is the trick of how two parties can connect in secret in the presence of an eavesdropper. The main handlers of this system are the military, the diplomatic, banks, commercial, and government services. With the increase of computers and communications systems, a huge demand from the private sectors came to protect the information in digital form and for providing security services. From the cryptographic standpoint, many

Table 3 Blockchain architecture

	Application layer	Network layer	Contract layer	Consensus layer	Data layer
Bitcoin	Bitcoin trading	TCP-based P2P	Script	PoW	Merkle tree
Ethereum	Ethereum trading	TCP-based P2P	Solidity/Script EVM	PoW/PoS	Merkle patriciatree
Hyperledger	Enterprise blockchain	HTTP/2-based P2P	Go/Java Docker	PBFT/SBFT	Merkle Bocket tree

cryptographic techniques have already been revealed and profoundly active in blockchain platforms and use-cases [34].

The dimensions of data across the internet are amplified a lot. Providing security and privacy to this huge amount of data is a critical problem. However, continuous research activities are going on in this field to achieve complete security. The security facets of a network are:

- *Confidentiality*
- *Data integrity*
- *Authentication*
- *Availability*
- *Non-repudiation.*

Confidentiality and Integrity are accomplished by different cryptographic algorithms. A number of authentication algorithms are available to prove the legitimacy of the sender. Availability guarantees that systems, application and data are available to users when they need them. The most common attack that impacts availability is DoS, in which the attacker interrupts access to information, system, devices or other network resources. Non-repudiation means a sender cannot deny sending a message and a receiver cannot deny receiving a message at a later time. This is achieved by means of a digital signature.

Cryptography entirely depends upon mathematical computation. The primary desires of cryptography are:

- *Customer information must be secret from enemies*
- *Protection of user data*
- *Increased protection at the time e-transaction through Internet.*

Quantum cryptography is another new technique to achieve security. Quantum computers always use quantum algorithms for the advantageous achievement of quantum parallelism. These algorithms simplify the mathematical problems which are found difficult in the classical computing model. Homomorphic cryptography is a fresh methodology for ensuring data privacy on the cloud and increasing the prospective cloud computing techniques [35].

The conventional encryption algorithms and Message Digest or Hash algorithms are combined to generate authentication and digital signature algorithms. Several techniques have evolved in recent times such as digital watermarking and steganography to achieve digital signatures. Major developments are noticed in the field of authentication using biometric data.

A digital signature is a mathematical outline on public-key cryptosystem that produces codes termed as signatures of digital messages with the help of a private key. These signatures are verifiable with the help of the corresponding public key. Hence, digital signatures act as protectors against tampering and forgeries of digital messages [36].

In banking sectors, customer's consent is the key to success. All the data of customers must remain private and secured. But these data are also shared with third-party providers. However, the customer must remain confident about sharing their data. Some of the banking sector requirements are:

- *Customer Banking Experience*
- *Open Banking Perspective*
- *Technology as an Enabler.*

Nowadays, many customers are. taking part in open banking initiatives. As a result, each and every individual customer gets profit. The transparent process makes the customer confident about participating in the whole process. As more and more customers provide their consent to share their data, the framework will be able to generate and show the customer the rating of the bank. Blockchain-based rating for every bank increases the trustworthiness and transparency effects in the system. Consortium blockchain makes the framework highly robust. This also increases high transaction output and a profitable solution.

Presently data mining is used with embedded systems. IoT has already made a significant impact by providing a competitive advantage. Many people think that Technology and Business are two different domains but there is a symbiotic relationship between them. As a result, both important strategic elements contribute to business success.

Our smartphones, tabs, laptops, wearable and other handheld devices, which are capable of interacting with each other, are examples of IoT devices. Many IoT devices are built using sensors, actuators and communication systems which are cost-effective and can communicate with each other and transfer data to a centralized system. This collected information is processed and many times sent to other destinations which can use the information for their own benefit. But still, IoT is well thought-out as an enabler for the next industrial revolution by bridging the gap between the physical, digital, cyber and virtual space. IoT has subdivisions such as smart homes, personal wellness applications and wearable devices, smart cities, smart mobility, smart manufacturing, smart energy, smart farming to name a few. Nowadays the rates of deployment of IoT devices have been enhanced. So, device connection and security issues are also increased. IoT would change the way we live our lives.

IoT devices are targeted by attackers. Therefore, an effective mechanism needs to be created to ensure the security of IoT devices. Hence, with hardware device security and blockchain-based encryption, end to end encrypted communication between the different devices and the server against hackers and intruders is ensured. In the coming days more secured IoT devices using a hybrid technique with cloud realization is going to be implemented.

The invention of quantum computers has enhanced the computing ability and strengthened the network security. But it also increases susceptibility within the system. The obligation of lightweight cryptography also improves the IoT-based smart technologies.

3.3 *Impact of Artificial Intelligence and Machine Learning on Security Aspects*

In the twenty-first century, Artificial Intelligence (AI), a branch of computer science, is well thought-out to be one of the three most superior technologies (Genetic Engineering, Nano Science and Artificial Intelligence) [37]. AI is used for building machines that imitate human cognitive functions. They also perform well turned-out or "intelligent" things on their own without any human intervention. With the indisputably noteworthy potentials of AI, a cybercriminal also starts weaponizing it and uses it for boosting and expanding the horizon of threat. One of the biggest concerns is that hackers misuse AI to computerize cyber-attacks on a massive scale. Cybercrime and cyber security settings are going to be changed by using AI and machine learning to do the dirty work [38]. Computer scientists deploy AI and machine learning to harmonize the scarcity in human resources and to put aside the cost in cyber security. But criminals also use it for the same purpose. The investments and resources needed for initiation and coordinating attacks will mitigate extremely. Machine learning facilitates computers to figure out things from the dataset and uses it to deliver AI-based applications. In machine learning, we feed a lot of data to an algorithm to analyze things out on its own. Deep learning, a subset of machine learning uses ANN with supplementary neurons, layers and interconnectivity to enable computers to solve more complex problems. These cataloged technologies can help to solve a lot of security-related problems. Machine learning is that part of AI which has proven to be extremely helpful when it comes to detecting cyber threats based on analyzed data and identifying a threat before it exploits a vulnerability in the information systems. So, it is used for behavioral analytics.

Cyberspace is the implicit computer world, where the electronic medium is used to form a global computer network to assist online communication. So, cyber security is a very important area as it takes in everything that relates to protect all sensitive data from burglary and damage attempted. Cybercrime has become a regular phenomenon nowadays. For every organization, it is essential to preserve its data, information using security tools. Providing secured Cloud services to both the business and personal users is the need of the hour as everybody wants their information and messages to be safe and secured. Data storage must ensure proper security for business-related activities as they have added legal obligations to keep data secure.

Almost the major part of cyber-security researchers employs machine learning techniques to transform a reactive to a prognostic approach for threat exposure. The trend of implementing machine learning in cyber-security is influenced because of the increasing amount of data. So, both cyber-threats and cyber-crimes turn out to be essential parts of our daily lives.

The fundamental rules of AI are: learning, reasoning, and problem-solving. Here learning, works on information gathered with principles to use data instances, reasoning finds the suitable information from learning, and problem-solving selects instant response on the basis of learning plus reasoning. Because of the powerful ability of machine learning, it is also applied to cyber-security and recent research has shown that it notably reforms the cyber-security view. Machine learning in the cyber world is used to improve malware detection, triage actions, identify infringes and alert organizations to security issues. Deep learning, which is the subset of machine learning, provides a simple or combinational explanation to a mixture of cyber-attack exposure troubles such as malware and intrusion detections. In the recent past, billions of US dollars were invested for cyber security startups especially those that specialize in AI and machine learning. Those were the top investment areas. Therefore, the collaboration of AI and cyber security is important to save our systems as well as AI-based projects in the digital world from upcoming cyber threats automatically [39]. With the prolonged research, an AI-based IoT device can be developed which will have an automatic intrusion and bug detection system, and will block malicious activities, create logs and will inform the user about the activities on a regular basis.

Contributing security measures to stop cyber-attacks has also materialized as a giant profit business. Government and many organizations are investing ample of money to secure personal and financial information. Despite bountiful uses, AI can be a threat to the human life itself. Hence, human interference is required to monitor all actions of AI. Without strong security trials, AI may become valueless as very easily it may go to the wrong hand. Consequently, government, banks, healthcare system and multinational companies will face threats from online hackers. Lots of personal and commercial information may be leaked and exploited by hackers. It has been seen that AI technology can be used by the adversaries for different sorts of Phishing attacks over the Internet. At the same time, AI is also used for malicious URL detection. So, in recent times, rigorous research in the area of cyber security with AI has grown importance.

3.4 Security Networking

Recent progress in Software Defined Networking (SDN) and network programmability can be used to simplify operations, enhance agility, and meet new mission requirements for security networking. SDN is an architecture that controls the network intelligently and centrally. It is controlled by software applications. SDN and network programmability have been materialized to deal with recent trends in communication to provide superior automation and synchronization of the network by permitting dynamic, application-based configuration and services. To meet all these criteria, networks must be open, programmable and application aware. To meet the requirements, there should not be any compromise for resiliency, service or security.

Conventional IP-based networks are very complex and difficult to manage. Here each network elements are configured separately, using vendor-specific commands for maintaining high-level network policies. In the current network architectures, automatic reconfiguration and reaction are also not present. The architecture is also built in a vertical direction. The control plane is having the responsibility of handling traffic and the data plane forwards that traffic. Both the planes are placed together inside the networking elements. Hence, the flexibility of the network architecture is reduced. SDN proposes a promising prototype where control and management parts are disconnected from network traffic forwarding functionalities and disobeys the vertical integration. In this case, switches and routers play the role of simple traffic forwarding devices. Traffic forwarding is not destination-based, rather alternative flow-based forwarding is adopted in SDN. A centralized controller is present for providing all control functionalities. SDN is programmable, thus reconfiguration becomes easier. Policy incorporation is also simple in the case of SDN. Therefore, the management of the network is done constantly regardless of the underlying network technology. SDN layer basically operates as a virtual software switch or router for the physical network devices. So instead of software embedded in the routers and switches managing the traffic, software from outside the devices takes over the job. SDN security is built within the architecture and also brought as a service to establish confidentiality, integrity and authenticity of all connected resources and information. Figure 15 shows SDN planes.

OpenFlow is the programming interface for SDN. Switches for this network protocol have one or more flow tables of traffic forwarding rules. Actions such as forwarding, modifying or dropping of packets can be performed only when the incoming traffic matches flow table rules. It makes network-wise intrusion detection and malicious switch detection easier. As an effect, the forensic analysis also becomes easier. Presently, SDN and *OpenFlow* have gained noteworthy popularity in the industry.

Another type of security network architecture is Named Data Networking (NDN). Previously Internet was meant for a packet data network where the client and server used to communicate with specific IP addresses over a pre-established communication link. Nowadays this client–server data transmission model has advanced into a

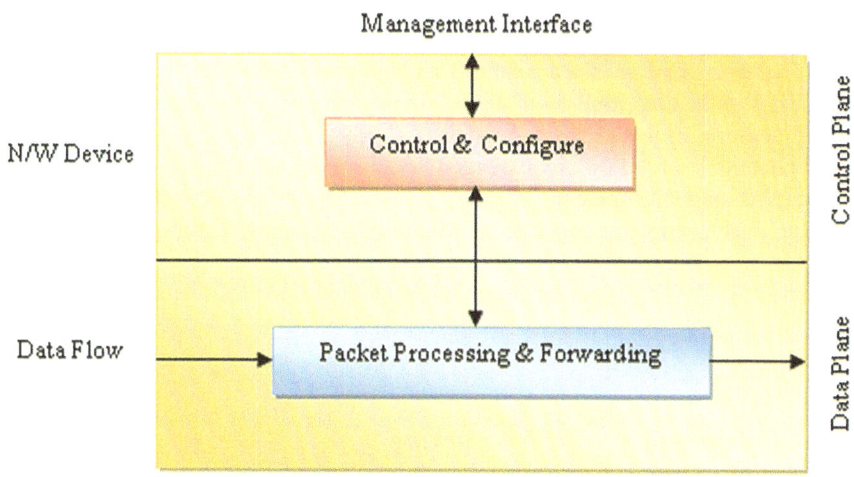

Fig. 15 Control and data plane of *SDN*

peer-to-peer model for communication purpose. Different applications such as social networks, YouTube, Instagram have modernized their idea of communication on the basis of user-generated contents. Present users are only bothered about a specific data item irrespective of its source. Thus, using IP addresses for the identification of servers hosting a meticulous content is no more in demand. NDN names the data instead of locations for network packet forwarding. Hence communications in NDN are now consumer-driven [40]. NDN architecture is designed to set connections in the world of computing devices, ranging from IoT sensors to cloud servers, by naming each and every data bit. Each piece of sensitive data is cryptographically signed. As a result, NDN communications are secured in a data-centric manner.

In the current era, we use the Internet, more often for content distribution and retrieval. However, the Internet is still lying on the conventional IP-based architecture. The variance between the architecture and the handling has given rise to many problems in IP, together with safety measures. The security model in TCP-IP is channel oriented. Once the data stream goes out of the channel, there is no assurance about the contents of the packets. Though many security measures are followed, the number of network attacks is still increasing.

The tremendous growth of secured applications such as online banking, a healthcare application has forced NDN researchers to consider "security as primary design goal" from the very beginning. The data coming through NDN carries a digital signature signed by the novel producer. After receiving the data, it is verified before utilization. If the verification is successful, then there is no need for the origin of data and the retrieval method. If the application requires confidentiality, the original producer can encrypt the content so that only the intended recipient can decrypt the data. A digital signature is also added with NDN for boosting its security.

The architecture of NDN helps in achieving security with mobility and efficient content distribution. Still, opponents can take advantage of some features such as in-network caching and may launch various kinds of attacks. It is also seen that an entirely new set of attack vectors are now employed with different impacts. Most of them are targeted toward the availability and integrity of data. So, the application of different hashing algorithms with digital signature may be one solution to fulfill the safety measures of network traveled data.

Applications of NDN are extended to many application domains, such as Smart System Development, Wireless Sensor Networks, Educational systems, Gaming, Vehicular Communication, Multimedia systems, Online Conferencing, entertainment, etc.

Networks (WSN), vehicular communication, multimedia systems, conferencing and.

3.5 *Anonymous Traffic Networking*

The most important concern of today's world is exchanging information securely. It can only be achieved if the right information is sent by the right sender to the right receiver. But to maintain the secrecy of information, to a greater extent details are captured for both sender and receiver. Hence, a new concern comes out on Privacy. Users demand privacy as well as anonymity, especially when sensitive information is interchanged. With the invention of the Internet and its incursion to our daily lives, the issue of anonymity is becoming pertinent.

In the recent era, many techniques have come into the market to hide the identity of users. Pseudonym Servers can communicate through e-mails and do not reveal the actual identity of the sender or receiver [41]. The Internet is now an inevitable part of our lives. So, there is a growing need for using the Internet with anonymity to protect the privacy of users.

ATN is also known as Onion Router or TOR directs Internet traffic through an overlay network consisting of several thousand relays to hide a user's location and usage. Things are hidden from the adversaries who conduct network surveillance for analyzing the traffic network. In this type of routers, the data together with the next destination address is also encrypted several times. This is done at the application layer of the connecting protocol. The processing is done similar to the layers of an onion. After that, it is sent through a virtual circuit with successive and arbitrarily selected TOR relays. Each relay decrypts a layer of encryption to disclose the next relay in the circuit to pass the remaining encrypted data on to it. The ultimate relay decrypts the innermost layer of encryption and sends the original data to its destination without revealing the source IP address. As the routing of information was partially hidden at every hop in the TOR circuit, this technique removes any single point at which the communicating peers can be determined via packet scanning and surveillance. By this methodology, the user may avoid any kind of censorship from government. Apart from private information which is vital for authenticating user's identity in a secured environment, they may also have their own mindset, emotion and attitudes toward social issues, that they do not want to incarnate. Such growing need evolved into the development and popularizing ATN. Anonymizing networks such as TOR provide such facilities to remain anonymous. TOR uses a chosen set of proxy servers that start with an entry node and end up to the final destination through an exit node. TOR also works in combination with end-to-end encryption of data through HTTPS.

The social recognition of the ongoing use of these sorts of the network is still considered as doubtful and the moral use of ATN is still questionable. The government and law enforcement departments are perhaps ready to give justifiable space for freedom of speech, until it stops creating offences and confusion. Harmonizing the welfare with more trust on ATN can open up huge opportunities in the Internet domain. The blockchain technology with pseudo-anonymous public ledger can be well thought-out as a step toward constructing an inclusive system with an extension of ATN. In this era, the practice of an anonymous network is a prime factor and the effort to fight threats on these networks should be seriously considered.

4 Conclusion

This chapter has explained the meaning of network security attacks and threats. It has discussed different types of network attacks in depth. The chapter also introduced the various cutting edge technologies, mechanisms and services for network security. It has touched upon the artificial intelligence, machine learning and deep learning concepts to combat security attacks. Review of different digital forensics and their use cases are highlighted. It has also explained the use of blockchain and modern cryptography which are used for cloud security, IoT security and cyber security. Last but not the least the use of different types of next-generation networking technologies, their vulnerabilities and solutions such as SDN, NDN and ATN are discussed in detail.

References

1. Dharmarajan R, Thiagarasu V (2019) A literature survey on the network security and intrusion detection system using data mining techniques. Asian J Comput Sci Technol 8(1):7–12. ISSN: 2249-0701
2. https://www.em360tech.com/continuity/tech-news/opinion-piece/disgruntled-employee-cybersecurity/
3. https://www.solarwindsmsp.com/blog/types-of-network-security
4. https://geek-university.com/ccna-security/confidentiality-integrity-and-availability-cia-triad/
5. Pawar MV, Anuradha J (2015) Network security and types of attacks in network. In: Proceedings of international conference on intelligent computing, communication & convergence (ICCC-2014), procedia computer science, vol 48, pp 503–506
6. Sinha P, Kumar Rai A, Bhushan B (2019) Information security threats and attacks with conceivable counteraction. In: 2019 2nd international conference on intelligent computing, instrumentation and control technologies (ICICICT). https://doi.org/10.1109/ICICICT46008.2019.8993384. IEEE Explore 13 February 2020.
7. Laurent M, Bouzefrane S (2015) Digital identity management, book. Sciencedirect. https://www.sciencedirect.com/book/9781785480041/digital-identity-management
8. Singh GD, Vinod M, Anandh V CCNA security 210-260 certification guide. Oreilly Publication
9. Dhende SL, Shirbahadurkar SD, Musale SS, Galande SK (2018) A survey on black hole attack in mobile ad hoc networks. In: 2018 4th international conference on recent advances in information technology (RAIT), Dhanbad, pp 1–7. https://doi.org/10.1109/RAIT.2018.8389073
10. Ghoreishi S, Abd Razak S, Isnin IF, Chizari H (2014) Rushing attack against routing protocols in Mobile Ad-Hoc Networks. In: 2014 international symposium on biometrics and security technologies (ISBAST), Kuala Lumpur, pp 220—224. https://doi.org/10.1109/ISBAST.2014.7013125
11. Wu Z, Gao S, Cling ES, Li H (2014) A study on replay attack and anti-spoofing for text-dependent speaker verification. In: Signal and information processing association annual summit and conference (APSIPA), 2014 Asia-Pacific Siem, pp 1–5. https://doi.org/10.1109/APSIPA.2014.7041636
12. Saini SK, Singh P (2016) Analysis and detection of Byzantine attack in wireless sensor network. In: 2016 3rd international conference on computing for sustainable global development (INDIACom), New Delhi, pp 3189–3191
13. Aliyua F, Sheltamia T, Shakshukib EM (2018) A detection and prevention technique for man in the middle attack in fog computing. Proc Comput Sci 141:24–33. https://doi.org/10.1016/j.procs.2018.10.125

14. https://www.uscert.gov/sites/default/files/publications/DDoS%20Quick%20Guide.pdf
15. Namanya AP, Cullen A, Awan IU, Disso JP (2018) The world of malware: an overview. In: 2018 IEEE 6th international conference on future internet of things and cloud (FiCloud), Barcelona, pp 420–427. https://doi.org/10.1109/FiCloud.2018.00067
16. Asish MS, Aishwarya R (2019) Cyber security at a glance. In: 2019 fifth international conference on science technology engineering and mathematics (ICONSTEM), Chennai, India, pp 240–245. https://doi.org/10.1109/ICONSTEM.2019.8918889.
17. Xiao K, Forte D, Jin Y, Karri R, Mohammad SKB, Tehranipoor M (2016) Hardware Trojans: lessons learned after one decade of research. ACM Trans Des Autom Electron Syst May 2016. Article No.: 6 https://doi.org/10.1145/2906147.
18. Boukerche A, Zhang Q (2019) Countermeasures against worm spreading: a new challenge for vehicular networks. ACM Comput Surv May 2019. Article No.: 34 https://doi.org/10.1145/3284748
19. Robbes R, Lanza M (2008) SpyWare: a change-aware development toolset. In: ICSE '08: proceedings of the 30th international conference on software engineering, May 2008, pp 847–850. https://doi.org/10.1145/1368088.1368219
20. Schwarz M, Lipp M, Moghimi D, Van Bulck J, Stecklina J, Prescher T, Gruss D (2019) ZombieLoad: cross-privilege-boundary data sampling. In: CCS '19: proceedings of the 2019 ACM SIGSAC conference on computer and communications security. November 2019, pp 753–768. https://doi.org/10.1145/3319535.3354252
21. Chaudhry JA, Rittenhouse RG (2015) Phishing: classification and countermeasures. In: 2015 7th international conference on multimedia, computer graphics and broadcasting (MulGraB), Jeju, pp 28–31. https://doi.org/10.1109/MulGraB.2015.17
22. Patel K, Dubey SK (2016) To recognize and analyze spam domains from spam emails by data mining. In: 2016 3rd international conference on computing for sustainable global development (INDIACom), New Delhi, pp 4030–4035
23. Gao J, Li L, Kong P, Bissyandé TF, Klein J (2019) Should you consider adware as malware in your study? In: 2019 IEEE 26th international conference on software analysis, evolution and reengineering (SANER), Hangzhou, China, 2019, pp 604–608. https://doi.org/10.1109/SANER.2019.8668010
24. Satheesh Kumar M, Ben-Othman J, Srinivasagan KG (2018) An investigation on wannacry ransomware and its detection. In: IEEE symposium on computers and communications (ISCC). Natal, pp 1–6. https://doi.org/10.1109/ISCC.2018.8538354
25. Chadha S, Kumar U (2017) Ransomware: let's fight back! In: 2017 International conference on computing, communication and automation (ICCCA), Greater Noida, pp 925–930. https://doi.org/10.1109/CCAA.2017.8229926.
26. Kuncoro AP, Kusuma BA (2018) Keylogger is a hacking technique that allows threatening information on mobile banking user. In: 2018 3rd international conference on information technology, information system and electrical engineering (ICITISEE), Yogyakarta, Indonesia, pp 141–145. https://doi.org/10.1109/ICITISEE.2018.8721028.
27. https://www.firstlegal.com/what-is-digital-forensics-and-why-is-it-important/
28. Rahim N et al (2014) Digital forensics: an overview of the current trends. Int J Cryptol Res 4(2)
29. https://resources.infosecinstitute.com/category/computerforensics/
30. Ruan K, Carthy J, Kechadi T, Crosbie M (2011) Cloud fforensics. In: Peterson G, Shenoi S (eds) Advances in digital forensics VII. Digital forensics 2011. IFIP advances in information and communication technology, vol 361. Springer, Berlin, Heidelberg
31. Stoyanova M et al (2020) A survey on the internet of things (IoT) forensics: challenges, approaches, and open issues. IEEE Commun Surv Tutor 22(2)
32. Nakamoto S (2008) Bitcoin: a peer-to-peer electronic cash system. Consulted. 165:55–61
33. Zhai S et al (2019) Research on the application of cryptography on the blockchain. IOP Conf Ser J Phys Conf Ser 1168. 032077 IOP Publishing https://doi.org/10.1088/1742-6596/1168/3/032077.

34. Wang L, Shen X, Li J, Shao J, Yang Y (2019) Cryptographic primitives in blockchains. J Netw Comput Appl 127:43–58
35. Bhatt AP, Sharma A Quantum cryptography for internet of things security. J Electron Sci Technol 17(3):213–220
36. Alvarez R, Martínez F, Vicent J-F, Zamora A (2008) A matricial public key cryptosystem with digital signature. WSEAS Trans Math 7:195–204
37. Rashmi BH (2018) Impact of artificial intelligence on cyber security. Int J Comput Sci Eng 6(12):341–343. https://doi.org/10.26438/ijcse/v6i12.341343
38. https://www.entrepreneur.com/article/339509
39. https://def.camp/artificial-intelligence-cybersecurity
40. Saxena D (2016) Named data networking: a survey. Elsevier Comput Sci Rev. https://doi.org/10.1016/j.cosrev.2016.01.001, January 2016.
41. Gayathri S (2014) A survey of anonymous networks and related menaces. Indian J Comput Sci Eng (IJCSE) 4(6):505–509. ISSN: 0976-5166, Dec 2013–Jan 2014

Part II
Review of Recent Trends in Forensics

A Systematic Review of Digital, Cloud and IoT Forensics

Atonu Ghosh, Koushik Majumder, and Debashis De

Abstract Our world is proliferated with high-end devices that have multiple capabilities. These devices have an array of sensors embedded into them that collect an enormous amount of data. Because of rapid development in technology, these devices are capable of rendering state-of-the-art services that are based on the cloud. Assuredly, the proliferation of ultra-modern devices with high-end services, have modernized and eased our lives. But, some entities or adversaries of our society are exploiting the same technological innovation for their benefits. Such criminal activities have become even easier to be carried out with modern devices and other technologies. Fortunately, we can examine the confiscated digital devices such as android smartphones, laptops, smartwatches, etc., from a crime scene and gain insight into the criminal activities carried out by the adversaries. It is also possible for us to recover the data the adversaries have collected or transmitted. Since modern devices utilize the services provided by cloud computing, it naturally becomes essential for the analysis of the cloud in case of the investigation of a crime scene. IoT forensics is interdisciplinary, as the data to be investigated may be collected from sensors, smart devices, etc., connected to a crime scene and the cloud too. Digital Forensics is an easier task to be performed when compared to Cloud Forensics. The segregation and collection of evidence in a cloud forensic investigation is a mammoth task due to the distributed and multi-tenant nature of cloud computing. But there is no such problem associated with digital forensic investigation. In this chapter, we have reviewed the work on Digital, Cloud and IoT Forensics by the scientific community. We have compared their effort and the outcome of their work, and summarized the state-of-the-art in the field of Digital, Cloud and IoT Forensics. Our aim of this chapter is to create a thorough review that will help fellow researchers in thought association for getting a clearer picture of the current state of research on Digital, Cloud and IoT Forensics and find the research gaps.

A. Ghosh · K. Majumder (✉) · D. De
Department of Computer Science and Engineering, Maulana Abul Kalam Azad University of Technology, Kolkata, West Bengal, India
e-mail: koushikzone@yahoo.com

M. Chakraborty et al. (eds.), *The "Essence" of Network Security: An End-to-End Panorama*, Lecture Notes in Networks and Systems 163,
https://doi.org/10.1007/978-981-15-9317-8_2

Keywords Digital forensics · Android forensics · Cloud forensics · Internet of Things forensics · Blockchain · Blockchain in forensics · IoT forensics

1 Introduction

The modern human life is unimaginable without electronic devices and has become so dependent on various electronic devices that we no more can efficiently manage our day to day life without using them. According to a report by Statista [1], the global market share of the consumer electronics industry alone is worth $426,238 million in 2020 with a user penetration rate of 26.9% and is expected to have a volume of $565,345 million by 2024 with a user penetration rate of 38.5. The exponential growth in the global revenue share of the segment proves the consumers' inclination towards the electronic devices.

We find the application of electronic devices in almost every sphere of our modern-day lives. Most modern devices be it a smartphone, tablet, pen drives, laptops, portable music players, etc., have storage of some form in them. These devices become the nodes in the chain of our day to day activities. Also, these devices store a great deal of data which might be extremely confidential and sensitive. In addition to the in-device functions, these ultra-modern devices exploit the capabilities of cloud computing to render state-of-the-art services to consumers. Lately, with the advent of the Internet of Things (IoT) technology, these devices' capabilities and service range have got even enhanced. Devices such as smartwatches, smart home systems, smart cars, etc., have broadened the data collection and storage endpoints of the existing systems.

The confidential data stored in such electronic devices, cloud environment or IoT infrastructure which can be of monetary gain or can serve any other purpose of the adversary, attracts them to break into such systems. Once the adversary gets to access these systems containing valuable data for them, the adversary can have the data and use it for their benefit. Such criminal activities have been on the rise ever since. Thus, there is an urgent need for law enforcement agencies to get hold of these adversaries and frame them. To facilitate the process of entrapping the suspect and prosecuting them, the branch of science has evolved. This branch of science is called the Forensic science. The branch of forensics that deals with criminal activities in the digital environment are called the Digital Forensics. It is an ever-evolving field of forensic science. It encompasses many sub-disciplines such as Computer Forensics, IoT Forensics, Network Forensics, Cloud Forensics, etc. Each of these subfields essentially includes the application of digital forensic principles and techniques to carry out the forensic process in each of the subfield's environments such as in an IoT infrastructure, computer network infrastructure, cloud computing environment, etc.

The field of Cloud Forensics is more challenging than that of Digital Forensics due to the distributed nature of cloud computing. By design, cloud computing systems are geographically spread across the globe. This brings in the multijurisdictional issues.

The multijurisdictional issues make it extremely hard or sometimes impossible to identify and seize the appropriate digital devices that might provide significant shreds of evidence in digital crime. Due to these issues, most of which is still an open research area, there is an enormous number of pending cases involving the cloud computing environment. Apart from the multijurisdictional issues, the multi-tenancy of cloud computing makes the evidence segregation a challenging task. In cloud computing systems, multiple clients share the same hardware resource. Due to the sharing of resources, it becomes hard to identify the devices that need to be seized for forensic investigation. This often leads to a privacy breach of other tenants in the cloud computing environment. To avoid this, the cloud service providers are reluctant to provide access to the cloud infrastructure by the forensic investigators as such a privacy breach, in turn, jeopardizes the cloud service provider's reputation and it makes the cloud service provider violate the service level agreement with his client. Once the forensic investigator can get access to the cloud computing infrastructure and identify the devices to be seized, the forensic process to be followed next is essentially digital forensics as the devices used in the cloud infrastructure are all digital devices.

Adding to the complexity of cloud forensics, the introduction of connected things i.e. Internet of Things has led to the explosion of data. We live in an era of connected things where everything is connected be it the refrigerator, air conditioner, coffee maker, car, drone or garage door. Each of these connected devices is exchanging data. Thus, each of these devices can prove to be a piece of evidence. In a crime scene, it becomes a challenging task to identify the device rather than the "thing" that needs to be forensically investigated. The introduction of IoT has lead to exponential data collection endpoints. Handling and analyzing such huge volumes of data to gain insight into a crime scene is another challenge for the forensic investigator. Once the forensic investigator identifies the devices or the "things" to be investigated, the forensic process may include cloud forensics if the implementation of IoT in the crime scene involves cloud services which is mostly the case. If not, then the way forward is simple digital forensics of the digital devices from the crime scene. IoT forensics is at the intersection of Digital Forensics and Digital Forensics. It is complex but it promises to produce more granular data about the crime scene.

In this work, we review the work on Digital, Cloud and IoT Forensics by the scientific community. We compare and contrast their effort and the outcome of their work; we summarize the state-of-the-art in the field of Digital, Cloud and IoT Forensics. Our aim of this work is to create a thorough review that will help fellow researchers in thought association for getting a clearer picture of the current state of research on Digital, Cloud and IoT Forensics and find the research gaps.

Organization: In Sect. 2 we have reviewed Digital Forensics. We reviewed Cloud Forensics in Sect. 3. Section 4 contains a detailed review of IoT Forensics. In Sect. 5 we have identified the open areas of research in the field of digital, cloud and IoT forensics. Finally, we conclude the chapter in Sect. 6.

2 Digital Forensics and Its Evolution

Digital forensics is the branch of science that deals with the examination of the digital pieces of evidence collected from a crime scene. The purpose of the evidence collection is to correlate to evidence with the happenings of the crime scene and try to recreate the crime scene. Digital forensics is a broader term that includes the branches of forensics such as cloud forensics, network forensics, etc., as essentially these include digital devices but of special types and purposes. Palmer [2] defined digital forensics as—

> The use of scientifically derived and proven methods toward the preservation, collection, validation, identification, analysis, interpretation, documentation and presentation of digital evidence derived from digital sources for the purpose of facilitating or furthering the reconstruction of events found to be criminal, or helping to anticipate unauthorized actions shown to be disruptive to planned operations

In general, Digital Forensics deals with the globally accepted set of procedures and norms which are carried out for investigating a digital device. It consists of five steps such as Identification, Collection, Preservation, Analysis, and Reporting as depicted in Fig. 1.

Identification: In this phase, the digital objects from the crime scene are identified which can be forensically investigated for leading the case to confidence.

Collection: In this phase, the objects that were identified for forensic investigation in the crime scene are seized from the crime scene in a forensic sound manner.

Preservation: In this phase, the shreds of evidence collected from the crime scene are preserved in a forensic sound manner.

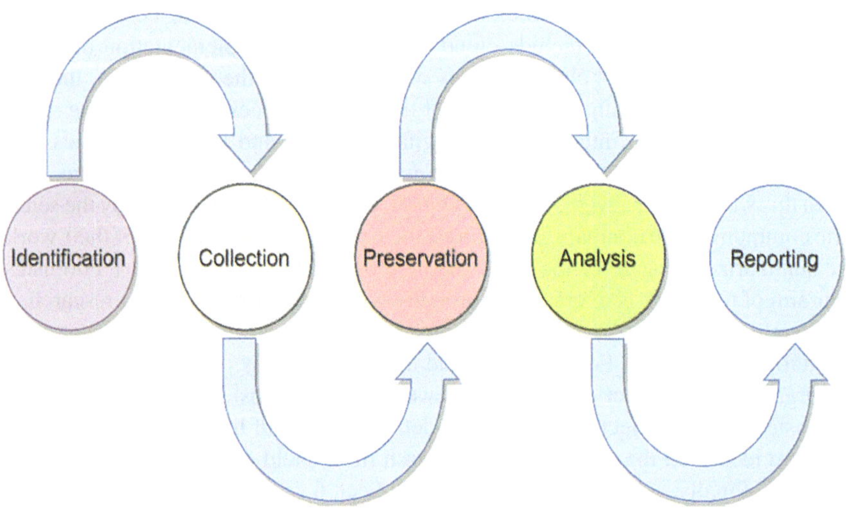

Fig. 1 Digital forensics process

Analysis: In this phase, the digital devices are analyzed for shreds of evidence. The tools and techniques used in this process are forensically sound and globally admissible.

Reporting: In this phase, the findings of the previous stage are documented and reported in the court of law as a shred of evidence for the proceeding.

Initially, digital forensics suffered from a lack of standardization. Though a few groups of people who were actively involved in digital forensics tried to produce some elementary guidelines for digital forensics these eventually turned out to be technology or device-specific rather than being a general guideline. These guidelines were appropriate but lacked abstraction to be applied for other technologies, for example, there were guidelines developed for UNIX in but not for Windows NT. Had there been some common guidelines or principles, it could have been applied for both. The first standardization was done by the DFRWS [2.2] which is an academic consortium. This was the first standardization done by such a body as all earlier standardization efforts were made by the law enforcement agencies. The DFRWS as a first designed and proposed framework for forensics which comprised of steps such as Identification, Preservation, Collection, Examination, Analysis, Presentation and Decision [2]. This framework can then be reworked and tuned by anyone for their applications. Based on the DFRWS forensics framework, in 2002 Reith et al. [3] proposed a digital forensics model to contain all future sub-processes comprised of nine steps such as Identification: It is the step where the crime incident is identified based on the identifiers of the crime, Preparation: Getting ready for forensics with appropriate tools, methods and warrants needed, Approach Strategy: Formulating a strategy to collect evidence, Preservation: Preserving the digital evidence, Collection: Collecting data from digital evidence using standard procedures, Examination: Thorough search of the evidence data to correlate the crime scene, Analysis: Conclude the evidence data, Presentation: Prepare and provide evidence report to the law enforcement body and Returning evidence: Returning the physical-digital evidence to the owner. Rogers et al. [4] in 2006 proposed a triage digital forensics model that was sensitive to the urgency of the need of the hour at the crime scene. It stressed the urgent course of action that needed to be taken. For instance, in cases of child abductions, pedophiles, missing or exploited persons, which are extremely time-sensitive, the model proposed to carry out the preliminary digital forensics at the crime scene itself to get crucial clues about the crime and then carry the digital pieces of evidence to the lab for a thorough investigation. This model has been applied in the past in practice and has proved to be of benefit in such scenarios. After all, technology is to serve mankind. Digital forensics has faced several challenges since its inception. The major challenges have been the sheer lack of tools and lack of corporate governance [5]. The digital devices for forensic investigation found at the crime scene were modern than the tools prevalent. This posed a major challenge for the forensic investigators. The modern digital forensic scenario is though not very different because the technology around the digital devices evolves rapidly compared to the tools. In the past, there have been many cases of major corporate shutdowns due to a lack of strong corporate governance as most of the corporate communications are conducted over electronic media. Keeping track of such transactions is of extreme urgency for future pieces of evidence. With the rapid increase in technology, there has been an explosion in the volume of data generated. The forensic investigators face this as a challenge. It is extremely difficult for them to analyze such a huge volume of data with the existing tools. They need more sophisticated tools and techniques to reduce the turnaround time of the cases. The simplest of the techniques that are commonly used by the forensic community to reduce the analysis time is to save the precious CPU cycles in the analysis tools [6]. This has proven to be highly beneficial, but the increasing volume of data has outpaced this gain over the CPU cycles. The better solution lies in speeding up the evidence data analysis using distributed computing. It is beneficial to harness the power of a grid of computers to handle such a volume of data and get a smaller turnaround of time provided the urgency of the data that is being computed. Marziale et al. [7] found that the modern Graphical Processing Units GPUs are equipped with a large number of general-purpose processors whereas in earlier designs of the GPUs special purpose hardware was used. These general-purpose processors in the GPUs can be used for parallel processing in the processor-intensive environments such as the data analysis tasks which are CPU intensive. They experimented with several GPUs and such CPU demanding software to attain a significant amount of development in the throughput. They threaded the software to take advantage of the multi-core GPU architecture. Another effective way is selective data extraction. In this method, the data that is being extracted from the digital evidence is selective and is arguably believed to have pieces of evidence. Data parts such as common files of operating system applications can be avoided and may reduce the whole forensic time significantly. But this method of data extraction and analysis is open to debate and legal challenges. Anti-Forensics [8] is another major challenge that digital forensics faces. It is the employing of various tools and techniques by the criminal entities to thwart the process of digital forensics. So, it becomes hard for forensic investigators to retrieve the data from the digital pieces of evidence and comprehend them. “Data Hiding”, “Artefact Wiping”, “Trail Obfuscation” and “Attacks on forensic tools” are the four major anti-forensic methods prevalent [9]. In Data Hiding, the data is hidden in numerous ways such as by the means of steganography [10], in the slack space, in the unallocated area of hard disk drives, in metadata of files, etc. In the past, people have also used servers of third-party to hide their data. Artifact Wiping tools aim at overwriting the entire drive repeatedly with garbage values which makes the process of data retrieval difficult. At a time after this wiping action, the data recovered is impractical and is of less significance. Trail Obfuscation is the alteration of the logs that trace back to a network activity. Modern technologies such as the Onion routing, anonymous Secure Shells make it very misleading for the forensic investigators. These mostly lead the cases in the wrong direction and eventually bring a case to a complete halt. Attacks on the forensic tools are the latest type of techniques. These try to present the exploitations of the forensic tools. They bring the forensic tools into question which in turn jeopardizes the whole process of forensics.

2.1 Document Metadata in Digital Forensics and Steganography

Document editors such as the Microsoft Office that use structured storage for encoding the metadata of the document file, always carry much more data than intended [11]. The metadata in the file gets added merely upon the creation of a new empty document file and while editing an existing file. This metadata includes sensitive information such as the author of the document, the modification dates, the institution who created or modified the document, the reviewers of the document, the system information on which the document was created and edited, the software that was used to create and edit the document, the deleted content of the file and also it includes the contents of files that are unrelated to it but have co-existed in the system while the document file was being edited in the system. Breach of sensitive information is not just due to the presence of metadata but also for the presence of undesirable data in the document files. Such data in the documents are referred to as "trash". Over time several tools have emerged with time for the removal of sensitive information from the document files but the efficiency of the tool is questionable. Some of the popular tools are Microsoft Remove Hidden Data Tool [12] and ezClean [13]. To prove the fact that none of these tools are effective and provides a false sense of document sanitization, Castiglione et al. [11] experimented with the document sanitization tools. They used the tools to clean the unwanted sensitive information contained in the documents and then used the "strings" tool in UNIX to print all the printable items of a document file. Astonishingly, the authors found that even after the sanitization of the documents, the "strings" tool printed out a lot of unwanted and unexpected sensitive information from the document file along with the contents of the file. The output content included the revision history of the document, the path of the file, the original title of the document that the document had, which suggested that the current author did not start with a new document file rather an existing document file was edited. The authors of the original organization were found along with the software that was used for the creation of the document.

The authors established that on an average about 6.53% of the document's size is allocated for the "trash" and most of which remains unused throughout the life of the document. This unused portion of the trash can be exploited for data hiding using steganography. Castiglione et al. [11] have proposed "StegOle," a tool that exploits the technique of steganography to hide data in the free space of the trash region of the Microsoft Office documents. "StegOle" hides the data as well as retrieves the hidden data from the trash section of the Microsoft Office documents. The authors have also implemented encryption and compression of messages in the "StegOle" tool developed by them for better efficiency of the tool. "StegOle" can operate in two modes, that is, in "Single-file-mode" and "Multi-file-mode." In single-file-mode, the tool embeds the secret messages in a single Microsoft Office document, whereas in Multi-file-mode the tool embeds the secret message hides the message in multiple documents. The data concealment by the tool is a four-step process as depicted in Fig. 2a.

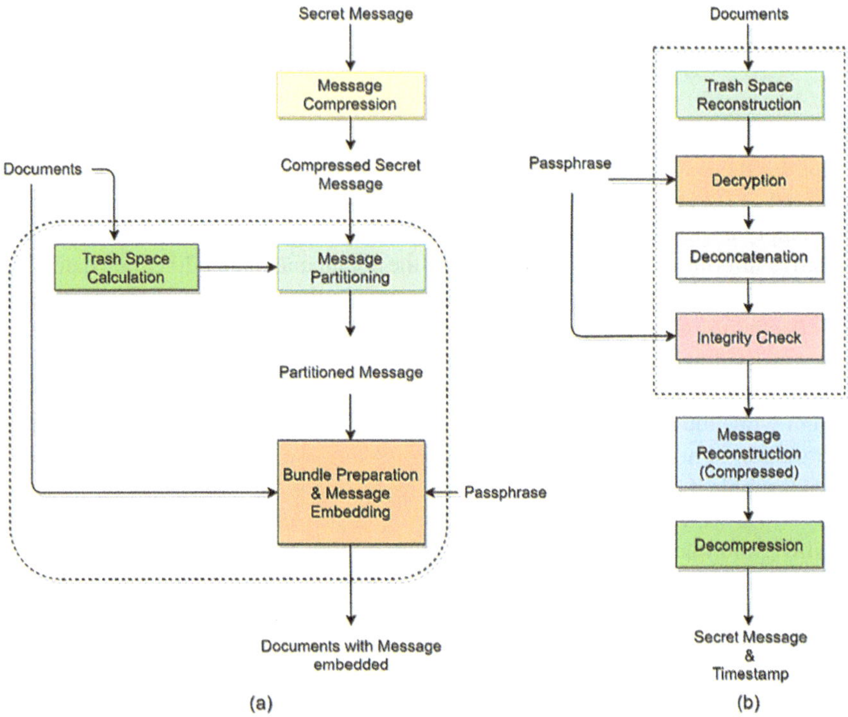

Fig. 2 **a** StegOle multi-file-mode message hiding process. **b** StegOle multi-file-mode message revealing

a. **Free Space Calculation**: In this step, the tool calculates the amount of available free space in the trash area of the document.
b. **Message Package Preparation**: It is further divided into subtasks that are carried out based on the file modes. For Single-file-mode, the subtasks are compressing the message, computing the header and encrypting the message. For Multi-file-mode, as the message needs to be embedded in multiple files, the already compressed message is chopped and then for each part, the header is calculated. For a proper reconstruction of the dismantled message, each segment of the message is assigned a sequence number. During compression, the tool uses a standard compression algorithm and produces a fixed number called the HMAC value along with the timestamp.
c. **Increasing Free Space (if needed)**: The tool frees up the trash area to hold the message to be concealed.
d. **Message Package Hiding**: The tool embeds the message in the free space of the trash area.

The process of message extraction is a two-step process as depicted in Fig. 2b.

a. **Free Space Reconstruction**: The tool takes a Microsoft Office document as input and rebuilds all the trash area free space.
b. **The Decryption of Message**: The message in each file undergoes decryption and the header from the compressed message is split. The HMAC is calculated and the newly calculated HMAC is matched against the old HMAC found in the header. Upon a match of the two HMACs, the message is decompressed in Single-file-mode else it indicates an error. Thus producing the message along with the timestamp. In Multi-file-mode, after the successful processing of all the parts of the message and integrity check using the HMAC, the whole message is reconstructed and decompressed and then is presented to the user along with the timestamp.

2.2 Digital Forensics and Full Disk Encryption

Lately, the systems are implementing operating system encryption which makes the task of Digital Forensics challenging. These methods of encryption have the capability of preventing data recovery from the digital devices of interest. Encrypted devices require a key or a password for the decryption of the encrypted device. Without the key or the password, access to the data becomes near impossible. Thus it is essential for forensic investigators to decide at the crime scene whether to power off the device or to perform a live forensic on the device. Casey et al. [14] studied the impact of full disk encryption in Digital Forensics. They have also demonstrated ways to recover data from fully encrypted devices in alternate ways other than which the data recovery from the fully encrypted device would not have been possible at all. Various programs have emerged with time for encryption of data in the hard disks. They evolved as data container encryption that makes it somewhat difficult for the data to be extracted. Several tools tried different strategies for the protection of the data in the disk such as the tools used more than one access paths and paraphrases for decryption of data, manipulation of the metadata in the disk, etc. For the better part of the forensic investigators, these tolls were not integrated into the operating system. As a result, despite the encryption used by these tools, there existed files in the system such as the page-files, temporary files, printer spool files, and deleted files that could be used for data extraction by the forensic examiners.

Recently, tools such as the Microsoft BitLocker have been introduced which provides full disk encryption in the computer systems. It stores the encryption keys protected in a removable media. The BitLocker is a tool that encrypts even the unallocated space in a hard drive and all parts of the operating system. The BitLocker handles the mechanism before the booting of the operating system. Thus making it impossible to get a logical view of the disk without unlocking the disk.

It is often found that the forensic investigators in a hasty, do not check the disks in a crime scene for full disk encryption and bring the drives to the laboratory for further examination when they find the full disk encryption and cannot find any logical information of the disk. In such a case acquiring a logical forensic duplicate is not possible although a physical forensic duplicate is possible. Thus the forensic investigator in the crime scene needs to decide whether to just pull the plug of a live system found in the crime scene only to find a fully encrypted disk later on or to perform live forensics at the crime scene itself. Though the full disk encryption makes it hard for the data extraction process but the recovery option in the disk encryption tools can be exploited to extract the data from the disk. Also as the encryption keys are the only way to recover the data; it must have been stored by the user on some safe device. So during a forensic collection stage, it is essential to thoughtfully collect such thumb devices from the crime scene for later disk decryption if the need arises.

Casey et al. [14] suggested that it is, in fact, possible to recover the data from a fully encrypted hard disk. They demonstrated that, if a disk is encrypted using BitLocker then it can be connected in read-only mode to a forensic system and then by providing the recovery password for the BitLocker would let the data to be acquired to be in the logical unencrypted form although the physical data will be still in encrypted in the device. In another example, the authors have demonstrated the process of acquiring a forensic duplicate of a disk. In such a scenario the digital evidence system whose data needs to be acquired is boot from a forensic boot media and then the contents of the evidentiary digital device are copied to some different storage device for further forensic investigation. This process has an added advantage over other methods of acquisition as it does not alter the metadata of the evidentiary digital device.

The authors further emphasized the "forensic soundness" in the field of Digital Forensics. They mentioned that the notion of "preserve everything but change nothing" needs to be reviewed. As this notion suggests that for the acceptability of evidence in digital forensics, no evidence can be altered rather they all have to be preserved. But this can hardly be implemented in real-world scenarios. Examining a system forensically is bound to alter some of the aspects of the system such as inserting a USB device alters the system state. For supporting their statement the authors argued that the process of DNA analysis is destructive. If the results of DNA analysis can be acceptable, then the notion of "preserve everything but change nothing" needs to be changed otherwise the future of digital forensics is unsure. They mentioned that it is rather essential to make it more stringent to document every minute detail of the forensic process.

2.3 A Solution to the Problem of Scale

The data generated by modern systems is huge. It is ever-growing and is growing exponentially. According to the FBI statistics, the average gigabytes of data examined per criminal case have risen from 83 in 2003–277 in 2007 [15]. The case turnaround time for Digital Forensics has been on a constant decline due to the massive amount of data that has to be analyzed by a human to find the tiny bits of a clue to correlate with the crime. It is a mammoth task for forensic investigators, given the volume of data they have to analyze for each case. As a result of this growing data to be analyzed multiple cases remain pending. Roussev [16] suggested that the analysis space of data of the forensic investigator can be drastically reduced if the objects that are known to exist in the acquired data for forensic analysis could be eliminated from the data space. All digital forensic data acquired from systems contain common files relating to operating systems, software, etc., that have proved to be of no use in the prior forensic investigations. If these data were intelligently eliminated from the data space of the forensic investigator, the remaining data space would be the pure evidence for the investigator to analyze. The author proposed to implement this heuristic using the hashing technique. Hashing is a technique where a function takes random data as input and produces unique output value for unique input. The output is spread over a fixed number of characters. This output is often called the hash digest. The hashing method is a one-way method which means providing the output, the input can never be determined. The idea proposed by the author is that several entities can maintain hash sets of different data commonly found in forensic data that do not contribute to a case's confidence. These hashes can then be compared with the hashes of the forensic evidence data. Thus the matched ones are the irrelevant data and the unmatched are the relevant data to the forensic investigator. One such hash set is hosted by the National Institute of Standards and Technology (NIST) called the National Software Reference Library [17].

2.4 *Digital Forensics Is a Real-Time Task*

Tasks that need to be completed within their stipulated deadlines are called real-time tasks. Further, real-time tasks are of two types: Soft real-time and hard real-time. Hard real-time tasks are those which if fail in turn fail the whole system. On the other hand, the soft real-time tasks are those in which some of the tasks may fail or may complete after the deadline. These are typical to a real-time system. Roussev et al. [18] argued that a digital forensic process ought to be viewed as a soft real-time task similar to an application in a soft real-time system that might miss some of its deadlines but meets the overall deadline requirement. The authors suggested that the performance lagging is because the whole process of digital forensics is an open-ended linear process. In this, all subsequent stages are dependent on the prior stages. The data acquisition rates are as slow as 123 MB/s for a 3 TB hard disk drive which takes almost 7 h. They expressed their "performance objectives" as—

a. Parallel acquisition and processing at a maximum throughput of a target device
b. Same start and end (approximate) time of acquisition and processing
c. Availability of partial results.

The authors suggest that the tasks of processing and acquisition should start at the same time with the full capacity of the device that is being used in the forensic investigation process thus maximizing the overall performance of the system. As the data is being acquired and processed at such a high rate at the same time, the partial results must be available to the users as and when the tasks begin. This will reduce the analysis time which will result in overlapping of tasks and speed up the overall digital forensic process.

Triage in Digital Forensics is the scenario when the Digital Forensics process needs to be quickly carried out outside the laboratory for fast decision making in the crime scene. The authors defined the triage as an optimization problem and have defined triage as "Digital forensic triage is a partial forensic examination conducted under (significant) time and resource constraints." In such a scenario, the digital forensic investigator has no option but is presented with limited resources and time. Within these constraints, the investigator needs to extract the most valuable part of data from the digital evidence. They experimented with various setups and evaluated the performances of the systems in triages. They found that for a 2 TB hard disk drive as a target device, with eight core workstation the tasks of file and metadata extraction, hashing, and the extraction of the registry meets the constraints. Whereas for a 48 core server the heavier tasks such as processing of bulk data, data indexing, data curving meet the constraints and are feasible on the server.

Thus it is evident from their work that to enhance the performance of the digital forensic process we need to consider it as a real-time task as proposed by Roussev et al. [18] because we aim at an overall performance improvement in the forensic process even if some of the sub-tasks of the forensic process fails to meet the deadlines. Designing forensic systems and tools based on this philosophy is surely going to speed up the already lagging forensic processes. Moreover, the authors have stressed the kind of systems that can handle the different kinds of workloads in the forensic environments. They have demonstrated that the traditional practice of a forensic workstation is no more viable in the modern-day forensic where the data volume is so high. The more powerful systems such as the servers should be used to cater to the computing-intensive tasks and the less computing-intensive tasks can be handled by the workstations. The choice of the systems for performing forensic processes plays a vital role in the turnaround time of the cases.

2.5 *Ramping up the Speed of Forensic Processes*

It is customary to carry out the digital forensics' tasks on a single workstation. This, in turn, bottlenecked the performance of the tools and has drastically reduced the case turnaround time with the exponentially growing data in the modern digital systems. The forensic investigators now have to dig into huge data piles to find the little hints in the cases in contrast to the smaller data volumes that had to be dealt with in the past. Analyzing such a huge volume of data is a time-consuming task. To minimize the case's turnaround times, the systems' capabilities are also pushed to the extreme limits where there is hardly any room for more gain in the performance. In such a scenario, harnessing the power of multiple computers seems the way forward to deal with the computing-intensive tasks in digital forensics. Roussev et al. [18] in their work proposed a "distributed digital forensics" system whose architecture is depicted in Fig. 3.

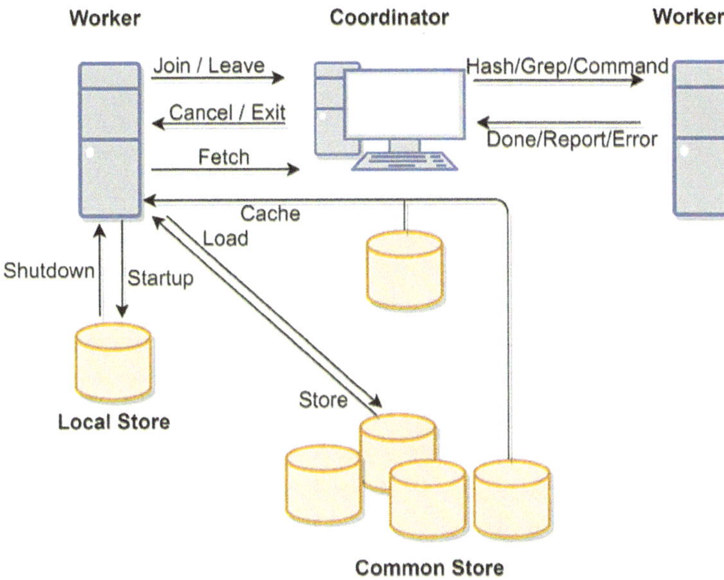

Fig. 3 Distributed digital forensics architecture

The proposed distributed system consists of "workers" and "coordinator" which are essential processes running on different systems. The coordinator is the main process that is responsible for accepting commands from a user. Upon receiving a command from a user, the coordinator process then distributes the tasks among the "worker" processes. The whole system uses a simple communication protocol for internal communication among the sub-systems. The communication protocol is a text-based simple protocol, and this is to minimize the complexity of the system as discussed by the authors in their work.

The authors have set up the prototype of the proposed system for the experiment using the C language and TCP/IP service. They have set up both the coordinator and the worker processes to be multithreaded for better performance. They experimented with FTK on Windows XP on a 3.0 GHz Pentium 4 processor, 2 GB RAM and 15,000 RPM hard disk drive versus their distributed system running on 8 nodes running on the eight nodes of the "72 node Gumbo-72 Beowulf cluster" each node with 2.4 GHz Intel P4 processor, 1 GB RAM, and Gigabit Ethernet port. A fileserver is attached to the cluster with dual CPU 1.4 GHz Intel Xeon, 2 GB RAM, 504 GB RAID5 storage.

As observed by the authors, the performance of the proposed system was impressive with the workload that has to be dealt with in the forensic scenario. A forensic image of 6 GB was handled by the system. Graphs 1 and 2 depict the initialization times are taken and search times were taken by the FTK system and the 8 node system, respectively. The performance comparisons justify the use of multi-noded systems in forensic applications. There has been a dramatic change in the performance with the introduction of the multi-node system. The preprocessing was reduced from hours

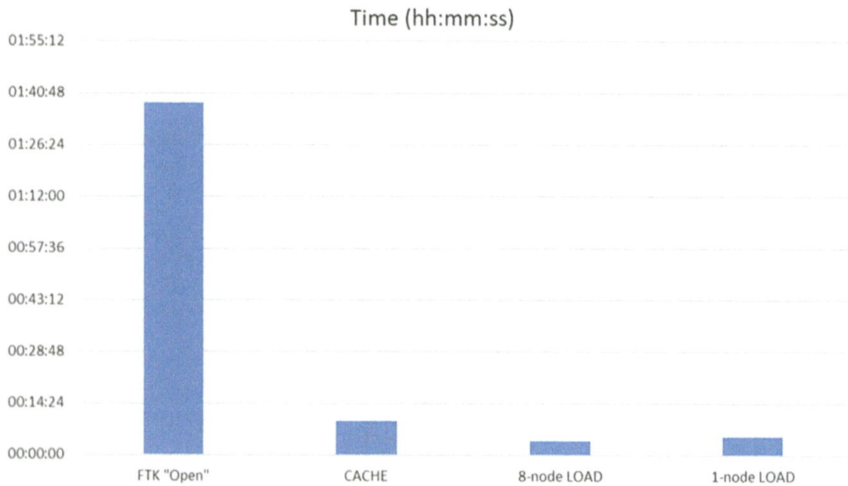

Graph 1 Initialization time by FTK

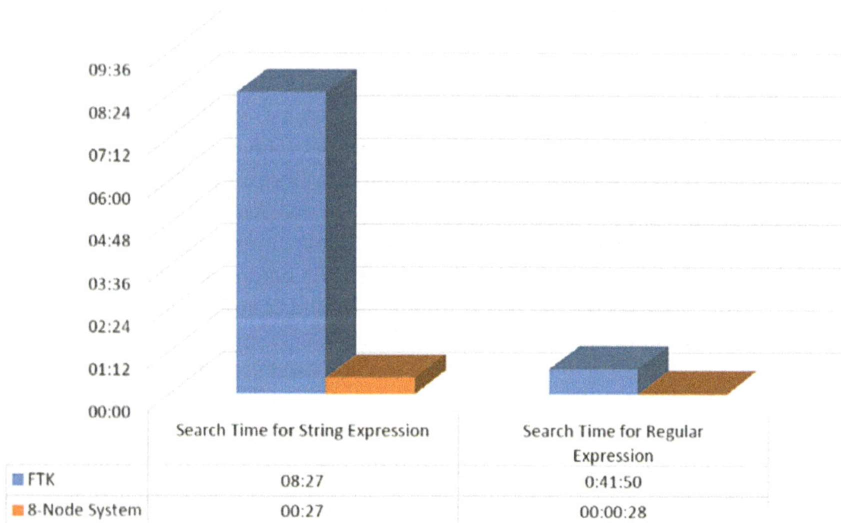

Graph 2 Search times on 8-node system

to just a few minutes and the search time got reduced by a factor of 18–89 times to that of the time taken in a single workstation earlier. Such systems are the need of the hour. Only by harnessing the power of such systems, the forensic processes can be completed faster and their turnaround time can be brought down significantly. Table 1 gives a comparison of digital forensics literature.

Table 1 Comparison of digital forensics literature

	Castiglione et al. [11]	Casey et al. [14]	Roussev [16]	Roussev et al. [18]	Roussev et al. [19]
Contribution	Showed how metadata exposes sensitive data	Stressed the effect of full disk encryption in digital forensics	Proposed the use of hashing to reduce data scale in forensics	Argued that forensic tasks are real-time tasks	Proposed distributed model for a forensic process
Performance enhancement mechanism	None	None	Used hashing to eliminate common data	Proposed to increase the overall performance of the forensic process	Showed how a distributed system can speed up processing in forensics
Proposed tool	A tool to embed a message in "trash" portion of documents	None	None	None	Model was proposed
Triage in forensics	No	No	No	Yes	Can help in Triage in Forensics

The effect of "trash" has been outlined by the authors Castiglione et al. [11]. They have also demonstrated that how critical it is to be mindful of the files before sharing them as a lot of sensitive information is shared in them. Also, it is for the benefit of the forensic investigators who intend to find clues in such type of files discovered in digital devices. The authors have also successfully demonstrated the tools designed by them and exploited the "trash" space to steganographically conceal data in them and retrieve the concealed data. Their tool is capable of error checking and concealing data in multiple files. But the authors have not contributed in any form in the performance optimization of the forensic process as done by Roussev [16], Roussev et al. [18] and Roussev et al. [19]. Roussev [16] proposed the use of hashing to eliminate the huge amount of known data found in the forensic data acquired from digital devices. This drastically reduced the data analysis time. On the other hand, the authors Roussev et al. [18] proposed that the tasks in the forensic process are all real-time tasks and hence they must follow a deadline. They also suggested that these tasks are soft real-time tasks meaning that some sub-task could miss the deadline, but the overall task meets the deadline. Thus, the systems designed for the forensic process should follow this principle. Again, Roussev et al. [19] in their work proposed a distributed model for computing the forensic tasks. Their proposed system when put under test with the forensic data, proved to be extremely beneficial as it demonstrated drastically reduced times for operations in the forensic scenario as compared to a traditional workstation. Casey et al. [14] in their work showed how disastrous full disk encryption can be and how it can pose a challenge for the forensic investigators. They have demonstrated the two possible ways to recover the data from a fully encrypted disk in their work.

The field of Digital Forensics has witnessed major advancements concerning its capability enhancements. There have been numerous efforts by law enforcement agencies and academia in tandem for the maturity of the field of digital forensics. Nevertheless, the forensic investigators regularly face a set of challenges that are either existing and have not been addressed or have emerged as a result of the rapid technological advancement. Heterogeneity of digital devices is one of the major challenges that the forensic investigators face in an investigation [20]. On average, a modern human being uses five digital devices daily [21] which is an amalgamation of a smartphone, computer, tablet, PDA, smartwatch, etc. Each of these devices is manufacture by different manufactures and has different architecture and different operating systems. Not just different devices, smartphones alone in the markets are manufactured by several manufactures and come with different software and operating systems in them. This, in turn, poses a challenge to the forensic investigators to carry out an investigation. Due to this reason, a forensic investigation often involves several forensic investigators working as a team in the same case. As a result, the human there is a high requirement of a greater number of forensic investigators personal to cater to the cases which have been on the rise due to the increasing rate of criminal activities involving the digital devices. Thus, there is a huge demand for such professionals in the forensic industry. But the positions in the Government and private sectors remain vacant due to a shortage of such professionals.

The ever-increasing data volume poses another major challenge for the forensic investigators. With billions of email and text messages being exchanges and the mobile traffic being of 2.5 exabytes/month, the data scale can be imagined well. Also, billions of IoT sensors that collect data are adding to the already huge volume of data. Dealing with such volume is a major challenge and is certainly going to drastically diminish the turnaround time of the digital forensic investigations wherein each case terabytes of data have to be analyzed. To cope with such data volumes various suggestions have been made such as distributed computing [18], data reduction [22], increasing the processing power [23], data mining [24] by various contributors.

It is completely unknown to the forensic investigators where the clue for a crime lies. So, in a crime scene, they confiscate all possible sources of information. This leads to the collection of far more information than that is required. This exposes the private and irrelevant data private to the user of the digital device. The privacy of the suspect is jeopardized in such a scenario. To protect the right to privacy activists have advocated in the past and to a major extent, their advocacy stands true. But the forensic investigators have no option other than seizing all possible devices at the crime scene when their only aim is to lead the forensic case to confidence. To deal with such a sensitive scenario it is to the forensic investigators' judgment and expertise to decide which devices lead to minimizing exposure of the privacy of the suspect and at the same time to help lead the forensic investigation to confidence. The decision is critical but finding the balance is essential.

In this section, we discussed the evolution of digital forensics, the issues and the solutions that have been proposed by the scientific community to address the issues. It is worth noting that digital forensics finds its applications in a wider range of forensics applications such as cloud forensics. In cloud forensics, the underlying infrastructure is comprised of the devices that are essentially digital. Hence cloud forensics can be thought of as the juncture where the principles of digital forensics are applied in a cloud-based computing system. In the next section, we move our discussion to cloud forensics.

3 Cloud Forensics Till Date

Lately, there has been a shift in the paradigm of computing. The traditional in-house computing resources are being rapidly replaced by their counterparts in the cloud. This trend of growth is exponential, and the revenue generated by the public cloud is predicted to grow 17.5%, that is, from $182.4 billion in 2018 to $214.3 billion in 2019 [25]. Individual and business entities are opting for the cloud computing alternative because of the lucrative characteristics of cloud computing.

Characteristics of cloud computing [26, 27]—

a. **Resource Pooling**: The resource in the cloud is an aggregation of underlying infrastructure which appears to the users as a large pool of resources. To utilize every possible amount of hardware resource and to get the most out of each hardware resource the cloud implements the multi-tenant model, where a single hardware accommodates several cloud tenants. The resources are dynamically provisioned as per the needs of the tenants.
b. **On-Demand Self Service**: In the cloud computing system a client does not need the intervention of the cloud service provider to provide her with resources. She can order a resource to be provisioned as per her needs and the resources are at her disposal almost instantaneously.
c. **Rapid Elasticity**: To a cloud user, the cloud provides an illusion of endless resources. Because of this feature of cloud computing users' applications can scale rapidly with the capability to handle massive loads. When the need arises the resources are provisioned automatically and deallocated when there is no need for large resources.
d. **Pay-As-You-Go**: The cloud computing billing model is extremely flexible for its users. It only charges for the measured amount of resources that have been consumed by a client.
e. **Broad Network Access**: Cloud computing resources are accessible and manageable from anywhere on the globe over the internet. It supports the access and management of resources over a variety of devices such as laptops, desktops, tablets, smartphones, etc.

It is clear from the discussion of the characteristics of cloud computing, that it is an efficient alternative to the in-house model of computing. Given the rise in demand for cloud computing, there has been an agitating amount of research and development in this field. Technologies around the cloud computing have grown leaps and bounds. To our misfortune, the technological advancements in the field of cloud computing are being continuously exploited by some entities in our society. These adversely motivated entities are up to their gains either for monitory or for their ideology. Cloud computing, which potentially seems infinite in the computing capabilities are being exploited by these entities to launch attacks such as the Distributed Denial of Service (DDoS). Let alone, the cloud is a gigantic warehouse of data critical to industries and private to the human being. The adversaries are targeting these data stores to dig either financial information or to use this private information to blackmail the information owner. Assuredly, the cloud computing is the emerging technology that will flourish in the future. Such entities who intend to exploit the capabilities of the cloud computing for ill-purposes need to be framed. This need for the conviction and framing of these adversaries has led to the inception of the field of cloud forensics.

The National Institute of Standards and Technology (NIST) defines Cloud Forensics as—"the application of scientific principles, technological practices and derived and proven methods to reconstruct past cloud computing events through identification, collection, preservation, examination, interpretation, and reporting of digital evidence" [28]. Cloud forensics essentially aims at the reconstruction of the chain of criminal activities in the cloud from the pieces of evidence that are gathered. This helps the forensic investigators to gain insight into the crime scene. There exists no single forensic process model. Over time several forensic process models have been proposed to overcome the demerits of the preceding models. Four standard processes are followed in the forensic process, the standards are—"Digital Forensics Research Workshop (DFRW)", "National Institute of Justice (NIJ)", "National Institute of Standards and Technology (NIST)" and "Integrated Digital Investigation Process (IDIP)" [29]. In simple terms, cloud forensics is the application of digital forensics in the cloud environment. Cloud forensics is essentially digital forensics as all the devices in the cloud computing environment are digital apart from a few special kinds of digital devices such as the networking devices. The only way in which cloud and digital forensics differ is the applicability of the forensic practices on a distributed platform. This distributed nature of cloud computing makes the identification of pieces of evidence and data collection a challenging task as discussed later.

Manral et al. [30] have categorized cloud forensics based on the response to the incident. They have categorized the forensics activities as—a. "Pre-Incident" which has been termed by them as "Continuous Forensics" and b. "Post-Incident" which has been termed as "Post Incident Forensics". In essence, the pre-incident forensics aims at the forensic readiness of the cloud service providers should a breach happen. In such a forensic scenario, the cloud service providers continuously keep gathering the shreds of evidence that might turn out to be substantial clues in leading a criminal case in the cloud to confidence. Marco et al. [31] proposed an OVF (Open Virtualization Format)-based system for forensic readiness of the cloud. They proposed an OVF-based architecture where the OVF modules were deployed between the cloud architecture and the forensic readiness system. As the OVF applications can be executed on the virtual machines, they used the OVFs to convert the cloud data to an appropriate XML data format as required by them. They successfully collected various logs and monitored data from the cloud computing environment with their proposed architecture. This facet of forensics has led to the emergence of the notion of forensic friendly cloud that is referred to as "forensic-by-design" [32]. Post-incident forensics, on the other hand, is the one which deals with the forensic activities after the criminal activities have taken place on the cloud. It tries to gather the shreds of evidence from the artifacts in the cloud and tries to reconstruct the crime scene to gain insight into what exactly might have happened during a breach in the cloud. In such a forensic activity the logs generated in the cloud systems play a pivotal role as they provide key clues to the activities that have happened in the cloud plane.

The collection of the logs is a major challenge in cloud forensics. The very nature of cloud computing makes it to be globally distributed. The data centers can be located anywhere in the world at a given time. It is impossible for anyone but the cloud service provider to pinpoint the location of the logs or the data of a cloud client. Adding to the complexity of the distributed nature of cloud computing, different states in the same country may have different jurisdictions. When it comes to different countries, it becomes extremely difficult to coordinate the multi jurisdictions involved. In a forensic investigation even if the location of the logs or the data of the parties involved in the investigation is known, it is unknown whether the country where the data center is situated permits the export of such data for investigation purposes. This kind of multijurisdictional issue has brought many cases in the past to a complete halt. Moreover, the Service Level Agreement (SLA) of most cloud service providers prohibits cloud service providers from exposing the private data of their clients. Even if forensic investigators seek such data from the cloud service providers are not always obliged to comply with the requests. Such issues of multi-jurisdiction have led to rigorous research and development in the field of cloud forensics especially in the field of Log-Based Cloud Forensics. It is worth mentioning that almost all facets of cloud forensics would require logs to help reconstruct the trail of past activities that might have taken place in the crime scene. So the logs are of utmost importance. In the next sections, we discuss some of the major works that have been done by various researchers in the field of cloud forensics.

3.1 Tools Under Test

Cloud forensics and traditional forensics comprises the same set of activities. The only thing that differentiates cloud forensics from traditional forensics is the data collection. The distributed nature of cloud computing makes this an extremely difficult task for forensic investigators to get hold of the evidence data. Apart from the open-source alternatives, EnCase [33] by Guidance Software and FTK [34] by

Access Data, have been the seemingly most trustworthy tools for traditional forensics. These tools in addition to the capabilities of traditional forensics allow data acquisition from remote sources. Both EnCase and FTK includes the "client-server" feature, where an executable is deployed on the machines to forensically investigate known as the clients. The forensic software can then request data from the client machine over a communication channel. These tools have been around for a long time and have gained significant trust among the forensic community due to their ability to stand criticisms in the past. But there has been no study in the past that proves their effectiveness and correctness when applied in the cloud forensic scenario. Also, the Computer Forensic Tool Testing (CFTT) project of the National Institute of Standards and Technology (NIST) is responsible for testing and certifying forensic tools. The enterprise versions of EnCase and FTK have not been evaluated and certified by them.

To establish the effectiveness and correctness of the commercial tools such as EnCase and FTK in Cloud Forensics, Dykstra et al. [35] performed a set of three experiments. They performed the experiments on Amazon Web Sevices's IaaS setup. For the sake of the experiment and simplicity, it was assumed by the authors that the cloud service provider produced correct and unbiased results when their services were being used. The authors extracted the data from the cloud using the tools and then compared the remotely extracted data with the data that was generated from their control machine. Their control machine was a Dell workstation with 32-bit Windows 2008 R2, a single 30 GB hard disk drive and 2 GB of RAM. The forensic workstation of the authors was a single system with Windows 7 Enterprise 64—bit with EnCase Enterprise 6.11, SAFE (Secure Authentication for EnCase) installed as per the manufacturer's instructions and FTK 3.2. The forensic workstation was connected to the internet through a proxy.

For the first experiment, the authors provisioned an Amazon EC2 virtual machine. The virtual machine was a Windows 2008 R2 32-bit with 30 GB hard disk and 1.7 GB RAM. The Amazon firewall was configured by the authors to allow Remote Desktop Protocol only to the virtual machine so that the authors could interact with the EC2 instance they have deployed. Using the Remote Desktop Protocol, the authors installed Apache Web Server on the Amazon EC2 instance and create few web pages. Then they deleted a few web pages to imitate the activities of a malicious user on the webserver. They further accommodated the webserver with be-based vulnerabilities to make the machine appear more realistically compromised by an adversary. Once the remote machine on Amazon Web Services imitated a compromised system, they went to extract the system image remotely using the EnCase and FTK client executables. They transferred and executed the files using the Remote Desktop Protocol on the EC2 instance. The image transfer was successful. Post the analysis and comparison of the image with the control machine's image, the authors concluded that the tools EnCase and FTK work excellent. The images divulged all crucial detail of activities in the system that has taken place such as the webserver installation, web page creation, modification, and deletion. The authors also do not question the integrity of the data. The acquisition of the data took 12 h for EnCase and FTK for 30 GB image over the OC-12 connection.

In the second experiment, the authors experimented with Virtual Machine Introspection [36] which enables live forensic of virtual machines by injecting agents into the virtual machines. In this experiment, the authors set up their test system on a workstation with Eucalyptus [37] installed on Ubuntu. As the Eucalyptus system uses the Xen hypervisor, the authors used the LibVMI [38] library to inject the EnCase and FTK agents directly into the guest OS. Once the agents were introduced into the guest OS, the agents communicated with the EnCase and FTK tools over a communication channel. The authors found that in this experiment the images extracted were complete and the images provided a complete timeline of the activities in the system.

For the third and the last experiment, the authors tried testing Amazon's data shipping method that mostly in use in forensic cases presently where a warrant is issued for the data to be shipped and the order is complied with by the cloud service provider. Amazon only ships data that is stored in S3 buckets. But the authors had to ship the data that was attached to an EC2 instance which is an EBS volume. To the shipping possible, the authors attached the EBS volume to a Linux VM and used the dd command to create an image of the volume and stored the image created into the S3 bucket. Then they requested Amazon to ship the data stored in the S3 bucket. For the data to be sent to the authors, they sent a "Seagate FreeAgente SATA external hard drive" to Amazon. When the hard drive with the data was received the authors did not find any difficulties in loading the data to EnCase and FTK. Also, they verified the contents and concluded that the contents were untouched and completely intact. Rather to their advantage, the data shipped by Amazon came with a report of export in which the details of each file was present. Each file's entry contained the size, time-stamp, time of transfer, location on the device and MD5 checksum.

From the three experiments, the authors concluded that the EnCase and FTK tools stand the claims they make to be able to retrieve data in the field of traditional as well as cloud forensics. They do maintain data integrity. Virtual Machine Introspection is an extremely beneficial tool and the cloud service providers should look into it for implementation. Also, they concluded that the data shipped by Amazon uphold the integrity and can be trusted as far as their experiment is concerned. But considering the cost of the methods, the authors only suggest using Amazon's data shipping methods when the data to be transferred using the commercial tools exceeds 240 GB. Table 2 shows the summary of tools under test.

Table 2 Summary of tools under test

Experiment	Tool/method	Data acquired	Data integrity held
1	EnCase	Success	Yes
1	FTK	Success	Yes
2	EnCase	Success	Yes
2	FTK	Success	Yes
3	Amazon Data Export	Success	Yes

3.2 Forensic Readiness: A Proactive Measure

The over piling of cases in the forensic investigation has led the community to look for measures that would help reduce the pending cases as well as help upcoming cases to be solved in a minimum time frame. Most of the cases take time in data acquisition and data analysis phases. When it comes to cloud forensic, data collection is a critical thing. This scenario has led to the evolution of pro-active measures for forensic that would keep things ready to go for forensic investigations if not all, to a major extent. This concept of pro-activeness is decided by how forensic ready a given cloud computing infrastructure is. To our misfortune, none of the present-day cloud service providers comply with this critical concept of forensic readiness. The pro-active measure aims to continuously gather systematic data in the cloud computing environment. Should a criminal activity happen in the cloud, these gathered data then can be searched for crucial clues for the answers to the "who," "when," "how," "where" questions. Kebande et al. [39] proposed a Botnet-based [40] forensic readiness model that would collect data from cloud infrastructure. The author proposed a model where the botnet is used to infect the virtual machines in the cloud infrastructure and collect data from them. It is to be noted that the botnet here does not contain any malicious code instead of just mimic the behavior of botnets. Figure 4. It depicts the botnet model proposed by the authors. The authors propose that virtual services are provided by the cloud service providers. These virtual services are infected by the botnet which is "non-malicious" and the botnet harvests data from these digital services and stores them for future investigations. They have divided the model into two sections, the "front-end layer" and the "back-end layer." The "front-end layer" comprises the cloud services and it is where the botnet infects the services. From here the data is collected. The data collected by the botnet is then sent to the backend which comprises the IaaS (i.e. the servers, network and storage devices) and there the data is stored for forensic readiness.

The Digital Forensic Investigation (DFI) is the process that is carried out when a breach is detected in the cloud with the help of the data collected using the botnet that is stored in the backend.

The data collection mechanism proposed in this model has not been mentioned by the authors. They have not mentioned whether the cloud client is aware of the data collection or not. Data collection without the consent of the client will breach the privacy of the client at the same time the criminals need to be punished. So this is a topic of debate and more research is required in this area to what extent the stealth mode of data collection is appropriate.

3.3 Logging and Log Segregation in the Cloud

Logs are critical to cloud forensic investigations. They provide crucial clues for the reconstruction of the timeline of the activities in the cloud. In a crime scene investigation, it is of utmost importance for the forensic investigators to rebuild the

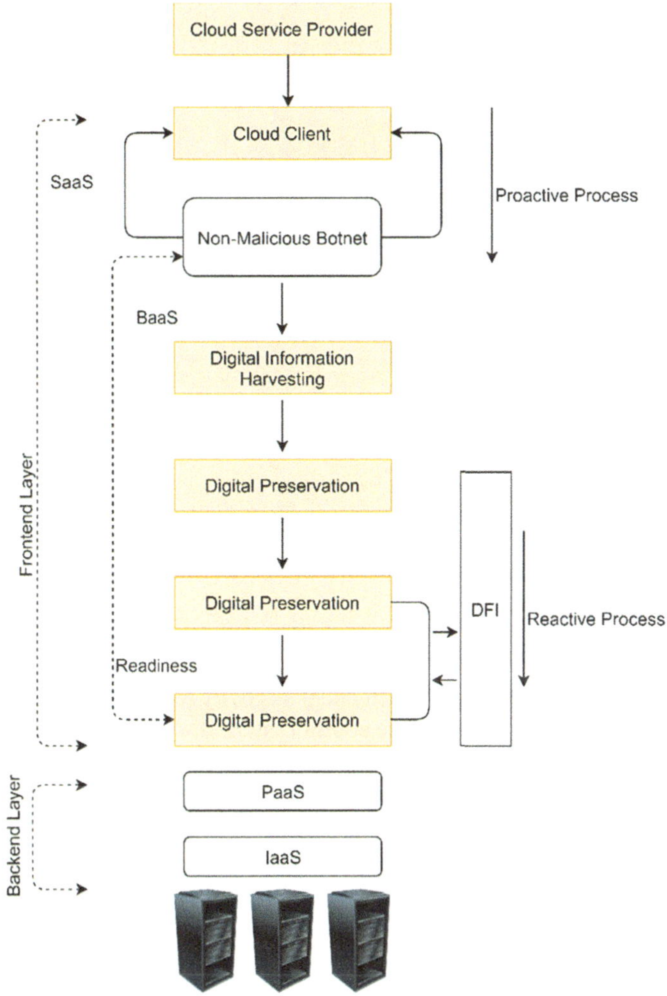

Fig. 4 BaaS: Botnet as a service model

crime scene to gain insight into the criminal activities in the cloud. Unfortunately, the logs being so crucial part of the investigation has not undergone any standardization of any kind. Neither they have undergone any standardization in their procurement, nor have they any standard of storage and representation. Because of this different cloud service providers adhere to their logging principles. This, in turn, makes it extremely difficult for the forensic investigators to comprehend different log formats if at all they get some. Moreover, logs often do not contain essential information about the activities that happened in a system. This is due to poor practices logging. In the cloud computing environment log segregation is another major challenge. In a multitenant environment, if logs are collectively collected for each client then it is

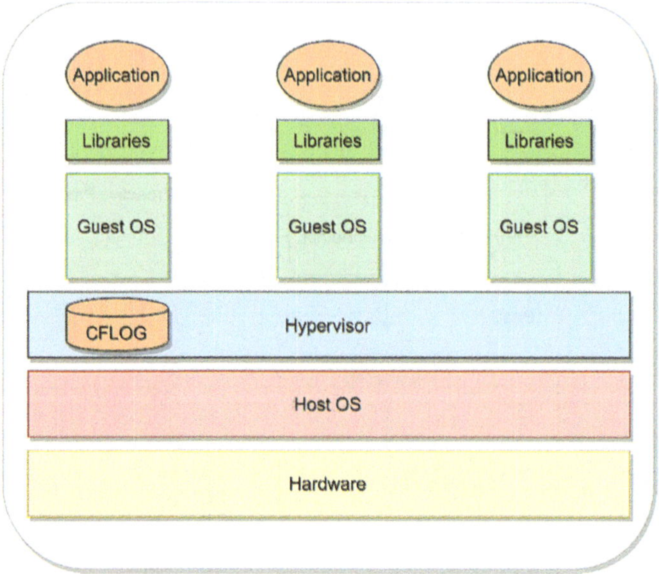

Fig. 5 CFLOG overview

extremely difficult to understand which client's activity led to what. Moreover, such tailored logs will not acceptable in the court of law and will be subject to suspicion.

To mitigate this problem of log handling, Pichan et al. [41] proposed a logging framework called the "CFLOG" depicted in Fig. 5. The designed system logs without the awareness of the cloud client thus increasing the integrity and trustworthiness of the logs generated by the proposed system.

As depicted in Fig. 5, the CFLOG in attached to the hypervisor of the cloud architecture. The hypervisor and the layers below the hypervisor are under the control of the cloud service provider whereas the layers above the hypervisor are under the control of the cloud clients. The authors claim that the proposed CFLOG system is highly scalable and can accommodate any number of logging parameters as might be needed in an implementation. The CFLOG system is capable of generating one or multiple log files per user per system. It also stores the log files in a pre-dedicated location in the cloud computing environment. The fact that CFLOG generates separate log files per user and application is highly beneficial because it makes it extremely time-saving and easy for the forensic investigators to grab the log files in a forensic triage. This significantly reduces the turnaround time of a forensic investigation and not to mention the effort needed in the investigation.

The authors further claim that they have designed CFLOG by keeping in mind the best practices of logging so that the logs generated by their proposed system can answer the six key parameters of an investigation viz. "who", "what", "when", "where", "how", and "why" things happened in the cloud environment.

The CFLOG system successfully addressed the problems of insufficient information capturing by the logs and log segregation in the cloud computing environment. The authors tested their proposed system using their own Cloud where they successfully gathered segregated logs from a multitenant cloud environment.

3.4 *Enhancing Reliability and Trust in Cloud*

The forensics of the cloud computing environment involves a great deal of trust. There is a need for putting trust in third parties for the forensic investigation to proceed. The logs that are collected from the cloud service provider itself has to be of great trust because there is no other option than putting trust in the cloud service provider as the cloud service provider is the only source of the cloud logs. Our aim is always to minimize this trust issue as much as possible. To help address this scenario Zawoad et al. [42] proposed the Open Cloud Forensics (OCF) model that enables the design of trustworthy clouds. The proposed model is depicted in Fig. 6.

The Open Cloud Forensics model enables the cloud of ceaseless forensics and at the same time converts the generated forensic data into electronically verifiable information. This information is would then later be used by law enforcement agencies to verify the integrity of the shreds of evidence presented to them.

The OCF compels the cloud service providers to store all data related to the services rendered by them. It is also mandatory for cloud service providers to securely store the data logged by them. They must be able to produce the data when they are asked for by the court. The cloud service providers expose the logged data through read-only APIs or web consoles. The forensic investigators use these APIs or the web consoles to collect data. They collect the required data and present it to the court in the specified format. The OCF ensures the segregation of the information

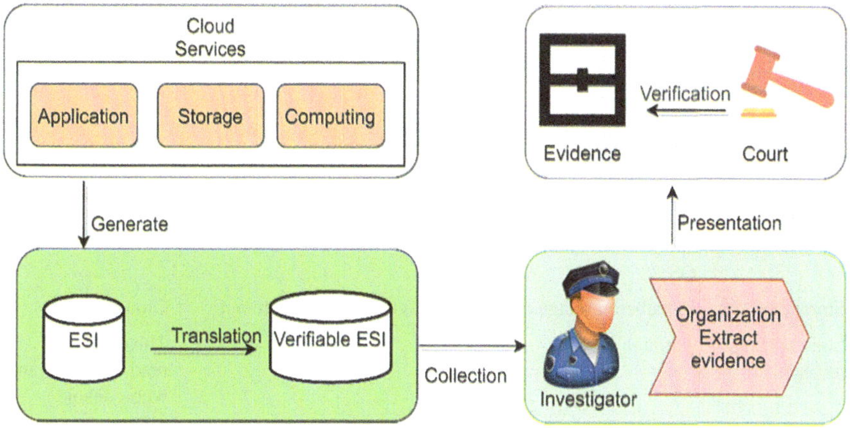

Fig. 6 Open cloud forensics model

collected and stored in the system. The host system determines which log belongs to which virtual machine and segregates the logs accordingly. It also ensures the integrity and security of the data stored in the OCF system. For the encryption of the logs, it enforces the public-private key cryptography thus protecting the crucial data from malicious insiders in the cloud service provider's organization.

To make the data published by the cloud service provider more reliable and trustworthy the OCF model implements a "Cryptographic proof" mechanism. In this mechanism, a cryptographic accumulator which is a One-way accumulator is used that stores a cryptographic value for the log files collected and stored by the OCF cloud system. As the cryptographic proof does not leak any data, at the end of each day, the cryptographic value of the accumulator is published on the internet which the cloud provider even cannot modify later. Table 3 gives a comparative analysis of cloud forensics literature.

The existing and trusted commercial forensic tools, the EnCase and FTK were thoroughly tested by Dykstra et al. [35] for their effectiveness and correctness when applied to the cloud forensics. The tools proved to stand the test and have demonstrated appreciable performance in the experiments performed by the authors. Kebande et al. [39] proposed a model where they tried to use the botnet to monitor the cloud services and gather data. But the botnet instead of being malicious is harmless. They are deployed on the front-end, that is, SaaS part of the cloud computing environment. They continuously gather data and store the collected data into the database in the back-end which is the IaaS part of the cloud computing environment. A similar model has been demonstrated by Pichan et al. [41] where they collect data for pro-active cloud forensics. But their approach is a bit different in the sense that they have

Table 3 Comparison of cloud forensics literature

	Dykstra et al. [35]	Kebande et al. [39]	Pichan et al. [41]	Zawoad et al. [42]
Contribution	Experimented and tested the correctness of the commercial versions of EnCase and FTK	Proposed Botnet-based pro-active cloud monitoring method	Proposed "CFLOG" a logging mechanism that segregated the logs and was highly scalable	Proposed Open Cloud Forensics model
Logging mechanism	Only those provided by applications and OS	Continuous	Continuous	Continuous
Storage of logs	In gathered image	Cloud storage	Cloud storage	Cloud storage
Log exposing mechanism	Through tools	N/A	N/A	Through read-only API and web-console
Security of logs	None	None	None	Private-Public Key Cryptography

developed a system called "CFLOG" which is embedded in the hypervisor of the cloud computing architecture. The CFLOG then collects data and stores the data in the dedicated location of the cloud computing environment. The CFLOG is capable of log segregation and is highly scalable. It can accommodate as many logging parameters as might be needed in a particular implementation. Zawoad et al. [42] proposed the Open Cloud Forensics model which demonstrates promising results and seems to ease the issue of trust involved in the cloud forensics process. The Open Cloud Forensics model makes the logging a continuous process and ensures the secure storage of the logs. The stored logs are verifiable by the court and are exposed to the forensic investigators through read-only APIs or web consoles only. This system also publishes proof of logs at the end of every day which the cloud service providers cannot modify thus making the process somewhat more transparent and trustworthy.

In this section, we discussed cloud forensics, the issues of cloud forensics and reviewed the works by the scientific community that has been proposed to address the existing issues. In the next section, we discuss and review the works in the field of IoT forensics which is essentially a combination of cloud forensics and digital forensics.

4 IoT Forensics: The Budding Field of Forensics

The Internet of Things (IoT) is the relatively new and emerging paradigm of computing by the virtue of which the objects/things of daily life are interconnected over a communication network and possess the capabilities to be controlled and monitored over the internet. This aggregation of interconnected devices over the internet is what we call the Internet of Things (IoT). The IoT devices typically comprise sensors that sense the environment, processes the sensed data based on predecided and stored algorithms and take actions that are applied to the environment for achieving a certain result. This capability of the IoT devices to sense the environment and act on the environment is what makes human life even easier. IoT has found its application in industrial automation, smart home systems, transportation systems, and agriculture, medical, to mention a few. The field of IoT is relatively new and is growing at a rapid pace. The market share of IoT is expected to grow to $14.4 trillion between 2013 and 2022 [43]. The sensors embedded in the objects/things that sense the environment collectively generate a massive volume of data. Moreover, the devices being of daily use such as smartwatches, smart home devices, etc., collect a lot of data that are private to the users. The data collected by the IoT devices attract the adversaries immensely. The IoT devices are designed to be cost-effective and hence they are usually small scale devices with limited onboard resources. These devices are not built on a security-first approach. Had they been built with the security-first approach, this would add up to the cost and bulk of the device [44]. This makes the IoT devices even more vulnerable to attacks by the adversaries. It is a research challenge to optimize the limited onboard resources to look after the IoT devices' security aspects too. The consequences of an attack can be imagined if an adversary breaks into a medical

IoT system network and gets control of the critical patient monitoring systems. Such an attack will lead to devastating effects let alone the private patient data that gets exposed due to the breach. Another such event can be imagined in a scenario where the IoT systems are being used for critical industrial system monitoring. If the systems are made to fail by the adversaries this will lead to loss of lives and result in a blow to the economy. Ling et al. [45] in their work performed extensive research on the Edimax smart plug and successfully found vulnerabilities in the smart plug that was being sold in the market. They successfully reverse-engineered the communication protocol that was being used in the Edimax smart plug and then could launch four attacks on the smart plug. They launched "device scanning attack", "brute force attack", "device spoofing attack" and "firmware attack". Then they implemented these attacks on the smart plug. They found that they were able to discover all online smart plugs through their device scanning attack. The passwords of the smart plugs were found by launching the device spoofing attack and the brute force attack on the smart plugs. Even root access of the smart plug was possible with the firmware attack. In their work, they demonstrated the attacks and provided guidelines to secure such devices. It is evident that as the IoT systems ease the processes in our daily walks of life, at the same time their misuse might bring a disaster. Hence their security is of utmost importance. Should a breach related to IoT systems happen a proactive investigation into the crime scene is essential to gain insight into the events to patch the vulnerabilities of the existing systems and to understand the level of planning of the adversaries. Like all disciplines of forensics, forensics of the IoT systems is required. In the next sections, we discuss the concepts of IoT forensics, next we propose our IoT forensics model, then we discuss the challenges in the field of IoT forensics and we discuss some solutions to the challenges in the field of IoT forensics.

The IoT devices are deployed in conjunction with other internet-enabled services such as cloud services. The IoT devices exchange data with these services through Application Programming Interface (API). The sensors embedded in the IoT devices sense the physical environment may or may not process the sensed data depending on the deployment and the capability of the IoT hardware and then the data is sent to the cloud. The cloud, which is virtually infinite in capabilities processes the data sent by the IoT device and then returns the result to the IoT device. The IoT device upon receiving the result from the cloud takes the appropriate action which might be specified in the returning result of the computation done by the cloud or an independent decision can be taken by a predefined algorithm stored in the IoT device. It takes a significant amount of time for a round trip of the data from the sensors of the IoT devices to the cloud and back to the IoT devices. For typical applications such as in agricultural IoT, smart homes, etc., this delay is perfectly acceptable. But consider the scenario where an IoT system is monitoring a patient and the data is being processed in the cloud. The decision taken in the cloud decides the next course of action to be taken by the patient monitoring system. In such a scenario a tiny deviation from the accepted delay might result in the loss of lives in the worst case. So it is evident that in such latency-sensitive scenarios a simple exchange of data between the IoT device and the cloud is not feasible due to the latency sensitivity of some applications. To deal with such latency-sensitive applications the concept of

Edge Computing [46] has been introduced that drastically brings down the latency by bringing the cloud to the edge of the network. So, unlike the traditional fields of forensics, the field of IoT forensics is an amalgamation of the enabling technologies behind it. IoT forensics essentially includes—Digital forensics, Network Forensics and Cloud forensics. In the next section, we discuss IoT forensics in greater detail.

4.1 Forensics of the Enabling Technologies of IoT

As discussed in the previous section, IoT devices alone are seldom used. IoT is the amalgamation of the enabling technologies acting behind it. The IoT technology comprises the IoT hardware, the network infrastructure over which the IoT devices communicate and gets connected to the internet, and the cloud computing services. So for the forensics of the IoT, we necessarily need to deal with the forensics of the underlying technologies too. So IoT forensics comprises—Digital forensics, Network forensics, and Cloud forensics [47] as depicted in Fig. 7.

Digital forensics in IoT forensics is required because the IoT devices used are all based on digital technology. These devices perform the sensing operations and limited computations using their limited local memory. The forensic investigator aims to collect as much information can be collected from the IoT devices. But the major problem is that since the IoT devices have very limited memory, they do not store information for longer durations. They usually overwrite the information in the memory in some time. So the crucial information for the forensic investigators is lost once they get overwritten. Forensic investigators need to get the non-overwritten data from the IoT device but that is a major challenge. One solution to this problem of overwriting is storing the data in some persistent storage such as the cloud. But this brings in even more challenges related to the integrity of the data.

Network forensics deals with the vitals logs generated by the network traffic generated due to the communication of the devices. The network logs have the potential to pronounce a network activity to be legitimate or otherwise. The IoT devices can be present on various networks such as the Wide Area Network (WAN), Metropolitan Area Network (MAN), Local Area Network (LAN), Personal Area Network (PAN), Hospital Area Network (HAN), etc.

Most of the critical activities of the IoT take place over the cloud. Hence it is where most of the vital clues of the IoT activates can be found. Since the IoT devices are low on resources they offload most of their computing tasks to the cloud via the network. Thus, the forensics of the cloud is a substantial part of IoT forensics.

As discussed in Sect. 4, IoT applications such as in medical IoT and applications in aerospace industries that are extremely sensitive to the latency cannot survive the latency of the cloud services. In such scenarios, edge infrastructures are used. When additional technology stack gets added to the IoT stack, the forensic of such stack becomes an overhead and needs to be carried out. In such latency, sensate IoT applications forensics of the edge devices needs to be carried out additionally. Rather it is the edge devices in such applications where most of the computations are done

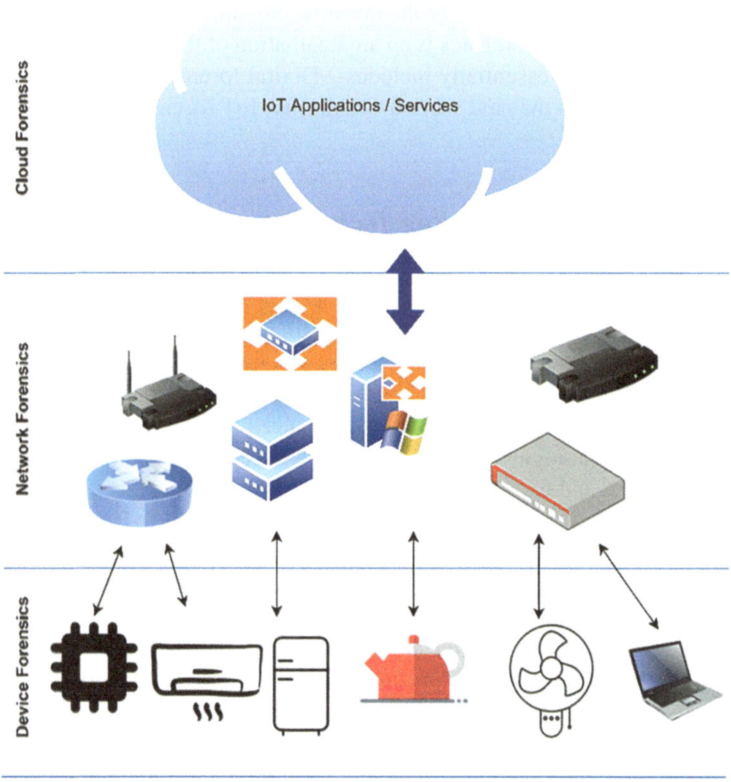

Fig. 7 IoT forensics

and the cloud receives on the filtered data that is sent by the edge system. Hence the forensic investigation of the edge system is essential even if it is an overhead. The edge device can vary depending on the application and deployment of the IoT infrastructure. In small scale applications, the edge devices can be simple routers and gateways and in industries, it can be edge servers.

4.2 Challenges in IoT Forensics

IoT devices being limited in resources make use of several enabling technologies to establish a working environment. This, in turn, results in the inheritance of the demerits of the enabling technologies along with their merits. As IoT forensics includes Digital forensics, Network forensics, and Cloud Forensics, it faces the existing challenges faced by the forensic investigators in these fields of forensics.

Volume: The explosion of IoT devices has led to tremendous data generation at endpoints. There has been an exponential growth in the data being generated by the IoT devices since its inception [48]. The vast amount of data needs to be handled efficiently to complete the IoT task which is a major challenge. In a digital forensic scenario, the forensic investigators are already dealing with an unprecedented number of pending cases due to the huge amount of data they need to investigate on an average per case basis. With the advent of IoT, this amount of data has grown exponentially and is going to worsen the forensic investigation scenario and lead to more pending cases with higher case turnaround times. The forensic investigation community is left with no option but to look for better alternatives to reduce this data volume and look for methods that increase the overall speed of the investigations.

Variety: IoT devices have undergone tremendous innovation in a short span of time. It is challenging for the investigators to identify the IoT devices in a crime scene as every possible object/thing in the crime scene can be an IoT device. IoT devices can be embedded in refrigerators, almirahs, and cupboards, to mention a few. It is crucial for the forensic investigator to carefully identify the IoT device in the crime scene for further investigations. Even if the devices are identified, it is a challenge for the forensic investigators to carry the device to the forensic laboratory. It is especially difficult to find devices that are hidden and are very small in size, such as hidden cameras. The variety of such devices is vast.

Moreover, devices from different manufacturers are of different architectures and have their firmware. Forensic investigators often find it difficult while dealing with such a huge variety of devices with different architectures and firmware. Thus often requiring a forensic investigation of a team of forensic personnel. Additionally, the data formats of these IoT devices are not standard across all devices. Each device follows its data and file formats. This brings in additional challenges. Moreover, that generated data in the IoT devices and that are stored in the cloud vary greatly. This is because the data in the cloud is processed upon receiving from the IoT counterparts. But for data in the cloud to be acceptable in the court of law has to be in the original format [49]. This again complicates the IoT forensics process.

Multi-Jurisdiction: As cloud services are a significant part of the IoT stack, the IoT forensics is also jeopardized by the existing issues of cloud forensics. The multi-jurisdictional issue in cloud forensics remains unaddressed. Since the IoT data is sent to the cloud, it gets stored in some storage service provided by the cloud service provider. The distributed nature of cloud computing makes it inevitable for the data to be stored somewhere in the globe in the data center of the cloud service provider. Should a forensic investigation happen on this data and the data is under some other jurisdiction, it will be very difficult for the forensic investigators to get hold of the data as the cloud service provider is not always obliged to provide access to the data. Moreover, even if a warrant is issued for the data and the cloud service provider is forced to provide the data, the integrity of the data solely depends on the cloud service provider. This, in turn, brings in the complexities of putting trust in a third party.

Short-Lived Data: The IoT devices which are built to be application-specific and are limited in resource, make use of the available resources optimally for their operation. One such optimization is in terms of their local data storage. IoT devices do not store much data locally. They store the data transiently and overwrite the data once the data is of no use. This happens over and again in the IoT device. The data in the IoT device that are meant to be stored are sent to the cloud for permanent storage. But for the forensic investigators, this local data is of much significance. This kind of volatile data can either be saved to the cloud or some other storage in the network. But this movement of the data to other storage poses the additional challenge of securing the chain of evidence and the proof that the moved data has not been altered in any form [50].

Poor Security of IoT Devices: As discussed in Sect. 4, the IoT devices are not designed on a security-first basis. They are designed to be cost-effective and to be light in terms of energy and form factor. This lightness of form factor and cost results in the limited onboard resources of IoT devices. The onboard IoT resources can get overburdened when they are made to handle the security aspects or sometimes, they are simply not capable of handling the security aspects of computing. Thus, making the IoT devices vulnerable and easy target for the adversaries. Moreover, most device manufacturers do not patch the existing bugs in their firmware and release updates. This lack of effort for IoT devices even leaves the legacy devices more vulnerable. Thus, there is a stringent need for regular bug fixes in the device firmware and regular release of updates for the IoT devices. On the hardware level, the devices need to be designed to utilize the onboard resources optimally and secure the device from vulnerabilities.

4.3 An IoT Forensics Model

IoT forensics involves not only the forensics of the IoT devices which is digital forensics but also involves forensics of the associated technologies such as the network and the cloud services. Typically, in digital forensics, the stages of forensics include identification, preservation, analysis, and presentation.

In the identification stage, the IoT devices such as the smart home systems, sensors, and other such related devices are identified for forensic investigation. In the preservation stage, the data from the devices are acquired from the identified devices using specialized tools and preserved in a forensic sound manner. The analysis stage is where sophisticated tools such as EnCase and FTK are used to analyze the acquired data from the devices. In the presentation stage, the forensic investigators present their findings in detail to the authority. Li et al. [51] in their work have proposed an IoT forensic model that is depicted in Fig. 8. The model begins with the classification of the IoT device into "IoT as a target", "IoT as a tool", and "IoT as a witness". Further, the devices are then forensically investigated along with their associated applications. Finally, the shreds of evidence are stored in a repository in an encrypted form.

The variety of IoT devices and networks can puzzle the forensic investigators and mislead them. It is essential for them to correctly identify the relevant sources of evidence from the crime scene for an optimal forensic investigation to proceed. The

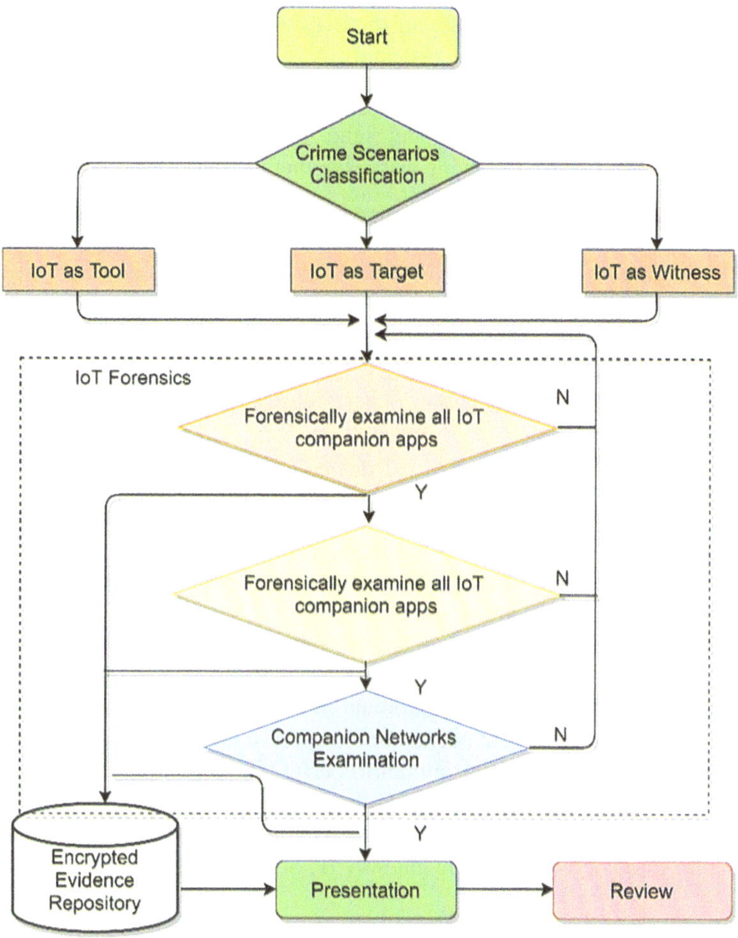

Fig. 8 IoT forensics model

authors have argued that it is essential for the type and severity of the crime to be decided before gathering the shreds of evidence for effectiveness as per their model of IoT forensics. They argued that if the forensic investigator knows the severity of the crime then she can decide on the resources to be spent on the forensic investigation, the methods to be used for data acquisition, the laws, and regulations. They have categorized the IoT devices in three categories such as—

a. "IoT device as a target": These are the devices that are limited in resources and capabilities that have less secure systems onboard. This limited or no security on them makes them vulnerable and an easy target for the adversaries.
b. "IoT device as a tool": IoT devices that are used for launching other malicious activities such as the botnet attacks.

c. "IoT device as a witness": These are the IoT devices whose stored data can serve as a witness for the forensic investigators. Amazon Echo is a very good example of such a device.

The next stage in the proposed model is the "IoT device Identification" where the IoT device is identified by following a six-step method that—

a. Identifies IoT devices associated with a crime
b. Identifies the time span for which the device needs to be forensically investigated to get clues related to a crime
c. Identifies the confidentiality, authentication, authorization, etc., for the identified device
d. Defines the data that the identified IoT device might produce
e. Identifies the network associations of the identified IoT device and isolate it from all network associations
f. Identifies access to the identified IoT device.

The next step in the proposed forensic model is "Evidence Preservation". It is where the data from the identified IoT device is acquired using a wide variety of tools. The primary source of vital information resource-limited devices such as the IoT devices are physical memory, page file, etc.

The final step in the proposed IoT forensics model is the "IoT Forensic Analysis and Presentation" step where the artifacts extracted from the devices are thoroughly investigated by the forensic investigator and their finding are reported in a detailed manner to the law enforcement agency or the court of law for further proceedings.

The authors further used their model to perform a forensic investigation of the Amazon Echo device. They were successful in their experiment and they successfully extracted artifacts such as Calenders, Lists, Alarms, Usage history, Books, Music, device connection details including gateway IP, MAC address, server address, etc.

4.4 A Decentralized IoT Forensics Framework

A centralized system that stores data is vulnerable. It is prone to critical system failure and as a result, may lose vital data permanently. Moreover, such data stores attract the adversaries for an attack on those systems. Additionally, the service that is managed by a third party and is opaque, poses problems of reliability. Such systems constantly are questioned about their privacy and integrity preserving principles. Applicability of such a central system in a forensic investigation is no exception. Rather it makes the whole forensic process even more contentious. Ryu et al. [52] proposed a Blockchain-based investigation framework for IoT forensics. The authors have identified the requirements of digital forensic investigation—

a. Integrity: The integrity of the shreds of evidence that are to be submitted to the court of law is of extreme importance. Without the integrity of the pieces of evidence the case can get mislead and the wrong person might get framed.

b. Non-repudiation: The forensic investigators are responsible for the pieces of evidence that they submit. The forensic investigators are liable for the investigation's outcome.
c. Mitigate single point-of-trust: The single-point-of-trust is to be mitigated to the most possible extent. There is an utmost need for transparency and reliability of each entity involved.
d. Forensics readiness: The IoT systems must have measures to store historical data that help in the forensic process.
e. Lightweight System: The IoT systems should be free from overheads.

The IoT systems in the proposed Blockchain-based system store the generated data that is being communicated in the Blockchain in the form of a transaction as depicted in Fig. 9. As depicted in the figure, sensors, smart cars, smart buildings, smart industry, smart homes, and smart grid are the components of the IoT environment. These components can get attacked by an adversary. Thus the forensic investigation of these components is essential. The authors argue that for the existing methods of IoT forensics it is difficult to handle the interaction among so many IoT components. They further state that their proposed Blockchain-based IoT forensic model can handle the complexity of the interactions as the Blockchain inherently provides integrity and chain of custody to the forensic data.

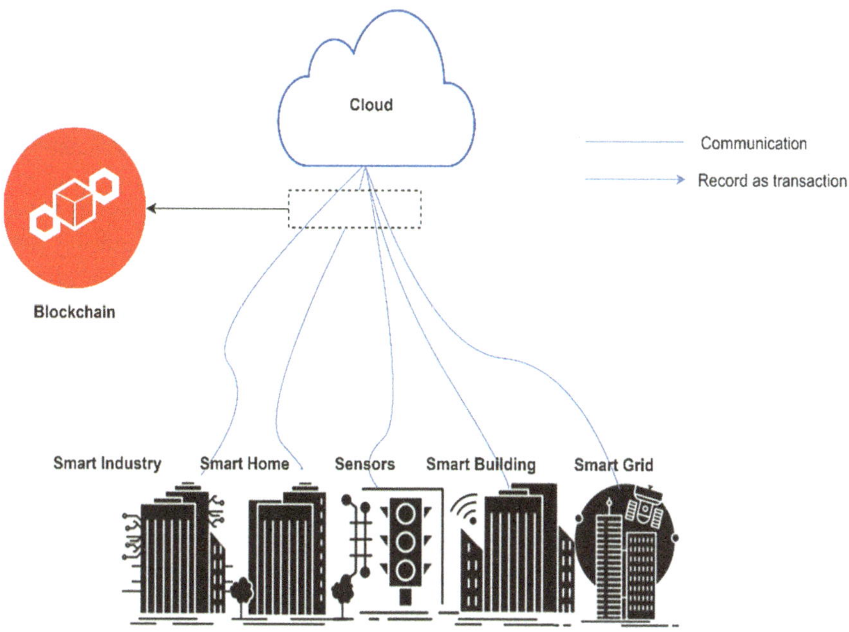

Fig. 9 Blockchain-based IoT forensics framework

The authors have developed their block structure for the proposed system which is different from that of a traditional block structure. The blocks are divided into two sections—a. Block Header and b. Transaction.

The Block Header is further divided into—Block Number: Used to number the blocks in a sequence, Merkle tree hash: Used to find transactions, and Timestamp. The timestamp records the block creation time.

The Transaction is further divided into—Transaction ID: Unique number to identify a transaction in the blockchain, Digital Signature: It is the signature generated using the combination of PUF ID and primary key of sender device, PUFsrc: PUF ID of the device sending the data, PUFdst: PUF ID of the device receiving the data, and data: the message that needs to be communicated.

The participants of the blockchain network are—User of the IoT device, Manufacturer of the IoT device, the forensic investigator and the cloud service provider. The blockchain data modification needs the consensus of all the participants in the blockchain network. The authors have addressed the issue of data integrity in this work and they have proposed to solve the problem of data privacy issues due to the blockchain's distributed ledger in their later work.

4.5 IoTFC: A Blockchain-Based IoT Forensic Framework

As discussed in Sect. 4.4, a distributed system such as the Blockchain system is preferred to a central system because the Blockchain-based distributed system has many advantages over the central system. The advantages include better asymmetric encryption, integrity, transparency, provenance, fault tolerance, immutability, consensus, and distributed trust. In any forensic investigation scenario, there is a pressing need for the integrity of the shreds of evidence and the elimination of trust in third-party by keeping the process as transparent as possible. IoT forensics is no exception. To fulfill these pressing needs in the field of IoT forensics Li et al. [53] proposed a Blockchain-based forensics framework for IoT. The proposed system was called "IoTFC" by the authors. The authors mentioned in their work that their proposed system achieved—

a. **Extensive Evidence Visibility**: Each evidence in the Blockchain-based system can disclose its source and co-relation with other pieces of evidence.
b. **Integrity**: Maintaining the integrity of the data being handled is a major challenge and it is one of the main objectives of the IoTFC system.
c. **Immutability**: The IoTFC system exploits the inherent immutable nature of the Blockchain system to ensure the immutability of the data stored in the system. Blockchain technology also enables the audit ability of the data being stored in the system. The data that is being stored in the Blockchain-based system is tamper-proof due to the immutable nature of Blockchain. Moreover, for any editing of the data, there needs to be a consensus among the participants of the Blockchain.

d. **Provenance**: In forensic investigations, the shreds of evidence have severe implications. The IoTFC system enables the forensic investigator to provide a full history of the pieces of evidence to the court of law for natural justice.

The data generated in the IoT environment is massive. It can easily overwhelm the forensic investigators and thwart the forensic investigation process significantly. The authors have used their previously proposed features-based devices fingerprinting methods [54] that identify the devices associated in a case and hence narrow down the devices to be investigated as a data source. The proposed system architecture is depicted in Fig. 10. The proposed system enables the user to store the forensic investigation events, findings and associated information in the Blockchain-based system. The recorded data is visible to all the participants of the Blockchain network. By nature, the Blockchain makes the proposed system cryptographically strong, immutable, resilient, provides shared trust, traceable data and timestamped data.

The proposed system consists of—

a. IoT Devices and Users: The users of the IoT systems can be the user, the forensic investigator, and the device owner. The devices include all possible IoT devices.
b. Merkle Tree: It is a tree of hash that allows the verification of the data stored in the Blockchain system. The Merkle tree generates a digital signature based on the contents of the system. Thus making it possible to verify the contents and decide whether to include a block in a transaction or not.
c. Block: It is the data structure where the actual data is stored in the Blockchain system. The IoTFC system includes the following in a block—"pre block hash", "version", "nonce", "timestamp", "block state", and "Merkle root". The "TE" is the list of pieces of evidence stored and the evidences is hashed into a Merkle tree.

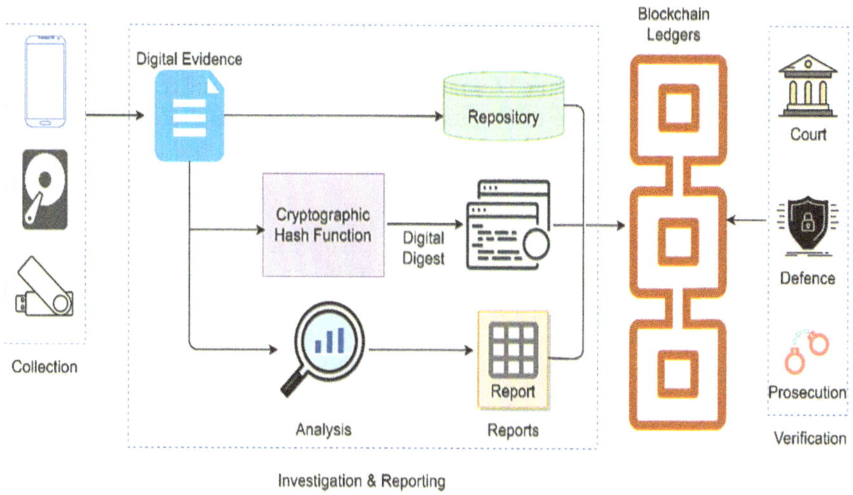

Fig. 10 IoTFC forensic framework

d. Smart Contract: It is a code that gets executed in the Blockchain when a pre-set condition gets fulfilled. Thus automating the processes in the Blockchain without the need for some third-party. The smart contract enables the IoTFC system to be autonomous, trustworthy, safe, fast, cost-effective and accurate.

The authors have further graded the shreds of evidence in a forensic investigation based on their difficulty as—

G1: Easily identified pieces of evidence such as plain text files, images, QR codes that are unencrypted.

G2: Deliberately hidden such as shreds of evidence for which the files extensions have been changed.

G3: Hardly identified pieces of evidence such as plain text in locations other than the files system and slack space.

G4: Pieces of evidence that are difficult to be identified such as data that is under encryption.

G5: Very difficult shreds of evidence such as encrypted files in a location other than the file system.

The authors have mentioned that all forensic investigation events, findings, and related information are stored in the Blockchain and is shared among all the participants of the Blockchain. The smart contracts manage the records of the transfer of pieces of evidence, the present state of the shreds of evidence, the permission level on them, the data and the time. All-access to the evidence also is recorded in the Blockchain-based system by the invocation of the smart contracts. Table 4 gives a comparative analysis of IoT forensics literature.

An IoT forensic model was proposed by Li et al. [52] that stored the acquired pieces of evidence in a central repository as compared to the works of Ryu et al. [53] and Li et al. [54] who implemented their IoT forensic frameworks using Blockchain that stored the pieces of evidence in a distributed fashion. None of the systems but the system proposed by Li et al. [54] recorded all access to the shreds of evidence stored in the Blockchain-based system using smart contracts. Only the systems proposed by the authors Ryu et al. [53] and Li et al. [54] provided forensic readiness, that is, they continuously logged the IoT systems and saved the crucial details that might prove to be beneficial in future forensic investigations.

Table 4 Comparison of IoT forensics literature

	Li et al. [51]	Ryu et al. [52]	Li et al. [53]
Contributions	Proposed an IoT forensic model	Proposed a Blockchain-based IoT forensic framework	Proposed a Blockchain-based IoT forensic framework
Logging mechanism	None	Continuous	Continuous
Storage	Central	Distributed	Distributed
Storage mechanism	Traditional	Blockchain	Blockchain
Record of evidence access	None	N/A	Yes

5 Open Areas of Research

From the review work undertaken by us, we found that there has been a tremendous development in the recent years in the field of Digital, Cloud and IoT Forensics but still there are several missing crucial pieces of research work that needs to be done to address the existing issues in the field of Digital, Cloud and IoT forensics. We found that Digital forensics has comparatively matured with time than cloud and IoT forensics. Whereas, there is a lot to be done in the field of IoT forensics. Cloud forensics occupies a position in the middle of very new and maturity.

5.1 Enhancement of Digital Forensics Tools

The current state-of-the-art tools of digital forensics such as the FTK and EnCase have been providing promising results in the analysis of the digital data. But they are incapable of detecting the atypical data that in the digital shreds of evidence as a result of the deliberate attempt of the suspect trying to hide information by altering file extensions and so on. Moreover, the timeline of activities in the crime scene needs to be plotted by the forensic investigator from the data analyzed by the tools. This timeline generation activity is critical and time-consuming. Had the tools helped in the timeline creation in any way, it would result in a significant reduction of the forensic investigation turnaround time. Thus, there is a lot to be enhanced and done in terms of tool development for digital forensics. There is an utmost need for these tools to be more autonomous and intelligent to be of more help to the heavy data-intensive task of digital forensics in the modern world.

5.2 Distributed Computing and GPU-Based Digital Forensics

Performing digital forensics activities in a workstation has been the practice so far in the forensics community. This, in turn, has restricted the growth of the digital forensics investigation capabilities and has been largely remained bound to the capabilities of the workstations on which the forensics are usually performed. The data to be investigated has been growing exponentially whereas the workstations' capabilities have not witnessed that much of technological advancement and have not been able to keep pace with the rapid growth of the data. To address this issue of single workstation's limiting capabilities; researchers have proposed and have demonstrated the use of distributed computing and the exploitation of the massive power of the graphical processing units in the computers. Undoubtedly this model of computing in the field of digital forensics will provide promising throughput. But the researchers demonstrated the use of not stressed the effectiveness of the modification of the tools to harness most of the computing power of the distributed computing. Not all tools are designed to work on multiple CPUs, that is, in a distributed system. Though the existing tools have proved to show exceptionally well results in the distributed and GPU-based digital forensics scenario but there is a lack of performance study of the specially designed tools for harnessing the power of such special architectures of computing. Moreover, the studies have been performed in a laboratory environment and hardware. So, there is a need for the more real-world implementation of the architectures with commodity hardware.

5.3 Enhancement of Physical Memory Forensics

This facet of digital forensics needs severe research and development. Digital forensics has been there for quite a long time now. This field of forensics has witnessed tremendous research and development by the academia and the industry. The commercial tools such as FTK and EnCase have there been for a very long time which provides promising results. Despite the research and development, the extraction of data from physical memory is still a challenge today. The forensic investigators still face numerous challenges while performing live forensics on digital systems. Efficient methods and tools are not at their disposal for the extraction of volatile information from the digital systems. There is a pressing need for efficient tools and methods for such data retrieval from digital systems as the volatile data in the digital systems might provide a breakthrough in the forensic investigation cases which the data in the file systems cannot. Moreover, proper development of guidelines is also essential for forensic investigators to help the forensic investigator take an informed decision in the crime scene whether to just pull the plug of a live system or not. Without these guidelines, it is obvious that crucial data will be lost.

5.4 Reduction of Cloud Data Acquisition Time in Cloud Forensics

The digital forensics tools such as EnCase and FTK have proved to be successful in extracting evidential data from the cloud devices and have demonstrated to preserve the integrity of the extracted data. But the time that was required for the data to be extracted even over a very fast internet connection is disappointing. This much time consumption is a setback for faster settlement of forensic investigations. It is an overhead for the forensic investigation process. Moreover, the data shipping method from Amazon took even more time and was high cost intensive. Thus, this field of remote data acquisition by the tools needs more research and development to find better ways to transfer the files over the internet. One possible research direction in this regard is whether the cloud data that is being acquired can be compressed in any form and then transferred. Again, the compression process brings in the challenge of evidence data alteration. So, this needs to be extensively researched and studied whether such schemes are possible or not and if alternatives are possible to reduce the remote data acquisition times.

5.5 Privacy Preservation in Blockchain-Based IoT Forensics

The systems studied in this review only takes care of the integrity of the data that are being stored in the Blockchain-based system. They have also taken care of the chain of custody of the shreds of evidence. But as mentioned by the authors, the blockchains share the data stored in them among all the participating members of the Blockchain network. This raises a potential issue. Not all data from the IoT system that is being generated needs to be shared among all the participants at all times. This will result in the privacy problems of the IoT system user. This issue of the Blockchain-based systems needs to be researched and enhanced further for a better and robust system that takes care of all the aspects of the data. Further, the Blockchain consensus algorithm is needed to be researched. Blockchain systems are one of the highest consumers of power. So, it needs to be researched whether the application of a Blockchain-based system is feasible in such an application or not. If applicable, then which consensus algorithms are applicable in such IoT forensics frameworks are applicable needs to be further researched.

5.6 Preservation of Overwritten Data in IoT Systems

The IoT systems have limited resources onboard. This compels them to optimally utilize the onboard resources. Because the local storage of the IoT is very small so they do not store much information locally for longer durations. They only store the local data for a certain time and overwrite the data once the data is no more required. This results in the loss of vital information from the point of view of the forensic investigator. As the data that is being overwritten can provide tremendous clues to what is being executed and performed by the IoT system. One solution to save the overwritten data is to save it to some remote data store or the cloud. But this relocation of the data from the IoT system brings in the challenge of integrity preservation. This facet of the IoT system needs to be researched extensively to make the IoT systems forensic ready and help the forensic investigators in solving the future cases.

5.7 Document File Meta Data Removal

The works reviewed in this work have proved that no tool has proven to completely remove the metadata associated with the document files such as the MS Word files.

The metadata poses a potential threat to unintentional data leakage by an uninformed user. The metadata is capable of leaking a considerable amount of data about a system and the user. This is a threat to the privacy of the user. There is a need for research in this field on how to limit and regulate the metadata in these files. A user should be able to control the data she exposes. More efficient and easy to use tools need to be developed that remove the metadata from these files as per the users' will.

6 Conclusion

We have performed a thorough and systematic review of the literature in the fields of Digital, Cloud and IoT forensics. We have compared and contrasted the efforts and the outcomes of the works of the scientific community; we have also summarized the state-of-the-art in the field of Digital, Cloud and IoT Forensics. We have been able to fulfill our aim of this work which was to create a thorough review that would help fellow researchers in thought association for getting a clearer picture of the current state of research on Digital, Cloud and IoT Forensics and find the research gaps.

References

1. https://www.statista.com/outlook/251/100/consumerelectronics/worldwide#marketglobalRevenue
2. dfrws digital forensics definition
3. Reith M, Carr C, Gunsc G (2002) An examination of digital forensic models. Int J Digit Evid 1(3)
4. Rogers MK, Goldman J, Mislan R, Wedge T, Debrota S (2006) Computer forensics field triage process model. J Digit Forensics. Secur Law 1(2)
5. Mohay G (2005) Technical challenges and directions for digital forensics. In: Proceedings of the first international workshop on systematic approaches to digital forensic engineering (SADFE'05). IEEE (2005)
6. Richard GG, Roussev V (2006) Next-generation digital forensics. Commun ACM 49(2)
7. Marziale L, Richard GG, Roussev V (2007) Massive threading: using GPUs to increase the performance of digital forensics tools. Digit Invest: Int J Dig Forensics Incid Response. ACM
8. Harris R (2006) Arriving at an anti-forensics consensus: examining how to define and control the anti-forensics problem. Elsevier
9. Kessler CG (2007) Anti-forensics and the digital investigator. Aust Digit Forensics Conf
10. Johnson NF (1998) Exploring steganography: seeing the unseen. IEEE
11. Castiglione A, Santis AD, Soriente C (2007) Taking advantages of a disadvantage: digital forensics and steganography using document metadata. J Syst Softw 80(5):750–764
12. https://www.microsoft.com/en-in/
13. https://www.techrepublic.com/article/clean-potentially-harmful-metadata-from-office-documents-with-ezclean/
14. Casey E, Stellatos JG (2008) The impact of full disk encryption on digital forensics. ACM SIGOPS Oper Syst Rev. ACM
15. Regional Computer Forensics Laboratory Program Annual Report FY2007 (2007) US Federal Bureau of Investigation. www.rcfl.gov/downloads/documents/RCFL_Nat_Annual07.pdf

16. Roussev V (2009) Hashing and data fingerprinting in digital forensics. IEEE Secur Priv 7(2)
17. https://www.nist.gov/itl/ssd/software-quality-group/national-software-reference-library-nsrl
18. Roussev V, Quates C, Martell R (2013) Real-time digital forensics and triage. Digit Invest 10(2)
19. Roussev V, Richard GG (2004) Breaking the performance wall: the case for distributed digital forensics. In: Proceedings of the 2004 digital forensics research workshop, vol 94
20. Vincze AE (2016) Challenges in digital forensics. Police Pract Res 17(7)
21. http://go.bloomberg.com/tech-blog/2012-08-29-average-household-has-5-connected-devices-while-somehave-15-plus/
22. Breitinger F, Rousseu V (2014) Automated evaluation of approximate matching algorithms on real data. Digit Invest
23. Lee W, Stolfo SJ (2000) Data-mining approaches for intrusion detection. Defense Technical Information Center, Fort Belvoir, VA
24. http://www.dfrws.org/2004/day1/Beebe_Obj_Framework_for_DI.pdf
25. https://www.gartner.com/en/newsroom/press-releases/2019-04-02-gartner-forecastsworldwide-public-cloud-revenue-to-g
26. Carlin S, Curran K (2012) Cloud computing technologies. Intelligent Systems Research Centre, University of Ulster. Derry. Northern Ireland. Int J Cloud Comput Serv Sci (IJ-CLOSER) 1(2)
27. Malathi M (2011) Cloud computing concepts. Department of Computer Science. T. John Engineering College, Bangalore, Karnataka, India, IEEE
28. NIST Cloud Computing Forensic Science Challenges (2014) NIST Cloud Computing Forensic Science Working Group Information Technology Laboratory
29. Ruan K (2013) Cybercrime and cloud forensics: applications for investigation processes. Inf Sci Ref
30. Manral B, Somani G, Choo RK, Conti M (2019) A systematic survey on cloud forensics challenges, solutions, and future directions. ACM Comput Surv
31. Marco L, Kechadi M, Ferrucci F (2013) Cloud forensic readiness: foundations. In: International conference on digital forensics and cyber crime. Digital forensics and cyber crime. Springer
32. Rahman N, Glisson W, Yang Y, Choo K (2016) Forensic-by-design framework for cyber-physical cloud systems. IEEE Cloud Comput 3(1)
33. https://www.guidancesoftware.com/encase-forensic
34. https://accessdata.com/products-services/forensic-toolkit-ftk
35. Dykstra J, Sherman A (2012) Acquiring forensic evidence from infrastructure-as-a-service cloud computing: Exploring and evaluating tools, trust, and techniques. Digit Invest 9(Supplement)
36. Suneja S, Isci C, Lara E, Bala V (2013) Exploring VM introspection: techniques and trade-offs. In: VEE'15: Proceedings of the 11th ACM SIGPLAN/SIGOPS international conference on virtual execution environments
37. Nurmi D, Wolski R, Grzegorczyk C, Obertelli G, Soman S, Youseff L (2009) The eucalyptus open-source cloud-computing system. IEEE
38. Payne B (2012) Simplifying virtual machine introspection using LibVMI. Sandia Report, Sandia National Laboratories
39. Kebande V, Venter HA (2014) Cloud forensic readiness model using a Botnet as a service. ICSA Research Group
40. Puri RB, Botnet (2008) An overview. GSEC practical assignment, Version 1.4. Option 1—Research on topics in information security
41. Pichan A, Lazarescu M, Soh S (2018) Towards a practical cloud forensics logging framework. J Inf Secur Appl
42. Zawoad S, Hasan R, Skjellum A (2015) OCF: an open cloud forensics model for reliable digital forensics. In: 8th International conference on cloud computing. IEEE (2015)
43. https://www.ironpaper.com/webintel/articles/internet-of-things-market-statistics/
44. Zhang Z, Cho M, Wang C, Hsu C, Chen C, Shieh S (2014) IoT security: ongoing challenges and research opportunities. In: 7th international conference on service-oriented computing and applications. IEEE (2014)

45. Ling Z, Luo J, Xu Y, Gao C, Wu K, Fu X (2017) Security vulnerabilities of internet of things: a case study of the smart plug system. Internet Things J, IEEE
46. Cao W, Zhang Q, Li Y, Xu L (2016) Edge computing: vision and challenges. Internet Things J. IEEE
47. Alenezi, A., Atlam, H., Alsagri, R., Alassafi, M., Wills, B. IoT Forensics: A State-of-the-Art Review, Challenges and Future Directions. The 4th International Conference on Complexity, Future Information Systems and Risk (COMPLEXIS 2019). (2019)
48. Oriwoh E, Jazani D, Epiphaniou G, Sant P (2013) Internet of Things forensics: challenges and approaches. In: 9th IEEE international conference on collaborative computing: networking, applications and worksharing. IEEE
49. Alabdulsalam S, Schaefer K, Kechadi T, Nhien-An Le-Khac (2018) Internet of Things forensics: challenges and case study. In: IFIP international conference on digital forensics. Advances in digital forensics XIV
50. Stoyanova M, Nikoloudakis Y, Panagiotakis S, Pallis E, Markakis E (2020) A survey on the Internet of Things (IoT) forensics: challenges, approaches and open issues. Commun Surv Tutor. IEEE
51. Li S, Choo K, Sun O, Buchanan W, Cao J (2019) IoT forensics: amazon echo as a use case. Internet Things J 6(4). IEEE
52. Ryu J, Sharma P, Jo J, Park J (2019) A blockchain based decentralized efficient investigation framework for IoT digital forensics. J Supercomput
53. Li S, Qin T, Min G (2019) Blockchain-based digital forensics investigation framework in the Internet of Things and social systems. Trans Comput Soc Syst 6(6). IEEE
54. Li S, Zhao S, Yuan Y, Sun Q, Zhang K (2018) Dynamic security risk evaluation via hybrid bayesian risk graph in cyber-physical social systems. Trans Comput Soc Syst, IEEE

Part III
Blockchain and Cryptography

Blockchain-Based Framework for Managing Customer Consent in Open Banking

Indraneel Mukhopadhyay and Abir Ghosh

Abstract Customer consent is the most important aspect in open banking. How it can be made transparent and effective, lots of areas have already been explored and also explorations are in progress. Bringing transparency to customers and to the whole process is not very easy to achieve especially when multiple parties are involved and they are competitors to some extent in their areas of operations and service offerings to customers. Blockchain technology can be an enabler to bring in the transparency in the process and can help to get the trust of customers which is absolutely essential to make the whole idea of open banking effective. Consortium Blockchain will be best fitted to have maximum efficiency. Regulators need to bring all the interested parties under one framework which will track each and every attribute of customer data and provide the required transparency to customers. The framework provides details to the customer about the rating of banks or any other Third-Party providers on a real-time basis which helps customer to take an instant informed decision about providing or withdrawing consent.

Keywords Open banking · Blockchain · Customer consent · Third-party providers · Regulator · Banks · Permissioned blockchain

1 Introduction

Customer consent is the key to success in fulfilling the opportunities that the open banking brings to the worldwide banking customers. It is the proven fact that the customer data must be private and secured in order to get customer consent. Several methodologies for providing security and privacy of customer data while sharing data with third-party providers have been explored. However, a different view is required

I. Mukhopadhyay (✉)
Institute of Engineering & Management, Kolkata, India
e-mail: imukhopadhyay@gmail.com

A. Ghosh
Larsen & Toubro Infotech Ltd., Pune, India

M. Chakraborty et al. (eds.), *The "Essence" of Network Security: An End-to-End Panorama*, Lecture Notes in Networks and Systems 163,
https://doi.org/10.1007/978-981-15-9317-8_3

on what can make customers confident about sharing their data and what can make them convinced about providing much-needed consent. Without keeping a focus on customer psychology how and what makes customer sure, a view is approached about what can make customers feel secured and confident about their own data. Whether this approach will actually make customers confident in data sharing or not that is a different aspect and will be justified in due course of time and need to be handled as a next step.

Customer Banking Experience

The customer banking experience is changing very fast in recent years. Brick and mortar to virtual branch experience is a thing of the past. Global penetration of mobile money to different other digitized services has opened a new world. Customer service is evolving into the customer experience now. Information is abundant, choices are temporary and time-conscious can be some of the identifiable characteristics of today's bank customers.

Open Banking Perspective

Open Banking has created an opportunity for banks to serve customers as per customers choice. Competitiveness between banks and service providers will help customers to get better service at a reduced cost. Open Banking provides the facility to customer to have a choice. Hence the customer can change their mindset and switch between service providers which brings altogether a new perspective about how the customer gets the banking service from banks or any other financial service providers. Customers are the key and at the center point of the whole concept of open banking. If the customer does not accept and provides consent, it cannot proceed further. On the other side, if the customers are happy and ready to share data and give consent, it opens up a new opportunity and new experience for them.

Technology as an Enabler

Technology to support customer expectations has evolved significantly in recent years. Blockchain technology is one such area. Recent advancements in Blockchain technology have actually made it possible to bring in the transparency in the process which will remove the bottlenecks in spreading open banking. Consortium blockchain can actually have different parties involved in the end-to-end process and regulatory bodies can act as a binding factor. Being decentralized, this also supports that no single organization controlling it for their benefit. Starting with this journey is not easy as it will require involvement from multiple parties who have their own priorities and work ethics. There are other factors like cost, resources and various other important aspects. We do not cover such aspects here in detail but keep our focus on the framework which needs to be a starting point in the journey.

1.1 Significance of Customer Consent Management in Open Banking

Banking still suffers from a poor reputation and relatively low levels of trust when compared to other industries [1]. An ingredient lies and remains important in the new financial system, that is trust (in banks, Fintechs, Techfins, regulators and users) and it is the keyword around which this revolution revolves [2]. Three quarters of consumer respondents in the United States and Canada say they are unlikely or very unlikely to allow their banks to share their account information, transaction history, funds overview and other data with third parties [3]. In this endeavor, "trust" is an essential ingredient for success [4]. There is a universal agreement that establishing and maintaining customer trust is essential to the success of open banking [5]. If an individual is clear about how their information will be used, then a level of trust for the online service will be gained. This is supported by research that finds that privacy concerns and the effect of the intention to share personal information were mediated by trust [6]. An important element of customer trust is provided by the regulatory environment within which data processing activities take place. This includes the existence of appropriate data protection laws coupled with effective, independent oversight (Greenleaf 2012) and a well-functioning legal system [7]. The informed consent is the most important dimension of the trust in open banking. A tick in a box saying that the consumer accepts terms and conditions of a document he will never read is not at all an informed, explicit and specific consent. As shared in The European Consumer Organization (BEUC) recommendation #2 - the consumer's consent should be explicit: "by ticking this box, I agree that company 'XXX' will have access to the following financial data (list data for which the access is being requested) managed by the ASPSPs (bank) 'YYY" [8].

1.2 How Technology Can Help

The latest technology can boost customer confidence in providing consent for data sharing with Third-Party Providers. Third-party providers can be any organization who provides financial services to customers. Technology can make customers informed and comfortable in making decisions to share data. Technology can actually help customer to take control of their own data. We have approached for a technology-based solution to boost customer confidence about data sharing. Before we go into details of different Blockchain-based solutions for consent management, let's understand about Blockchain technology first.

2 Blockchain Technology

Blockchain can be described as a distributed ledger that is immutable. It helps to build trust among the participants of the network. Some basic features of Blockchain Technology are:

- Immutable: Transactions cannot be altered or changed. It is one of the key aspects which help to build the trust.
- Distributed: No single party manages the network. By nature, every party in the network will have a view of every transaction.
- Enhanced Security: Blocks in the ledger contains a unique hash and also contains a hash of the previous block with the help of Merkel Tree, which makes tampering blocks almost impossible.
- Consensus: Blockchain technology thrives because of the consensus algorithm. It helps the network to make a decision.
- Smart Contracts: It is basically a set of rules which are stored on the Blockchain and executed automatically. It helps speed up transactions.

2.1 Types of Blockchain Networks

There are several ways to build a Blockchain network. They can be public, private, permissioned or built by a consortium [9]. Figure 1 shows the different types of Blockchain networks.

Considering the scope and nature of our research area, consortium Blockchain fits more into our solution approach. The below section covers some details about Consortium Blockchain [10].

Public blockchain networks	Private blockchain networks	Permissioned blockchain networks	Consortium blockchains
A public blockchain is one which is not restricted toany body i.e. anyone can participate e.g. Bitcoin. Some of the drawbacks are weak security, substantial computational power requirement, little privacy for transactions	A private blockchain network, similar to a public blockchain network, is a decentralized peer-to-peer network, with the significant difference that one organization governs the network. That organization controls who is allowed to participate in the network, execute a consensus protocol and maintain the shared ledger. Depending on the use case, this can significantly boost trust and confidence between participants. A private blockchain can be run behind a corporate firewall and even be hosted on-premises.	Businesses who set up a private blockchain, will generally set up a permissioned blockchain network. It is important to note that public blockchain networks can also be a permissioned. This places restrictions on who is allowed to participate in the network, and only in certain transactions. Participants need to obtain an invitation or permission to join	Multiple organizations can share the responsibilities of maintaining a blockchain. These pre-selected organizations determine who may submit transactions or access the data. A consortium blockchain is ideal for business when all participants need to be permissioned and have a shared responsibility for the blockchain

Fig. 1 Different types of blockchain network

2.2 Consortium Blockchain

The consortium Blockchain is a hybrid between the public Blockchain and the model of private Blockchain. Hyperledger, Quorum, Corda are examples of Consortium Blockchain.

Some of the characteristics of consortium Blockchain are:

- It needs permission for all participants, that is, for a predetermined group of enterprises;
- It is not fully decentralized. It works under the supervision of members of the limited group;
- It has a multi-party consensus where all the operations are verified by special pre-approved nodes.

Figure 2 shows how consortium Blockchain is placed in comparison to private and public Blockchain [11] when it is about managing consensus.

3 Related Work

Though Blockchain framework-based approach is not considered so far however to maintain the data privacy of the customer sensitive data, there are several approaches taken or being considered. The recent development in the technology has helped us to define these approaches. One such development is Blockchain-based data privacy management framework which consists of three components: (i) a data privacy classification method according to the characteristics of financial data, (ii) a new collaborative filtering-based model and (iii) a confirmation data disclosure scheme for customer strategies based on the Nudge Theory. Although Blockchains can effectively protect the privacy of financial data, there are still some deficiencies: firstly the privacy-preserving customer data and the granularity of customer data are not suited to the banks' business on the existing applications, and customers are unable to understand how their data is used and how their information is empowered; secondly the multi-dimensional data and numerous systems require banks to have dynamic and convenient methods to deal with customers' data authorization to avoid tedious operations; thirdly, due to the need for regulations, banks' operation and customer management, data should be hierarchically managed in Blockchains; fourthly, some personalized data need to be chained up and down dynamically according to the actual needs. In order to address these issues, a new Blockchain-based data privacy management framework is designed in combination with the Nudge theory. The

Public	Consortium	Private
Consensus managed by Public	Consensus Managed by set of participants	Consensus managed by single Owner

Fig. 2 Consensus management in different types of blockchain network

framework dissolves the default data privacy disclosure schemes of new customers with the customer strategies collaborative filtering model, and then the schemes are confirmed in the application scenario of banks. A concrete Blockchain-based financial customer data privacy management—technical implementation is fulfilled [12]. In another solution, a decentralized personal data management system is described where users are the owner of the same and they control their own data. Blockchain-based automated access control manager is implemented which is a protocol that does not require trust in a third party [13].

Another approach develops a General Data Protection Regulation (GDPR)—compliant data management platform. This platform is developed on Blockchain and smart contract technologies. Its goal is to provide decentralized mechanisms to service providers and data owners for processing personal data. This platform takes care of data usage consent given by data owners so that data is accessible to only designated parties. This platform also ensures that all data activities are logged in a distributed ledger which is immutable [14].

4 Working Principle of Blockchain-Based Solution

Open Banking directives are centered on customer benefits. Customers are the primary movers here. If customers decide not to share information, the purpose of open banking will not be met completely. Customers give consent to share their information for their own benefit—this is the main objective of open banking however it takes a lot to convince customers to give consent for their data sharing. As consumers gain greater exposure to Open Banking, it is likely that: (i) trust, in mechanisms, services and providers will become more established; (ii) that the specific benefits will become clearer and better understood and (iii) that the active concern about risks will decline [15].

The most important factor is what makes the customer confident about giving consent to share their account information: Primary factors are: 1. Trust Factor about the system handling the information. 2. Visibility about Future benefits. 3. Ease and comfort factor to provide the consent. It can be safely assumed that the trust factor plays the most important factor here. General Data Protection Regulation (GDPR) gives control of personal data back to the owners by appointing higher requirements and obligations on service providers (SPs) who manage and process personal data. As the verification of GDPR-compliance, handled by a supervisory authority, is irregularly conducted; it is challenging to certify that an SP has been continuously adhering to the GDPR. Furthermore, it is beyond the data owner's capability to perceive whether an SP complies with the GDPR and effectively protects their personal data [16]. It is important to give confidence to the customers about their data protection and to give that confidence instantly and continuously about the SPs how they will be handling their data in the future. There is no standard or framework now clearly visible and accessible to all customers how SPs will be handling their sensitive data. Before we try to visualize this standard or framework, it is necessary to analyze if there is any such standard or framework available now for the handling of customer-sensitive data in the open banking world. It is present for the cards industry where merchants need to adhere to the Payment Card Industry Data Security Standard (PCI DSS) standards. To maintain a secure environment, the PCI DSS is a set of security standards designed for all companies that accept or process credit card information. There are 12 PCI DSS requirements that companies need to adhere to. In the open banking world, clearly, there is a need to build PCI DSS like standards which SPs need to adhere to and maintain before they start access to customer sensitive data.

4.1 Defining Standards for Third-Party Providers

As already stated in earlier sections of this chapter, the Consortium Blockchain-based approach will make the process much more secure as it will help to build a framework and standard for service providers to adhere to. How would that be possible? In this Blockchain-based approach customer data is tracked end-to-end and a view is also available all the time. Here which SP is accessing what data at what time and whether they have the authorization or not can be tracked.

As per Fig. 3, the customer has provided consent to the five different banks to access the data as shown in the above diagram. Also, the diagram shows which bank has access to what kind of data of the customer. Figure 4 shows how one of the banks, that is, Bank1 have accessed the same customer data over some duration of time.

As it can be seen that bank1 has accessed the Income details of customer 1 which is not as per consent provided by customer 1 to the bank1. So this can be considered as a violation. Hence we propose to track all violations and also categorize the data for which breach of trust has happened. After categorizing violated data for all customers, the framework will formulate the rating for the bank which will be displayed to all the customers on a real-time basis whenever any customer tries to give consent to the bank1. This will ensure that customers are aware on a real-time basis, how the

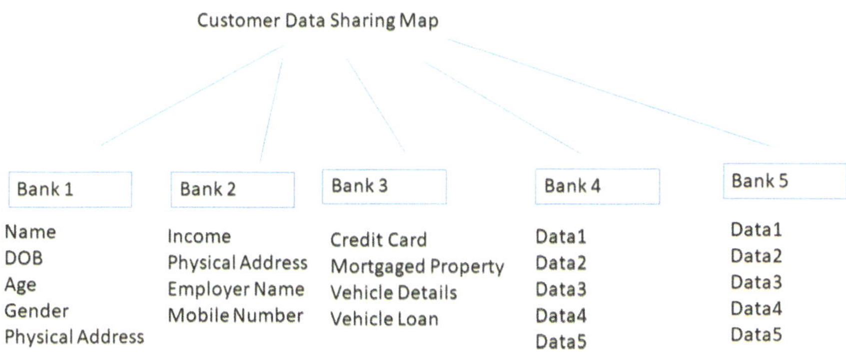

Fig. 3 Customer data sharing map

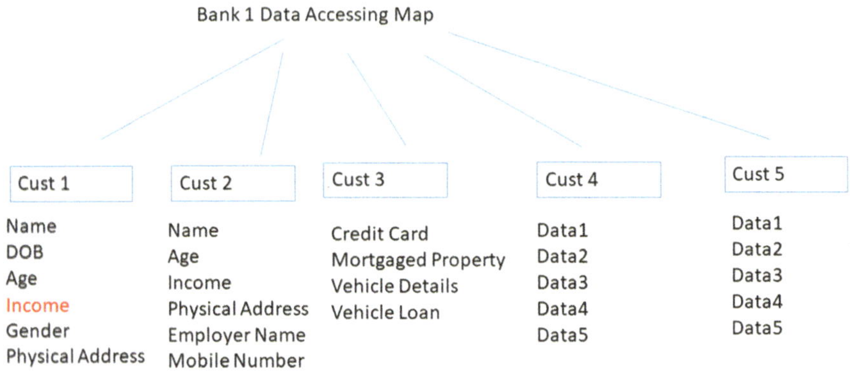

Fig. 4 Customer data accessing map by bank

bank is handling customer sensitive data and adhering to the standards. In order to achieve this rating generating framework, the first and foremost step is how banks agree to participate in it for their own benefit. Regulators will have a very crucial role to play here.

Figure 5 shows the steps from the bank and SP perspective what they need to follow:

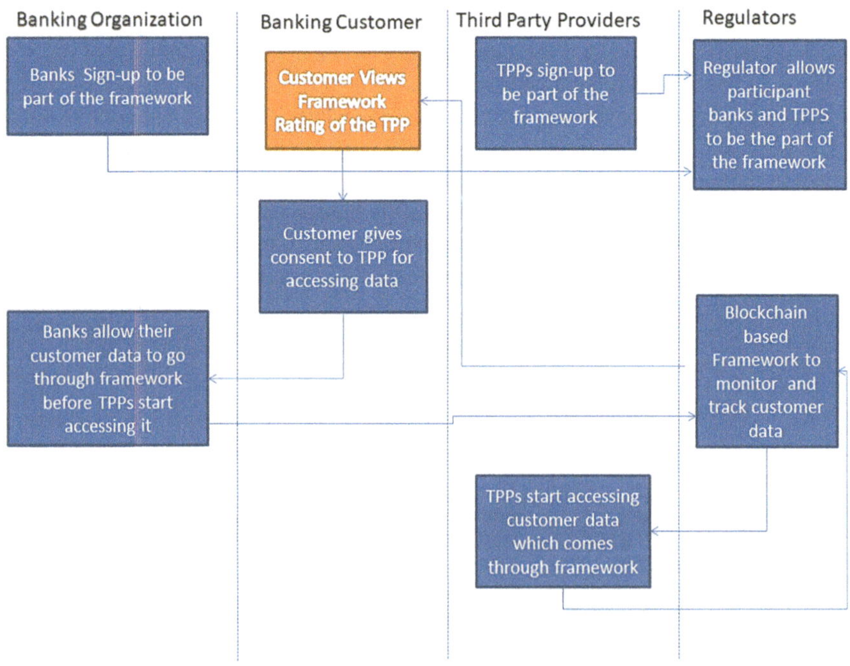

Fig. 5 Flow diagram to show the interactions by involved parties

Below are the sequential steps:

1. In the first-place banks sign-up with the regulatory body to be part of the framework.
2. Similarly Third-Party Providers (TPPs) also sign-up to be part of the framework.
3. Regulatory body does necessary review and allows banks and TPPs to be part of the framework.
4. As a next step, banks allow their customer data to go through the framework before TPPs start accessing it when the customer gives consent.
5. Customer accesses bank feature and view framework rating of the TPP.
6. Customer gives consent to TPP for accessing data.
7. TPPs start accessing customer data which comes through the framework.
8. Blockchain-based framework monitors and tracks it on a continuous basis of how TPPs are accessing customer data.
9. Banks refer to the framework and show customer framework rating of TPPs whenever the customer tries to give consent use of data to them.
10. Once sufficient transaction handling data of customer with consent details are accumulated at the framework regulatory body, more meaningful information can be passed on to the customer for taking the right decision.

5 Blockchain-Based Solution Architecture

Figure 6 shows the solution architecture. It indicates how customer data block will be maintained and accessed by different TPPs or banks. When the customer gives consent at the time of accessing bank1 then it creates one customer data block. The customer data block will contain all details about the customer which are maintained with bank1 which they want to share with other banks or TPPs. This customer data block is immutable and every time other banks try to access this customer data block, other bank details will get added to the customer data access log. This customer data access log will provide details about which all banks or TPPs have accessed customer data block. No TPPS or banks will be able to modify customer data block ever. This maintains the data integrity of the customer. From the access log, all types of customer data have been accessed by corresponding banks may be viewed. How do we ensure that when a customer withdraws the given consent, their data will no more be accessed by the banks or the TPPs? To handle this scenario, banks will be prevented from accessing customer data block whenever the customer withdraws given consent.

In Fig. 7, it is shown what can be typical components in the permissioned Blockchain architecture [17].

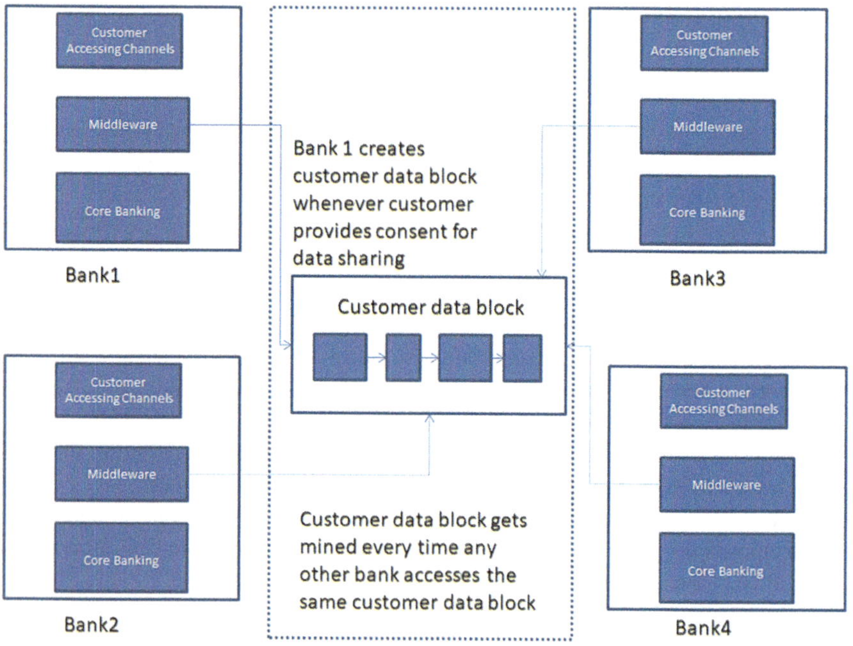

Fig. 6 Solution architecture

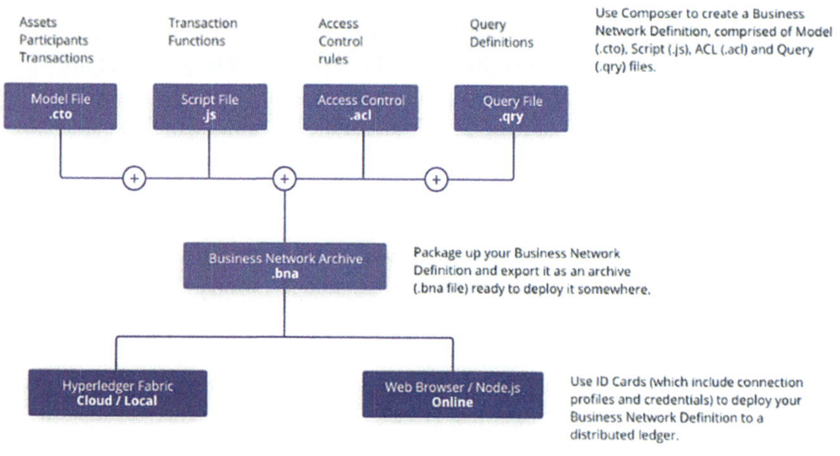

Fig. 7 Component in a typical permissioned blockchain architecture

6 Features of Blockchain-Based Solution

Some important features of our Blockchain-based solution are:

- Open source: The Blockchain-based framework is completely built on open source technology which makes it easy to adopt by different banking organizations, TPPs and supported by regulators. Licensed product will make the framework cost-sensitive which may be a hindrance to the wide acceptability.
- Cost-Effectiveness: Using open source technology makes the solution cost-effective hence attractive to every involved party of the framework.
- Less Hardware Intensive: Blockchain-based solution architecture is designed in such a way that it minimizes the need for using higher capacity hardware.
- Scalability: The solution is scalable in the sense that the whole design is done considering the participation from more and more parties.
- Secure: Permissioned Blockchain-based solution makes the framework utmost secure and this will help more customers to share their data.
- Compliance: Regulators taking the significant initiative in the proposed framework which in turn confirms the compliance on regulatory aspects.
- Distributed: Even though regulators take the significant initiative to bring the momentum into the framework, they do not become a controlling authority. Once the framework gains required momentum, every permissioned party will have equal rights.

7 Comparative Analysis

There are various approaches taken by the organization to manage customer consent. Below is the synopsis of some of the approaches (Fig. 8).

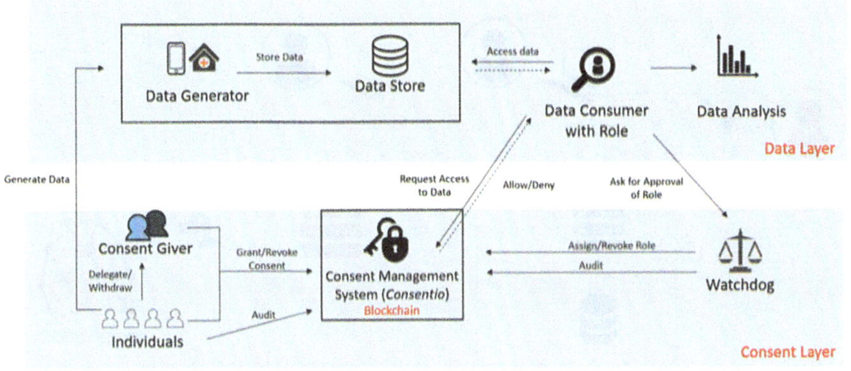

Fig. 8 Consent management using consentio

Hyperledger Fabric permissioned Blockchain system [17] —Consentio implementation:

Consentio is a consent management system using a permissioned Blockchain at the back end which is built on Hyperledger Fabric. No modification is required to the Fabric. To ensure high throughput and low latency of consent transactions given the fact that Fabric's key-value the world state ensures preserving compatibility with Fabric. The space of possible world state designs was analyzed. Using Fabric's key-value store, it showed that the world state for consent management can be implemented effectively. As per experimental results, 6,000 access requests per second running on a modest Fabric cluster can be handled in Consentio.

In another solution [18] smart contracts are used for consent management (Fig. 9). In this case, access modality is provided between two dates which can be extended if required. It can be done by integrating with other access modalities. When consent is given by the user, interaction with a consent smart contract happens and a new transaction is created. It is memorized and then recorded in a block. Maintaining the confidentiality and integrity of this block as well as the previous blocks, it is added into the ledger. This solution is hosted in Orange Flexible Engine cloud using Blockchain as a service infrastructure and can be integrated with Hyperledger nodes hosted elsewhere.

In a different solution [19] which deals with the design and implementation of a permissioned Blockchain third-party consent management system, a government agency decides the policy. A proof of concept is constructed using Fabric. It allows end-users to control and provide consent to SPs who are granted to manage their private information. The overall architecture consists of three main components: (i) the Blockchain network, (ii) the REST API consortium and (iii) the External Database. Both the user and the companies are required to be registered to the system.

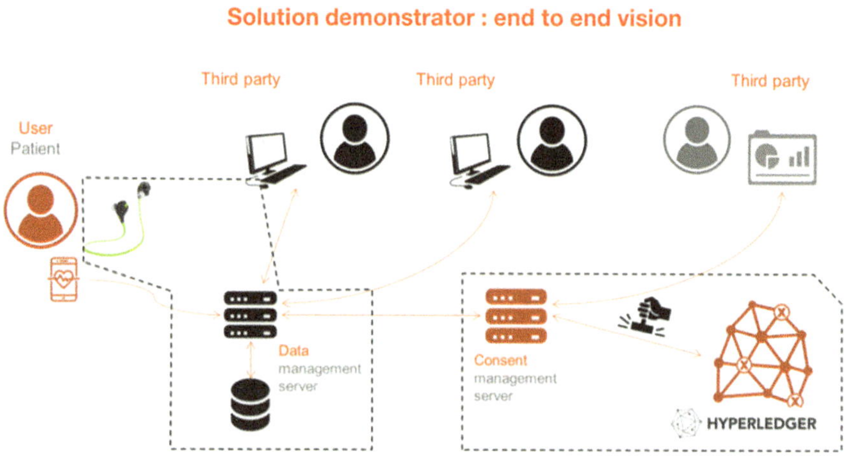

Fig. 9 Solution demonstrator

All the requests to Hyperledger are handled by the front-end. All requests for data are handled by the back-end.

Though none of these approaches are linked with open banking; however, these approaches can give us a fair idea about how permissioned Blockchain is effectively being used for managing customer consents.

8 Issues and Challenges

The participation of banks and third-party providers is the key to the success of this proposed Blockchain-based framework. Regulators have a key role to play by setting up the framework and by providing required governance to the whole system. Here comes the challenge of getting it to the priority of the regulators. Value propositions are there for banks, customers and third-party providers. However, unless it gains the required volume and momentum, benefits would not be clearly visible to all parties in the system. So the main challenge remains as initial set-up and then active participation from all parties. Secondly a lot of customer awareness would be required about the approach and objective so it needs to be started with few initial interested parties and then it needs to be scaled up. Some other challenges which might cause hindrance are:

Regulations: They vary from country to country from an open banking perspective. The core framework design needs to be flexible enough so that customization is possible.

Blockchain Technology: As a technology, it is evolving quite rapidly. The framework needs to be designed and implemented in such a way, so that it does not create any problem.

9 Conclusion

As more and more customers take part in the open banking initiatives, more will be the benefits to each individual customer. Transparency into the whole process will make the customer confident about participation. To bring customer confidence and trust, regulators need to play an active role. Regulators need to bring the banks and TPPs under one framework. At the initial level, customer needs to be encouraged by regulators to support open banking initiative. As more and more customers provide their consent to share the data, the framework will be able to generate and show customer upfront the rating of the bank or TPP. This rating can boost customer confidence about the bank or TPP before they can actually provide their consent. Blockchain-based framework which will generate a rating for each participating banking organization is one of the many alternative ways how trust and transparency can be effectively implemented in the system. This way information-based decisions will make the whole system very effective and customers will have lots of benefits that open banking can provide. Consortium Blockchain will be the best approach considering that it will make the framework secure. This will also help to have high transaction throughput and a cost-effective solution.

References

1. The future of banking is open—how to seize the Open Banking opportunity.https://retailbankinginnovation.fintecnet.com/uploads/2/4/3/8/24384857/the_future_of_banking_is_open.pdf
2. Omarini A (2018) Banks and fintechs: how to develop a digital open banking approach for the bank's future. Int Bus Res 11(9):23
3. https://thefinancialbrand.com/88283/open-banking-consumers-api-digital-data-privacy/
4. https://thepaypers.com/expert-opinion/open-banking-consumer-education-fear-of-fun--1239920
5. https://www.pinsentmasons.com/out-law/analysis/open-banking-opt-outs-and-revoking-consent
6. https://www.fs-cp.org.uk/sites/default/files/fscp_report_on_how_consumers_currently_consent_to_share_their_data.pdf
7. https://www.beuc.eu/publications/beuc-x-2018-082_consumer-friendly_open_banking.pdf
8. https://www.researchgate.net/publication/330940983_Nudging_Data_Privacy_Management_of_Open_Banking_Based_on_Blockchain
9. https://www.ibm.com/in-en/blockchain/what-is-blockchain
10. https://blockchain.intellectsoft.net/blog/how-the-consortium-blockchain-works/
11. Dib O, Brousmiche KL, Durand A, Thea E (2018) Consortium blockchains: overview, applications and challenges.https://www.researchgate.net/publication/328887130_Consortium_Blockchains_Overview_Applications_and_Challenges
12. https://blog.prototypr.io/a-non-architects-guide-to-blockchain-architecture-eb68360ad350
13. Zyskind G, Nathan O, Pentland AS (2015)Decentralizing privacy: using blockchain to protect personal data. In: 2015 IEEE CS security and privacy workshops
14. Truong NB, Sun K, Lee GM, Guo Y (2019)GDPR-compliant personal data management: a blockchain-based solution. IEEE Trans Inf Forensics Secur
15. https://www.ipsos.com/sites/default/files/ct/publication/documents/2018-04/open-banking-data-sharing-dilemmas.pdf
16. https://www.researchgate.net/publication/332263762_GDPR-Compliant_Personal_Data_Management_A_Blockchain-based_Solution
17. https://arxiv.org/pdf/1910.07110.pdf
18. https://hellofuture.orange.com/en/blockchain-for-consent-management-improved-privacy-and-user-control/
19. Aldred N, Baal L, Broda G, Trumble S, Mahmoud QHDesign and implementation of a blockchain-based consent management system. https://arxiv.org/ftp/arxiv/papers/1912/1912.09882.pdf

A Comprehensive Study of Pros and Cons on Implementation of Blockchain for IoT Device Security

Indraneel Mukhopadhyay

Abstract Internet of Things (IoT) one of the biggest buzz words is going around. The question arises what is it? Is it a group of computable devices and appliances connected using sensors, which use the Internet for data exchange? You might be right. We have to remember that if we are going for wireless transmission as we normally do in IoT solutions then the risk in these networks is very high. In this world of IoT 4.0 every device is connected with every device and these devices are used in all kinds of applications starting from the Military, Healthcare, Education, Home and list go on. One has to keep in mind that an IoT solution relies on wired, cellular, or adhoc communication networks. IoT is considered to play a very important role in the next decade or two. There is also an exponential rise of IoT-based devices so that there is a seamless exchange of data between the connected devices and hence decided to be taken based on the data that can be taken faster for the betterment of mankind. As the number of devices has increased it has given rise to various security concerns. We have to remember that confidentiality, authentication, access control, and integrity should never be compromised in the IoT environment. Currently, researchers are trying to develop efficient security and privacy protocols in IoT networks. IoT creates new security challenges for enterprises in scope and scale-like device identity, secure network scale, data security and physical security. This chapter deals with the issues and challenges of IoT devices and the prevention of security loopholes via Blockchain implementation.

Keywords Internet of things (IoT) · Security issues in IoT · Security · Blockchain · Privacy

---

I. Mukhopadhyay (✉)
Institute of Engineering & Management, Kolkata, India
e-mail: imukhopadhyay@gmail.com

M. Chakraborty et al. (eds.), *The "Essence" of Network Security: An End-to-End Panorama*, Lecture Notes in Networks and Systems 163,
https://doi.org/10.1007/978-981-15-9317-8_4

1 Introduction

As time has progressed from the Stone Age to the Digital Age, the method of doing business has also changed. Previously the business was done based on the exchange of surplus goods but nowadays we do business based on the exact need of the customer. And how do we know the exact need of the customer? We mine the data that has been collected via embedded systems called the Internet of Things (IoT). IoT has already made a significant impact by providing a competitive advantage. In days to come, IoT solutions will become the business block of Digital Business. Many people think Technology and Business are two different domains and each one is self-sustaining at its own level but in this VUCA (Volatility, Uncertainty, Complexity and Ambiguity) World that we live in is not true. There has to be a symbiotic relationship between technology and business in order to make the organization flourish as both are important strategic elements contributing to business success. Kevin Ashton in 1999 proposed the first concept of IoT [1].

With the recent technological advancement in communications technology, cloud computing and sensor technology, a lot of advancement has taken place in the field of IoT. With the advancement of the sensor technology and different types of identifiers, IoT devices have moved towards rapid hardware development. In recent times data storage has also become an advantage towards the implementation of IoT deployment. Nowadays, IoT devices can talk to each other either in a managed or unmanaged state. The question always arises about the different IoT devices. The answer is very simple from our smart mobile phones, tablets, laptops, wearables and other handheld devices, which are capable of interacting with each other are examples of IoT devices. Many IoT devices are built using sensors, actuators and communication systems which are very cost-effective but can communicate with each other and transfer data to one another or to a centralized system. As we have become too much dependent on technology hence, we see growth in IoT devices, which can make our life more comfortable. IoT devices collect, share and store the data in centralized servers. The information which is collected from these devices are processed and many times sent to other destinations that can use the information to their benefit. So the question arises "what data is shared and how is that data used for me or against me?". With time and growth of the service sector along with the technological advancement, our daily routine has got tangled with the digital and virtual world [1].

IoT is considered to be the enabler for the next Industrial revolution by bridging the gap between the physical, digital, cyber and virtual space. It is also a key enabler in the creation of Digital Single Market (DSM) which will have the potential to create jobs and economic growth. IoT has been divided into a number of ecosystems such as smart homes, personal wellness applications and wearable devices, smart cities, smart mobility, smart manufacturing, smart energy, smart farming to name a few. IoT makes a convergence of consumer, business and industry. We have to remember that we cannot automate every human activity [2]. Nowadays, the IoT density, that is, the rate at which the IoT devices are deployed has increased and as more devices get connected, issues and challenges towards security and privacy also increase. An appraisal discloses in a recent study: the global IoT market which was valued at US$190.0 Billion US$ in 2018 is projected to reach US$1,102.6 Billion by 2026 [3]. In another appraisal the forecast suggests that by 2030 around 50 billion IoT devices will be in use, creating a mesh of interconnected devices. Presently, that is, in the year 2020, there are around 25 billion connected devices compared to that of 22 billion in 2018. As the sophistication of both hardware and software is on the rise due to the use of advanced technology in the manufacturing industry the IoT industry is also seeing its boom. IoT will change the way we look at our lives and how we operate our business. The symbiotic relationship between both the technology and the industry will create the mesh of devices which will guide us through the Internet [4].

IoT creates new security challenges for enterprises in scope and scale-like device identity, secure network scale and data security and physical security. Currently, most IoT devices share security challenges such as brute-force attack and denial-of-service attack, malware, ransomware, botnets, host capture and communication privacy concerns to name a few. Most IoT devices are not designed to handle these security challenges and privacy attacks and thus unencrypted communication channel gives rise to a lot of security and privacy issues in the IoT domain keeping in mind confidentiality, authentication, data integrity, access control, secrecy [5]. IoT devices are targeted by attackers and intruders –70% of the IoT devices are very easy to attack due to the unavailability of hardware device security. Therefore, an effective mechanism needs to be created to secure the IoT devices; with the combination of hardware device security plus a communication framework that could secure end-to-end encrypted communication between the different devices and the server against hackers and intruders.

The chapter is organized in the following way. Section 2 talks about the architecture of the IoT device, followed by the classification of attacks in Sect. 3. This is followed by security concerns of the devices in Sect. 4. Section 5 deals with the issues and challenges of the IoT environment. In Sect. 6 we present the proposed Blockchain solution followed by other Blockchain solutions in Sect. 7 and finally conclude it in Sect. 8.

2 IoT Architecture

It is very important to understand that whenever we are talking of IoT devices we have to keep in mind that the devices work as independently and as smart nodes with the help of different types of sensors. They are in many cases self-conscious. The IoT devices are divided into a number of ecosystems such as smart homes, personal wellness applications and wearable devices, smart cities, smart mobility, smart manufacturing, smart energy, smart farming. Before we go into the ecosystem it is important for us to know the architecture of an IoT Device. Figure 1 describes the architecture of an IoT device.

The architecture of an IoT can be divided into four different blocks—communication block, hardware block, operating system block and network management

Fig. 1 Architecture of IoT

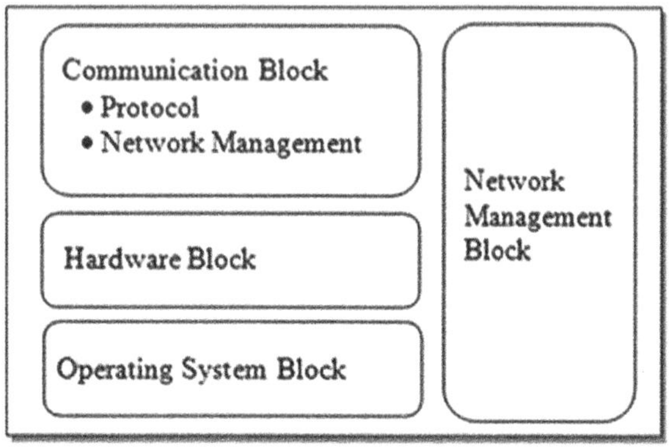

block [6]. The architecture is divided into different stages. In stage one sensor data that is collected by the device is used to convert the information obtained into data for analysis. Here we convert the input from the physical world to gain the necessary insights for further analysis. In stage two the data collected is sent via a communication medium to other IoT devices or to the centralized server. The huge amount of data collected in stage one is cleaned, formatted and made ready for further analysis. In stage three the prepared data from stage two is transferred to the virtual world. Data visualization and analytical processing is done in stage three so that the data collected in this stage may be sent to the next stage. In stage four the pre-processed data is analyzed, managed and stored and in due time sent to the intended destination. The data collected analyzed and stored is then used for the benefit of mankind.

3 Classification of Attacks on IoT Devices

Before we can jump into the broad classification of the attack types to an embedded system and its sub-types, we must understand that all IoT devices are also embedded devices. Hence threats to embedded devices are threats to IoT devices as well with additional threats related to a communications medium. Figure 2 summarizes the various classifications of attacks on IoT Systems [7].

Attacks on IoT Devices can be broadly classified into five different types.

A. Physical Attacks

These types of attacks happen when the attacker has the access to the physical IoT device. The attacker modifies the device components. This kind of attack is very hard to do as in most cases highly sophisticated equipment is required to modify and tamper with the hardware of the device. Some examples are micro-probing, reverse engineering.

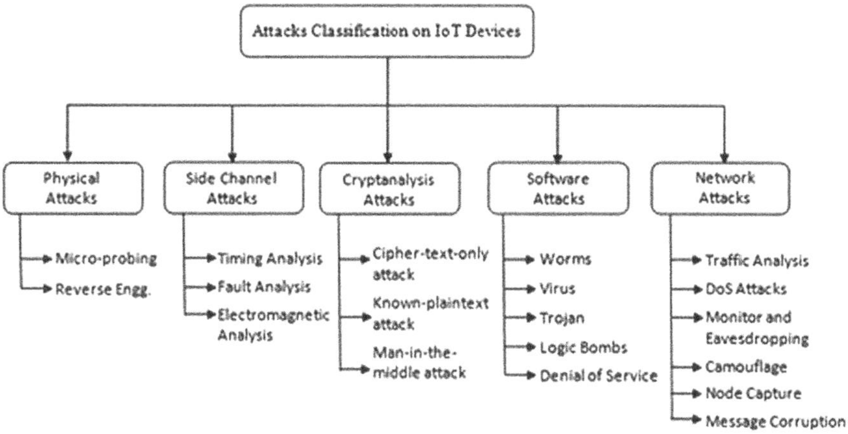

Fig. 2 Attack classification on IoT device

B. Side Channel attacks

These attacks are based on information that is produced during an encryption process; which can be retrieved and used against the IoT device. We all know encryption is the process of protection of data using different encryption algorithms. This encrypted data is stored so that only the intended party may have access to it. Now we have encryption devices that do the encryptions for us. But it comes with a cost; these devices produce different forms of signals which if used separately or in combination may give the attacker access to the device. The attacker can in turn gain control of the data we are trying to protect via these devices. Some examples are timing attacks, fault analysis attacks, electromagnetic attacks.

C. Cryptanalysis attacks

These attacks are focused on breaking the ciphertext to gain access to the encryption key and then get access to the original text which has been encrypted to protect from attackers. Some examples are man-in-middle attack, ciphertext-only, known-plaintext attack to name a few.

D. Software Attacks

Every IoT device that gets connected to the Internet is under attack within 5 min of it going online and most attacks that happen are cyber-attacks. The cyber-attacks that happen can be classified into simple denial-of-service attack to severe malware attack such as ransomware. All these attacks account for the software attacks as most of the IoT devices have little to no protection to these kinds of attacks. One of the main reasons being 92% of all data traffic in IoT devices go unencrypted. Some of the examples of software attacks are virus, trojan horse, worms, logic bombs, denial-of-service attacks to name a few.

E. Network Attacks

IoT works on data being transferred from sensors to actuators to the centralized server where the data is analyzed, processed, and sent to the intended destination. Most of the data is sent via wireless communications systems which are very vulnerable to network security attacks due to its transmission medium. In the network attack, the attacks can be passive attack such as monitor and eavesdropping, traffic analysis, camouflage whereas examples of active attacks denial-of-service attacks, node capture, message corruption to name a few.

4 Security Concern of IoT Devices

To protect the IoT devices from various types of attacks that were discussed in the previous section, we can follow some norms to make sure our devices secure. Few have been given below.

Device Security: IoT requires strong device security which is considered to be the root-of-trust. Hardware-based security is always the best solution but the cost is one of the important factors. Whenever you are using hardware-security implementation, it requires specialized equipment and also specialized people to run them. Once you have hardware security in place your device becomes tamper-resistant, and subsequently, it provides a secure environment for execution of the device operation.

Secure Network: For IoT device deployments, securing the point to point network connections becomes very important. Using the public key infrastructure (PKI) is the norm followed by most IoT developers but just using PKI is not enough to secure the network [7]. When we are transmitting data over an unencrypted network we have to couple PKI with security devices so that we have a secure transmission network. User/Node authentication is another very important feature that needs to be kept in mind before we can secure the network.

Data Storage Security and Physical Security: The basic function of IoT devices is to provide service at any time and at any place; hence we must remember to have inbuilt security of the data. The data can be in transit (data communication) or at rest (data storage), it is very valuable to the device. Most of the IoT devices lack physical security and the attacker can gain access to the device. So most organizations have device security implemented, and store the data which is collected in the cloud as to which they have subscribed. Many may argue cloud would bring security woes but nowadays cloud is much more secure than a physical server at organizational infrastructure [8].

5 Issues and Challenges

IoT devices have many challenges when it comes to security concerns [8, 9]. We have to remember here that we have to take a holistic approach that involves the IoT device, the network, the storage and also the application for which the device was built. There are many challenges for IoT devices as shown in Fig. 3.

Whenever we talk about IoT devices we are using an embedded system, which has a limited processing power and storage. If we really want to implement a very effective security algorithm then it may happen so due to the limited power it might not be enough for running the security algorithm for a long duration of time. Storage limitations are also hurdles for IoT device security features.

Privacy

IoT devices have a very high level of integration between the networks on which their data get transmitted and the services they render. So the devices need to protect the data so that attacks such as node capture or eavesdropping by an attacker do not happen. Here we need to remember many health care and personal wellness sensors are used to gather various personal and health care-related data which are very private. An IoT device which is collating such data has to be very careful in protecting the privacy of the users [10].

Integrity

When we use a centralized architecture to analyze process and store data it is very important to secure such storage. If an unauthorized person gets access to that original

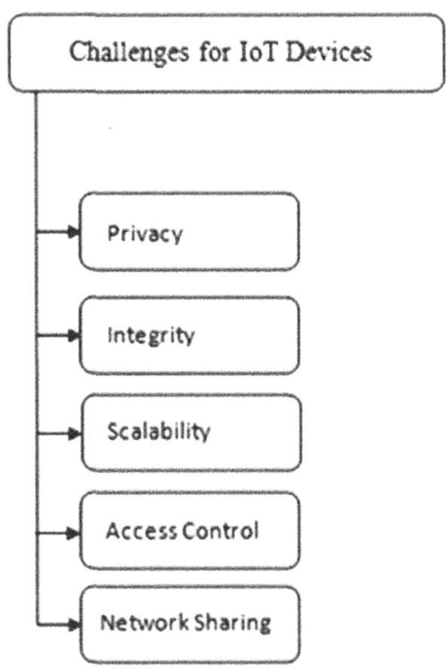

Fig. 3 Challenges for IoT devices

data then he/she may cause enough trouble. In several cases, many IoT implementations use a third-party environment which is always a cause of worry. When we discuss integrity we cannot forget the IoT environment from where the sensors are collecting the real-time data. Only authenticated IoT nodes should be able to send data thus ensuring data integrity [11].

Scalability

An IoT device is a mesh or swarm network consisting of a large number of sensors, actuators and other network and non-network devices using a large number of applications to capture, clean, format, process and store data all on the Internet. It is very important for those components to be scalable in nature. The biggest challenge that IoT environment faces currently is how to scale the rapid growth [12].

Access Control

Privacy and Access controls are two very important issues and challenges in an IoT device network. Based on the role each IoT device plays its authorization and data transfer is fully dependent on access control that each IoT node has. As IoTs are connected to one another it is difficult to give node-wise access right. Another reason being node failure. There has to be another node to take its place. Parallel channels with authentication need to be there with access control which is very difficult to perform for centralized architecture [9].

Network Sharing

The information gathered by IoT network devices is recorded distinctly for the purpose of analysis. Information sets may contain IoT devices network data load or their functioning logs. To confirm the efficiency of tools and tests, open accessibility of information plays a vital role. So, every time these information sets are openly shared their integrity becomes significant.

6 Blockchain Solutions for IoT Devices

In the previous section, we have discussed different types of attacks, issues and challenges that are present in IoT device networks. The solution to the above issues is the implementation of a Blockchain solution for the IoT devices. The reason for the effectiveness of Blockchain has been explained in this section.

Theoretically, the concept of Blockchain has a significant impact on the ways how organizations are doing business in the current scenario. The major advantage has been to shift to a decentralized form, with trustworthiness. All transactions that are happening can be very easily realized using different implementation mechanisms [13, 14].

With the technological growth related to Blockchain, the issues of IoT device is reduced as Blockchain bring its security feature which can be easily deployed to IoT devices.

Privacy

With the IoT device and its corresponding network put in place, the issue is data privacy. However, there are lots of advantages in implementing Blockchain technology. As Blockchain provides in-built cryptography support, data collection from IoT via the sensors up to the transmission of data over the communication medium are all protected upon the use of Blockchain. If an attacker who does not have proper authorization wants it to access any node of an IoT he/she will be prevented from unauthorized access. Illegal use of personal data can also be prohibited with the use of Blockchain. As Blockchain is a peer-to-peer network storing system it can verify and record all actions and transactions of data accomplished on IoT network [14].

Integrity

Blockchain uses a peer-to-peer network implementation strategy, where all blocks have the same copy of the records. Using the concept of private key infrastructure, the IoT node signs the transactions with its private key and sends it to other nodes for validation. We have to keep in mind that PKI may overload the node individually but when implemented over the whole network the overload is comparatively less. The IoT device block would then broadcast to all other nodes over the entire network. So, when this record is loaded under the implementation of Blockchain it becomes immutable [15].

Accountability

Records that are collected via the IoT devices are recorded (rather we can say audited) into the Blockchain network. This gives traceability to every node and data that is collected via those nodes. When an abnormal behavior is detected in an entity, Blockchain would be used for an additional investigation [15]. Fault Tolerant Decentralized devices are less likely to fail accidentally because they rely on many separate components. The Blockchain is a peer to peer decentralized network, where every device has the same identical copy of a record that is why the failure of a single node does not have any effect on the network. Hence Blockchain prevents a single point of failure of the IoT network.

Network Sharing

As the size of IoT network information sharing is increasing, thus the fundamental storage cost is also increasing. Required information is kept within an IoT device network and a centralized server is also kept separately. The connection between them is kept via a wireless connection. As the Blockchain has immutability feature hence accessibility with all IoT network devices in Blockchain ensures its integrity.

Trusted Data

Every IoT node in the network that implements the Blockchain is assigned a unique identification tag so that each of them can be tracked individually. When the sensors collect the data, the IoT device along with the hashed identification tag is sent.

Thus, we are able to remove transmission eavesdropping and also remove security loopholes due to the use of third-party applications. Thus, the device becomes capable of performing operations risk-free.

7 Other Blockchain Solutions

Blockchain can be integrated with IoT in three different ways. An important component of IoT and Blockchain convergence is cloud computing. With the advancement of cloud computing, the Blockchain network has been revolutionized as a new layer that is added between the cloud computing and IoT devices. The different solutions are as follows.

IoT–IoT

The basic type of data transfer from IoT device to another IoT device is the IoT to IoT connection via a transmission medium, with the routing mechanism being used. Here the data transfer rate is very high as we are not using any Blockchain mechanism. This is the normal IoT network that is normally used with its advantages and disadvantages.

IoT–Blockchain

This is the type of IoT network where IoT devices are connected via the Blockchain through the cloud. So IoT to IoT data transfer happens via the cloud. One must remember as this type of Blockchain implementation occurs in the cloud hence cloud security takes care of the data even for a traditional IoT network. A record that is stored in the cloud via the Blockchain becomes immutable, traceable and secured from third-party access. Security also increases as we are using the cloud-based approach.

Hybrid Approach

With the new approaches and technological advancement hybrid approach makes data transmission and interactions by IoT devices directly so that the data transmission rate is high, and Blockchain stores only control data. This approach brings both the benefits of Blockchain and IoT devices. The hybrid approach implements cloud computing to make up for the limitations of Blockchain and IoT devices.

Depending upon the application being developed and deployed we have to choose the right approach of Blockchain implementation. The hybrid approach may not always be the right solution for every deployment. Choosing the right approach will differ from a smart home to a smart car to a smart city. The data from these three different deployments will be different, will be differently mined and will be differently used by the designated users.

8 Conclusion

The main aim of the chapter was to highlight IoT device security issues and how we can use Blockchain to eliminate them. Many IoT devices still become the target of attacks even when all security measures are put in place. Considering the importance of security in IoT applications, the use of Blockchain could boost the security of the IoT device. The hybrid approach of implementation of IoT is currently used by most of the deployment. But other approaches can be used as they also have their own advantages and disadvantages. In days to come, we will see more and more secured IoT devices using the hybrid approach with cloud implementation so that security can be enhanced. But one should keep in mind that for every secure implementation that is done there will be attackers who will keep on attacking these IoT devices. So we have to keep on building secure IoT devices with respect to newer attacks.

References

1. Kumar JS, Patel DR (2014) A survey on internet of things: Security and privacy issues. Int J Comput Appl 90(11)
2. Abomhara M, Køien GM (2014) Security and privacy in the internet of things: current status and open issues. In: International conference on privacy and security in mobile systems (PRISMS). IEEE, pp 1–8
3. A web article published in Fortune Business Insight, Report ID: FBI100307 July 2019
4. Statista Research Department (2020) Number of internet of things (IoT) connected devices worldwide in 2018, 2025 and 2030, 19th Feb 2020
5. Hossain MM, Fotouhi M, Hasan R (2015) Towards an analysis of security issues, challenges, and open problems in the internet of things. In: 2015 IEEE world congress on services (SERVICES). IEEE, pp 21–28
6. Paul stokes, 4 Stages of IoT architecture explained in simple words December 2018
7. Ravi S, Raghunathan A, Kocher P, Hattangady S (2004), Security in embedded systems: design challenges. Trans Embed Comput Syst (TECS) (ACM) 3(3)
8. Paar C, Weimerskirch A (2007) Embedded security in a pervasive world. Inf Secur Tech Rep (Elsevier) 12(3):155–161
9. Eby M, Werner J, Karsai G, Ledeczi A (2007) Embedded systems security co-design. SIGBED Rev (ACM) 4(2)
10. Khan MA, Salah K (2017) IoT security: review, blockchain solutions, and open challenges. Future Gener Comput Syst. https://doi.org/10.1016/j.future.2017.11.022
11. Qureshi MA, Aziz A, Ahmed B, Khalid A, Munir H (2012) Comparative analysis and implementation of efficient digital image water-marking schemes. Int J Comput Electr Eng 4(4):558
12. AbdurRazzaq M, Sheikh RA, Baig A, Ahmad A (2017) Digital image security: fusion of encryption, steganography and watermarking. Int J Adv Comput Sci Appl (IJACSA) 8(5)
13. Conoscenti M, Torino D, Vetr A, Torino D, De Martin JC (2016) Blockchain for the Internet of Things: a systematic literature review. In: IEEE/ACS 13th international conference of computer systems and applications (AICCSA)
14. Zyskind G, Nathan O, Pentland A (2015) Enigma: decentralized computation platform with guaranteed privacy. https://arxiv.org/abs/1506.03471
15. Liang X, Zhao J, Shetty S, Li D (2017) Towards data assurance and resilience in IoT using blockchain. In: Conference Paper

Role of Cryptography in Network Security

Anindita Sarkar, Swagata Roy Chatterjee, and Mohuya Chakraborty

Abstract Network administrators employ several security mechanisms to protect data in the network from unauthorized access and various threats. The security mechanisms enhance the usability and integrity of the network. The design aspects of the network security mechanism involve both hardware and software technologies. The application domains of security mechanisms cover both public and private computer networks which are used in everyday jobs for conducting transactions and communications among business partners, government agencies, enterprises and individuals. The network security schemes vary depending on the types of the network, that is, public or private, wired or wireless. Data security includes encryption, tokenization, and key management practices in protecting data across all applications and platforms. The antivirus and antimalware software are also part of network security for protection from malware such as spyware, ransomware, trojans, worms, and viruses. Cryptography is an automated mathematical tool that plays a vital role in network security. It assures the confidentiality and integrity of data as well as provides authentication and non-repudiation to the users. This chapter primarily focuses on cryptography techniques and their role in preserving the network security. The cryptography technique consists of encryption and decryption algorithms. The encryption algorithms perform scrambling of ordinary text and generate an unreadable format for the third party known as ciphertext. The original data is restructured by the intended receiver using decryption algorithms. The cryptographic techniques are broadly classified into three categories namely symmetric-key cryptography, asymmetric-key cryptography and authentication. The cryptographic algorithms that are widely accepted are outlined with their relative advantages and disadvantages. Moreover, recent proficient cryptographic algorithms specific to cloud computing, wireless sensor networks and on-chip-networks are thoroughly discussed that provide a clear view about acquiring secure communication in the network using cryptography.

A. Sarkar · S. R. Chatterjee (✉)
Netaji Subhash Engineering College, Kolkata, India
e-mail: rcswagata@gmail.com

M. Chakraborty
Institute of Engineering & Management, Kolkata, India

M. Chakraborty et al. (eds.), *The "Essence" of Network Security: An End-to-End Panorama*, Lecture Notes in Networks and Systems 163,
https://doi.org/10.1007/978-981-15-9317-8_5

Keywords Network security · Data encryption · Wireless sensor network · Internet of things · Quantum computing · Network on chip

1 Introduction

In the last few decades, significant improvement in the field of computing is achieved by the researcher. Initially, the processors with large volumes were used which are replaced by the Personal Computers (PC) with much reduced physical size of both processor and storage memories. Many new devices have evolved from time to time which further increased the computing power and storage capability. PCs subsequently exhibit parallel processing with the development of multi-core chips. The usability of PC in modern society has extremely increased after the development of internet technology which makes a worldwide computer network. The PCs with Internet connection provide significant computing facility for the users. The advancement of technology provides the facility for the computer users to connect anywhere in the world, getting the required information, carrying out various financial transactions via the Internet. The advent of mobile computing further reduced the size of the computing devices. The handheld mobile devices currently offer significant computing facilities to the users through wireless interconnections. The volume of data across the Internet is increasing day by day. Providing security and privacy to this huge amount of data is a crucial issue that is to be addressed very carefully. There is a significant development in the area of network security to prevent vulnerabilities of losing security and privacy in the modern digital world. However, continuous research activities are required in this field to achieve complete security. The security aspects of a network achieved by different security mechanisms are follows:

- *Confidentiality*—Security mechanisms generate unreadable data for the third party.
- *Data integrity*—Security mechanisms restrict the alteration of data.
- *Authentication*—Security mechanisms assure that received data is only from the alleged sender.
- *Non-repudiation*—Security mechanisms restrict the sender from the refusal of sending data.

The first two aspects are achieved by different cryptographic algorithms. Several authentication algorithms are utilized to ensure the authentication of the sender. The non-repudiation is achieved by means of a digital signature.

Cryptography is the most accepted automated tool for network and communications security which entirely depends upon mathematical computation. The primary aspires of cryptography are:

- Keeping information secret from adversaries.
- Protecting user personal data such as military communications.
- Increasing protection at the time of financial transactions through electronic systems such as payment using a credit card over the Internet.

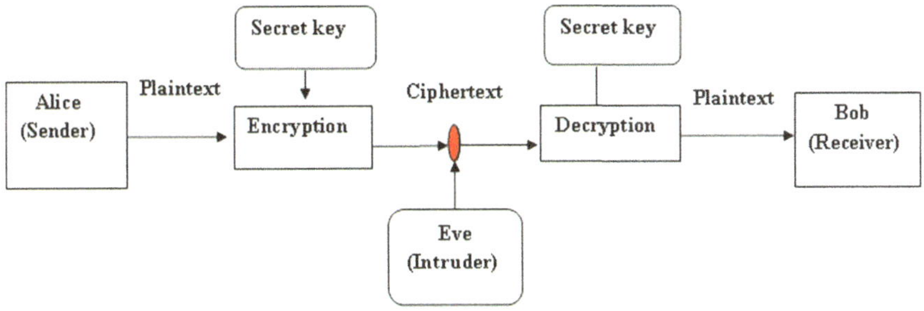

Fig. 1 Basic principle of cryptography

Cryptography is a collaborative study of mathematically designed protocols that prevent third parties from reading private messages. The basic principle of cryptography is shown in Fig. 1.

The original message is usually termed as plaintext and the scrambled message is called the ciphertext. By convention, Alice is considered as the sender, Bob is considered as the receiver and Eve is the intruder. The encryption algorithm converts the plaintext to the ciphertext and the decryption algorithm performs a reverse process to get back the original message. There are mainly two types of cryptography:

- Classical cryptosystems or symmetric ciphers.
- Public key cryptosystems or asymmetric ciphers.

The classical encryption system utilizes a single secret key for both encryption and decryption algorithms. However, public key cryptosystem uses a pair of a public key and private key. There are several cryptographic algorithms for both symmetric-key and asymmetric-key cryptosystems. Data Encryption Standard (DES), Advance Encryption Standard (AES), Blowfish, Rivest cipher5 (RC5), etc., are popular symmetric key algorithms. The well-known public key cryptography techniques are: Ron Rivest, Adi Shamir and Leonard Adleman (RSA), Elliptic Curve Cryptography (ECC) and Hyperelliptic Curve Cryptography (HECC).

With the advancement of quantum computers, many major improvements are observed in the field of quantum cryptography. This is due to the fact that quantum computers use quantum algorithms' which exploit the advantages of quantum parallelism. Such algorithms simplify the mathematical problems which are found difficult in the classical computing model. Many cryptographic algorithms are based on the difficulty of factorizing large numbers into their prime factors. Quantum algorithms make this task very easy and hence these classical cryptographic algorithms are vulnerable to many attacks in the quantum computing model. Hence strong cryptographic algorithms are to be developed to overcome these hurdles. Modern technologies such as Wireless Sensor Network (WSN), Internet of Things (IoT) and Network on a Chip (NoC) apply strong encryption techniques to assure security level in the networks. Lightweight encryption algorithms are perfect for these kinds of resource-constrained environments. Moreover, IoT generates a huge amount of data which gives rise to another area of research known as Cloud Computing and Big Data Analysis. Data security remains an important issue in the cloud particularly when a public cloud service is used to store data. Homomorphic cryptography is a newly developed scheme to ensure data privacy on the cloud and increase the prospective market for cloud computing. A NoC is a network-based communication subsystem on an integrated circuit and most typically used between the modules in a System on a Chip (SoC). In spite of lightweight cryptography, many hardware-based security systems are designed for NoC to provide a high level of security. Also, various security designs are proposed for NoC where coding theory has been applied in place of lightweight cryptographic algorithms.

Message authentication and digital signature protocols apply mathematical functions to produce an authenticator value that is used to authenticate a particular message. The conventional encryption algorithms and message digest or Hash algorithms are combined to generate authentication and digital signature algorithms. Several techniques have evolved in recent times such as digital watermarking and steganography to achieve digital signatures. Major developments are noticed in the field of authentication using biometric data.

Section 1 depicts the security aspects and the ways to achieve these aspects via different cryptographic applications. Sections 2 and 3 discuss classical cryptography systems with their relative advantages and disadvantages followed by the latest cryptographic algorithms such as Honey encryption, ECC and HECC. The algorithms used for authentication and digital signature are discussed in Sect. 4. Sections 5 and 6 consist of quantum computing and homomorphic cryptographic algorithms, respectively. Security schemes proposed for WSN are thoroughly explained in Sects. 7 and 8 discuss the security schemes for NoC. Section 9 concludes the chapter highlighting possible future research directions.

2 Classical Cryptosystems

The classical cryptosystems include both symmetric cipher and asymmetric cipher as mentioned in Sect. 1. In the case of a symmetric cipher single secret key is shared by both the sender and receiver. These kinds of cryptographic algorithms involve the use of substitution and transposition techniques. Here encryption algorithm is the reverse of the decryption algorithm. Common types of attacks on encryption algorithm are: cryptanalysis and brute force.

Among the most used encryption techniques, we have chosen five most popular algorithms to discuss here. The most widely used the encryption techniques so far are AES and Triple DES which are the advanced versions of DES. Along with these, we would discuss two algorithms viz., Blowfish and RC5 algorithms.

2.1 Data Encryption Standard

The DES algorithm was developed by IBM for the US Government. It was based on the Lucifer cipher [1] which used 64-bit data block and 128-bit key established in 1971. Lucifer cipher was modified with the help of the National Bureau of Standards (NSA) to 64-bit block size and 56-bit effective key size (the key length was initially considered as 64 bit). This modification was adopted in 1977 [2] as the name DES (FIPS PUB 46, ANSI X3.92) and reaffirmed in 1994 for five years more. Later on, it was replaced by AES.

In the beginning, the DES key was 56 bits, and the message characteristics were confusion and diffusion. However, distributed.net and the Electronic Frontier Foundation collaborated to break the DES key in 22 h 15 min using a brute force attack in 1999. DES is a 16-round algorithm. The whole data is segmented in 64-bit blocks and each block is used as the plaintext. After an initial permutation, the plaintext block is considered as the initial vector for the first round. The 64-bit input for a key is compressed of 56 bits and is considered as the effective key. The right-most bit in each byte is a parity bit, and should be set so that an odd parity is always maintained. These parity bits are ignored and only the seven most significant bits of each byte are considered, resulting in a key length of 56 bits. Each time this 56-bit key is used to go through some left circular shift and considered as the key for the next round. These keys are known as subkeys and a separate subkey is generated for each and every round. After round 16, the output of 64 bits goes through the inverse initial permutation and we get the desired ciphertext.

Next, we would discuss the key processing and subkey generation as well as the functions of each round. Sixteen different 48-bit subkeys, one for each round, are created from the 56-bit key. In order to do that, at first, the key is loaded according to the first Permuted Choice PC-1 and then halved. These permuted choices are defined by IBM and considered as the standard one. Then each half is rotated by 2 bits in every round except the first, second, ninth and last rounds. The reason for this is that it makes it secure against related-key cryptanalysis. Then 48 of the 56 bits are chosen according to a compression permutation PC-2. The subkey generation as well as the function of each DES round is shown in Fig. 2.

First, the 64-bit block is divided into two 32-bit blocks—Right Plain Text (RPT) and Left Plain Text (LPT). The right half first goes through an expansion permutation which expands it from 32 to 48 bits according to a given permutation table. The 32-bit right plain text is divided into 8 blocks, each block containing 4 bits. Each 4-bit block of RPT is then expanded to a 6-bit block using a substitution box or S-box. The 32-bit block obtained in this way is again passed through a permuted choice and then XORed with the 32-bit LPT to generate the RPT for the next round. The RPT directly becomes the LPT for the next round. The same process is repeated for the next 15 rounds to get the final ciphertext.

The DES uses a 56-bit key and hence there are 2^{56} possibilities of keys. So, Brute-force attack would take a very long time to match the correct key. Encryption and decryption take the same algorithm. Only that the function needs to be reversed and

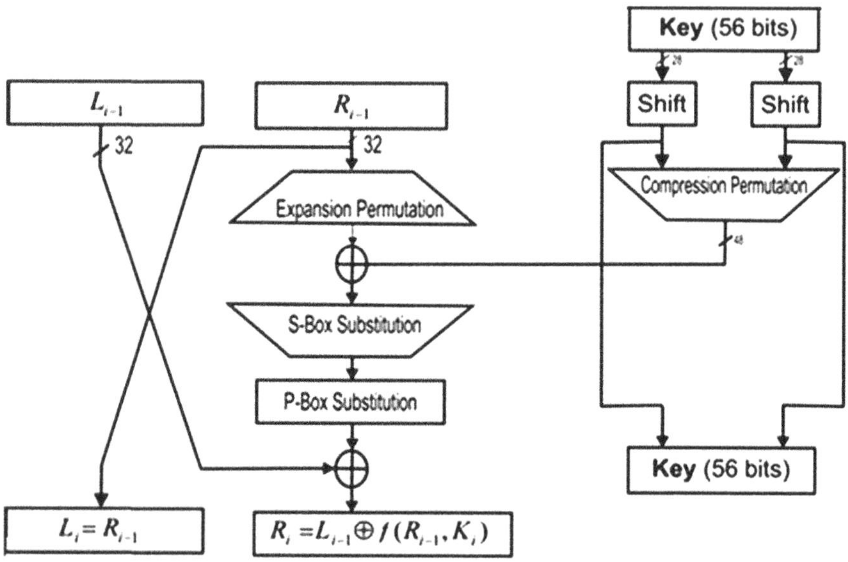

Fig. 2 Function of each DES round

the key should be taken in opposite order following the easier implementation. But DES fails in front of linear cryptanalysis because during its design this attack was not invented. Moreover, in the age of parallel computing, breaking DES has become easy with the help of a brute force attack, which was impossible during that time. So with time this algorithm is broken and has become obsolete itself. The National Institute of Standards and Technology (NIST) published that AES should replace DES and Triple-DES was published as an interim standard to be used until AES was made public. But AES is found as strong as Triple-DES and is significantly faster. Many security systems may support both Triple-DES and AES where AES is the default algorithm. Often Triple-DES is maintained for backward compatibility, but it is not recommended anymore.

2.2 *Triple Data Encryption Standard*

Triple Data Encryption Standard (3DES, 3-DES or TDES) is based on the DES algorithm; therefore it is found easy to convert the existing software to use Triple-DES from DES [3]. It has the advantage of having a longer key length (192 bit) that eliminates many of the known attacks as well as reduce the amount of time it requires to break DES. However, even this more powerful and reliable version of DES is only used for the small block size as it is not considered strong enough to protect much longer data.

Triple-DES takes three 64-bit keys or an overall key length of 192 bits. We may put the entire (24-character) key rather than entering each of the three keys individually. This 192-bit user-provided key is thus segmented into three subkeys, if necessary padding would be done to make each of the subkeys 64-bit long. The procedure for encryption is the same as the regular DES, but is repeated three times. That's why it is named Triple-DES. The data is encrypted with the first key, decrypted with the second key, and finally encrypted again with the third key.

Triple-DES is much more secure if used properly and also the implementation is easy in hardware as well as in software. It is ubiquitous, that is, most of the systems, libraries, and protocols include support for it. But 3DES runs three times slower than DES. The decryption procedure is the same as the encryption procedure, except it is executed in reverse. Like DES, it uses data in 64-bit chunks for both encryption and decryption. Although the input key for DES is 64-bit long, the effective key used by DES is only 56 bits long. So the effective key strength for Triple-DES reduces to 168 bits as each of the three keys excludes 8 parity bits during the encryption process. The block diagram of the TDES algorithm is shown in Fig. 3.

Advantages of 3DES:

- Implementation of 3DES is easy in both hardware and software approach.
- 3DES is ubiquitous in nature: most systems, libraries, and protocols include support for it.

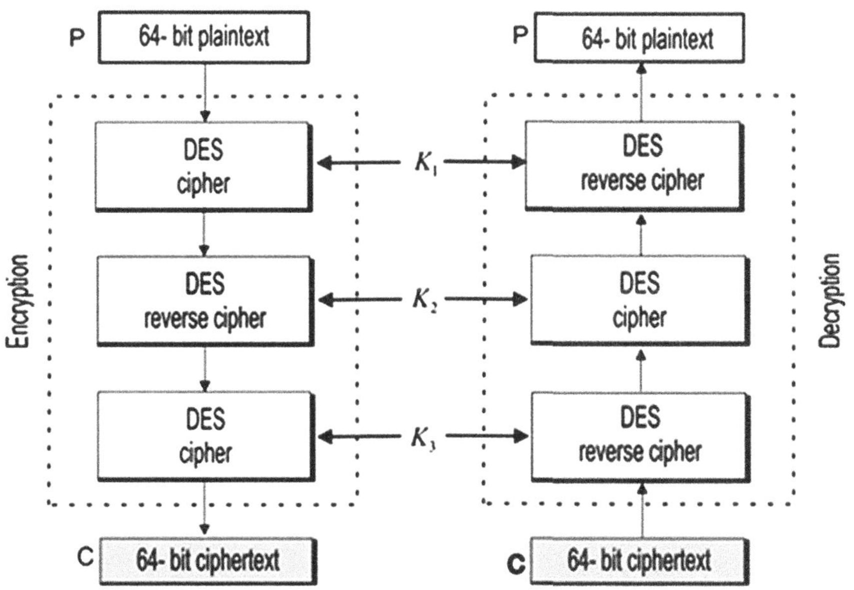

Fig. 3 DES algorithm

2.3 Advanced Data Encryption Standard

The AES is a well-designed cryptographic algorithm with a strong mathematical structure [4]. However, its main strength is the variation of key lengths. AES offers key lengths of 128-bit, 192-bit or 256-bit key, making it exponentially stronger than the 56-bit key of DES. The following diagram shows the steps of an AES round as well as the entire algorithm of 10 rounds. Here input data is considered as 256 bits or 16 bytes structured in a 4 × 4 array. Each byte is seen as an element of GF (2^8) which is a finite field of 256 elements.

There are mainly four steps involved in each of the AES round. Let us summarize the algorithm as follows:

Step 1: Initialize state ⟶ X (XOR) round key

Step 2: Following operations are involved in each of the nine rounds:

SubBytes: byte substitution works on bytes using a substitution table of 256 entries.
ShiftRows: It is nothing but simple byte shifting
MixColumns: Matrix multiplication in GF (2^8) which works on byte values and may be simplified by using a look-up table.
AddRound Key: works on byte XOR

Step 3: For the last round:

SubBytes
ShiftRows
AddRound Key

Step4: Output Y ⟵ State

Advantages of AES over 3DES:

- The security level of AES is high; it is less vulnerable to cryptanalysis than 3DES.
- The key sizes offered by AES are larger than 112 or 168 bits of 3DES.
- AES is faster than 3DES while implemented in both hardware and software.
- The 128-bit block size of AES makes it less susceptible to the birthday attacks than 3DES with 64-bit block size.
- AES is considered adequate by the latest international standards.
- It is resistant to known attacks.

2.4 Blowfish Algorithm

Blowfish encryption scheme was designed by Bruce Schneier in 1993 [5] as an alternative to the DES Encryption scheme. It is considerably faster than DES and provides a good encryption rate with no effective cryptanalysis technique found so far. It is one of those secure block ciphers which are not subject to any patents and hence freely available for anyone to use. The block size used is 64-bits. The key size

Table 1 Hexadecimal value of subkeys

P[0]	243f6a88	P[9]	38d01377
P[1]	85a308d3	P[10]	be5466cf
P[2]	13198a2e	P[11]	34e90c6c
P[3]	03707344	P[12]	c0ac29b7
P[4]	a4093822	P[13]	C97c50dd
P[5]	299f31d0	P[14]	3f84d5b5
P[6]	082efa98	P[15]	b5470917
P[7]	ec4e6c89	P[16]	9216d5d9
P[8]	452821e6	P[17]	8979fb1b

is variable and may vary from 32-bits to 448-bits. The number of subkeys used is 18. The number of rounds used is 16 and 4 substitution boxes are used each having 12 entries of 32 bits each. The workflow of the algorithm is explained step by step.

Step 1: Generation of subkeys

The algorithm utilizes 18 subkeys from P[0] to P[17] for both encryption and decryption process. These 18 subkeys are stored in a P-array with each array element being a 32-bit entry. It is initialized with the digits of P[i]. The hexadecimal representation of each of the subkeys is presented in Table 1.

Each of the subkey is changed with respect to the input key as:

P[i] = P[i] xor (i + 1)th 32 bits of input key.

The resultant P-array holds 18 subkeys that are used during the entire encryption process.

Step 2: Initialize Substitution Boxes:

The algorithm requires four Substitution boxes (S-boxes) from S[0] to S[4]. Each box having 256 entries S[0]…S[255], where each entry is 32-bit. It is initialized with the digits of P_i after initializing the P-array.

Step 3: *E*ncryption:

The encryption function consists of two parts:

- Rounds: The encryption consists of 16 rounds with each round (R_i) taking inputs the plaintext (P.T.) from the previous round and the corresponding subkey (P_i). The description of each round is shown in Fig. 4.

To calculate the F-function: The first byte of the 32-bits input is to fed as an entry in the first S-box, the second byte finds an entry in the second S-box, and so on. The value of the F-function is (X1(SB1) + X2(SB2)) XOR X3(SB3)) + X4(SB4) where addition is performed modulo 2^{32}.

Unlike the first 16 rounds the output after the 16 rounds is processed in a different way. The illustration of the last round is shown in Fig. 5.

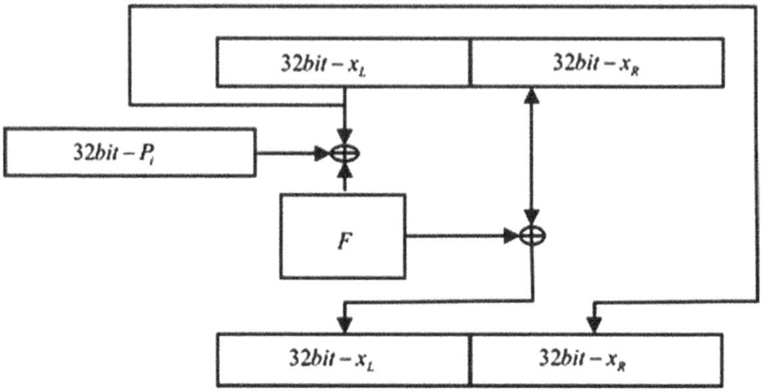

Fig. 4 Flow diagram of each blowfish round

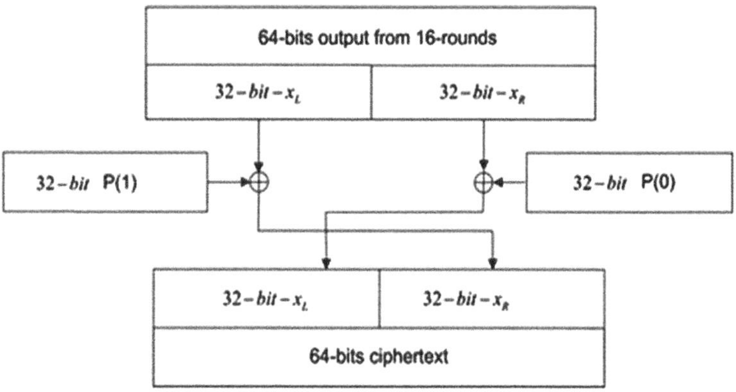

Fig. 5 Flow diagram of post-processing steps

The decryption process is similar to that of encryption only the subkeys are used in reverse, that is, P[17]—P[0]. The generation process of subkeys is the same as the encryption. The same S-boxes are used in decryption also.

Advantages and Disadvantages of Blowfish Algorithm:

- Blowfish is faster and much better than DES, but becomes slow while changing keys. This happens because each new key requires a pre-processing step which is equivalent to 4 KB of text.
- Blowfish uses a block size of 64 bits which makes it susceptible to birthday attacks.
- The version of blowfish which uses a reduced number of rounds is found to be vulnerable to known plain text attacks.

2.5 Rivest Cipher or Ron's Code Version 5 Algorithm

Rivest Cipher or Ron's Code version 5 (RC5) is a symmetric key block encryption algorithm designed by Ron Rivest in 1994 [6]. The algorithm is simple, fast (as uses only simple mathematical operations such as XOR, shift, etc.) and consumes less memory. It addresses two-word blocks at a time. The block size, number of rounds and key size use to vary according to the input plain text size. Various instances of RC5 can be defined depending on the values of word size (s in bits), number of rounds (r) and key size (b in bytes). Allowed values are shown in Table 2.

Since at a time, RC5 uses 2-word blocks, the plain text block size can be 32, 64 or 128 bits.

Step-1: Initialization of constants P and Q.

RC5 makes use of 2 magic constants P and Q whose values are defined by the word size s. Different values of P and Q according to the word size is shown in Table 3.

For any other word size, P and Q can be determined as:

$P = \mathrm{Odd}((e - 2)2^s)$

$Q = \mathrm{Odd}((\varphi - 2)2^s)$

Here, Odd(s) is the odd integer nearest to s, e is the base of natural logarithms and φ is the golden ratio.

Step-2: Converting the key (K) from bytes to words.

The secret key K of size b bytes is used to initialize array L consisting of s words where $s = b/u$, $u = s/8$ and $s =$ word size used for that particular instance of RC5.

Step-3: Initializing subkey

Subkey S is of size $t = 2(r + 1)$ and is initialized using magic constants P and Q.

Step-4: Subkey mixing.

The RC5 encryption algorithm uses Subkey S. L is actually a temporary array formed on the basis of user-entered secret key. Mix in user's secret key with S and L.

Table 2 RC5 attributes

Block/word size (bits)	16, 32, 64
Number of rounds	0–255
Key size (bytes)	0–255

Table 3 Values of constants *P* and *Q*

Word size (bits)	P (hexadecimal)	Q (hexadecimal)
16	b7e1	9e37
32	b7e15163	9e3779b9
64	b7e151628aed2a6b	9e3779b97f4a7c15

Step-5: Encryption.

The input plain text block is divided into two registers A and B each of size s bits. After going through the whole encryption process the result of A and B together forms the ciphertext block.

Decryption is simply the straight forward reverse of the encryption process.

2.6 Honey Encryption

Honey Encryption (HE) is a special type of Password-Based Encryption (PBE) scheme [7]. A PBE is a scheme for generating a symmetric-key that transforms an input string (a password) into a binary encryption key using various data-scrambling techniques. Typically, PBE accepts a low-entropy, user-given password, adds some entropy to it to increase its strength before turning it into a key. This key is then used for symmetric encryption. The problem is that the user's password often has very little entropy to start with. If an attacker may predict the salt, digest method and quantity of iterations, he may drive to guess the pre-strengthened password. He would try to match the related PBE, and then use the resulting key to match up with ciphertext that was created with that key. If the hacker is intelligent enough then he may get success in guessing the correct key with this kind of brute force attacks. HE is designed to develop such a ciphertext which when attempted to decrypt with any from a number of incorrect keys, yields plain texts, which look reasonable but actually bogus, known as honey messages. HE provides security against brute-force attacks that try every key, especially in those cases where the entropy is too little. In this way, HE provides security beyond conventional brute-force bounds. The heart of the conventional HE is Distribution Transforming Encoder (DTE).

A DTE is designed with an efficient decoder which produces the correct message from a given seed. So the estimate of the message distribution p_m is kept in mind while designing the DTE which makes it conceptually similar to arithmetic/Huffman coding [8]. If the seeds are uniformly sampled then the message produced by them is approximately distributed under p_m. A DTE is considered secure if it is able to produce enough confusion between these two distributions so that the hacker would not be able to distinguish these: (1) a message-seed pair generated by selecting the message from p_m and encoding it to obtain the seed and (2) a message-seed pair generated by selecting a seed uniformly at random and decoding it to obtain the message. DTE follows a two-step procedure to encrypt a message M under HE and the process is called *DTE-then-encrypt*. The DTE is at first used to obtain a seed S from a message M. Secondly, the seed S is encrypted with a conventional encryption scheme using a secret key K, which generates an HE ciphertext C. The message space must have to be equal to the seed space in this conventional encryption scheme and all ciphertexts must be decrypted to a valid seed under any key K. As the outputs of DTE require no padding (e.g., for Cyclic Block Chain-mode encryption) proper care must be taken to design it. A DTE is designed to estimate the target message distribution

p_m. If the DTE is only approximately good, we may prove that the message-recovery security is far beyond the brute-force-barrier. If the DTE is based on a poor estimation of p_m, we have to drop down to the normal security up to the brute-force barrier. This implies that it is never worse than prior PBE schemes, and, in particular, hackers must have to first execute the work of offline brute-force attacks before HE security becomes relevant.

There are significant examples of common DTE-then-encrypt construction schemes which found essential applications in the network security. There are HE schemes that are found to be very useful for RSA secret keys. Such schemes are used to develop a DTE for uniformly chosen pairs of prime numbers and make it possible to apply HE to RSA secret keys. Such systems are used by common tools such as OpenSSL which is a toolkit for Secure Sockets Layer (SSL). These DTEs are designed to implement a prime number generation algorithm. This scheme slows down the decryption process significantly but sometimes is found to be advantageous in PBE arrangement.

HE schemes are also crafted for password-based encryption of credit card numbers with associated Card Verification Values (CVVs), and user-selected PINs. A special DTE is required for encryption of PINs that deals with the problem of non-uniform distribution over messages. Several pragmatic studies show that users are heavily biased regarding PIN selection [9]. The consequential investigation results in a balls-and-bins game with non-uniform bin capacities, which is somewhat unusual association in the literature.

There are a few limitations associated with HE:

1. HE security is not strong enough when the adversary has side information about the target message.
2. As the decryption of an HE ciphertext using a wrong key produces fake but real-looking honey messages, a typographical error in passwords might confuse legitimate users.

3 Public Key Cryptosystem

The public key encryption or asymmetric encryption uses different keys for both encryption & decryption. One is called a public key which is known to everyone and the other is called a private key as it is private to the user. The advantage of these techniques is that it may be used for confidentiality, authentication or both. Here we would discuss a few well-known public key cryptography techniques namely RSA Algorithm, Elliptic Curve Cryptography and Hyperelliptic Curve Cryptography. Among these RSA is the widely used asymmetric algorithm.

3.1 RSA Algorithm

This algorithm was named after Ron Rivest, Adi Shamir and Leonard Adleman [10]. It was invented in 1977. RSA involves three steps: First is Key generation second is encryption and the third is decryption. These steps are explained one by one:

Step 1: Key Generation

a. Select r, s (primes)
b. Calculate $n = r \times s$
c. Calculate $\phi(n) = (r-1) \times (s-1)$
d. Select integer e such that gcd $(\phi(n), e) = 1$; $1 < e < \phi(n)$
e. Calculate p such that $p = e^{-1} \pmod{\phi(n)}$
f. Public key: PU $= \{e, n\}$
g. Private key: PR $= \{p, n\}$

Step 2: Encryption

a. Plaintext: $M < n$
b. Ciphertext: $C = M^e \bmod n$

Step 3: Decryption

a. Ciphertext: C
b. Plaintext: $M = C^p \bmod n$

The complexity of attacking RSA is based on the difficulty of finding the prime factors of a complex number. The complex mathematics makes it safe and secure for users.

The possible attacks on RSA are Brute force (involves trying all possible private keys), Mathematical attacks (factoring the product of two primes), Timing attacks (depend on the running time of the decryption algorithm), Chosen ciphertext attacks (exploits properties of RSA algorithm).

For any cryptosystem, two parties require to agree on a pair of keys (either the same key for the private key system or different keys for the public key system). Exchanging these keys over the Internet may be risky as the keys may be hacked by any third party. To eliminate this key transmission problem, key exchange protocols are used. Among the key exchange protocols, the most popular one is Diffie-Hellman key exchange protocol.

3.2 Diffie-Hellman Key Exchange Technique

It is one of the oldest examples of key exchange algorithm implemented in the field of cryptography [11]. The Diffie–Hellman algorithm allows both sender and receiver, who have no prior knowledge of each other, to mutually exchange a shared secret key over an insecure transmission channel. Diffie–Hellman key exchange is a symmetric key cryptosystem because the shared secret key and session key are used for both encryption and decryption. It is used by many protocols, such as Secure Socket Layer (SSL), Secure Shell and IPSec. Steps of Diffie-Hellman key exchange are as follows:

Step 1: Choosing Global Public Elements
q: Prime number
α: $\alpha < q$ and α is a primitive root of q
Step 2: User A Key Generation
Select private key X_A; $X_A < q$
Calculate public key Y_A; $Y_A = \alpha^X A \bmod q$
Step 3: User B Key Generation
Select private key X_B; $X_B < q$
Calculate public key Y_B; $Y_B = \alpha^X B \bmod q$.

The strength of this algorithm solely depends on the difficulty of computing discrete logarithms which are based on the extended Galois field GF(p^q) mathematics.

3.3 Elliptic Curve Cryptography Algorithm

Elliptic Curve Cryptography (ECC) was introduced in the mid of 80s by Victor Miller and Neal Koblitz independently [12, 13]. The elliptic curve cryptography algorithm mainly depends on the algebraic structure of the elliptic curves. An elliptic curve $E_{(a,b)}$(GF(p)), which is defined over a Galois field GF(p), is a set of points $(x, y) \in$ GF(p) * GF(p) which satisfies the Eq. (1) together with a special point, O, called the point at infinity. Each value of the "a" and "b" gives a different elliptic curve.

$$y^2 \equiv x^3 + ax + b \pmod{p} \tag{1}$$

The security of ECC depends on the difficulty of the elliptic curve discrete logarithm problem. If P and Q be two points on an elliptic curve such that $kP = Q$, where k is a scalar, then it is computationally infeasible to obtain k, if k is sufficiently large. k is the discrete logarithm of Q to the base P. Hence the main operation involved in ECC is point multiplication, that is, multiplication of a scalar k with any point P on the curve to obtain another point Q on the curve.

ECC includes three steps in its operation, that is, key agreement, key generation and authenticated encryption algorithm. The first step is used to share a secret key, the second step is the encryption algorithm that ensures confidential communication, and last is the digital signature algorithm which is used to authenticate the sender (or signer) and validate the integrity of the message.

Step 1: Key Agreement Process

In spite of the curve parameters a and b, there are other parameters that must be agreed by both the parties involved in a safe and reliable communication using ECC.

These are known as domain parameters. ECC domain parameters over GF(q), are a sextuple:

$T = (q, a, b, G, n, h)$
$q = p$ or $q = 2^m$
a and $b \in$ GF(q) such that

$$y^2 \equiv x^3 + ax + b\,(\text{mod } p)\text{for } q = p > 3 \tag{2}$$

$$y^2 + xy = x^3 + ax^2 + b \text{ for } q = 2^m \geq 1 \tag{3}$$

- a base point $G = (x_G, y_G)$ on $E_{(a,b)}$ (GF(q)),
- a prime n which is the order of G

(The order of a point P on an elliptic curve is the smallest positive integer r such that $rP = O$.)

- $h = \#E/n$, where $\#E$ represents the number of points on the elliptic curve and is called the curve order.

Step 2: Key generation for encryption

A public key $P_u = (x_u, y_u)$ is associated with a sextuple domain parameter (q, a, b, G, n, h) is generated for an entity A using the following procedure:

- Select a random or pseudo-random integer P_q in the interval [1, $n-1$].
- Compute $P_u = P_q\, G$.
- A's public key is P_u; A's private key is P_q.

A public key $P_u = (x_u, y_u)$ associated with a domain parameter (q, a, b, G, n, h) is validated for an entity A using the following procedure:

- Check that $P_u^{!}\ O$
- Check that x_u and y_u are properly represented elements of GF(q).
- Check that P_u lies on the elliptic curve defined by a and b.
- Check that $n\, P_u = \text{O}$.

Step 3: Elliptic Curve Authenticated Encryption Scheme (ECAES)

To encrypt a message m for B, A performs:

- Select a random integer d from [1, $n-1$].
- Compute $R = d\text{G}$.
- Compute $K = hdQ_B = (K_X, K_Y)$. Check that $K^{!}\ O$:
- Compute $k_1 \| k_2 = \text{KDF}(K_X)$.
- Compute $c = (k_1, m)$. Compute $t = \text{MAC}(k_2, c)$.
- Send (R; c; t) to B.

Here, MAC stands for Message Authentication Code, KDF stands for Key Derivation Function.

3.4 *Hyperelliptic Curve Cryptography*

The Hyperelliptic Curve Cryptography (HECC) cryptosystem is a subclass of ECC cryptosystem. As ECC creates smaller key sizes, it is considered better than other the public key cryptosystems. The minimum key size for ECC should be 132 bits versus 952 bits for RSA. Hyperelliptic curves have implementation advantages as compared to RSA or ECC because it contains a smaller key size than ECC. The operand bit length for RSA is 1024–2048 bit, for ECC it is 160–256 bit, but for HECC it is between 50 and 80 bits. That's why it is absolutely suited for constraint environments such as clouds, phones, smart cards, etc. Moreover, it offers greater speed along with less storage space which makes end users not worry about the required resources such as memory, processor, bandwidth, etc., for which he has to pay in the cloud. This makes HECC desirable in the cloud.

Hyperelliptic curves are special types of algebraic curves that may be considered as a generalized form of elliptic curves [14]. Generally, the curves defined with *genus* = *1* is known as the elliptic curve. Hyperelliptic curves have genus ≥ 1. Genus of a curve is a number of nonintersecting simple closed curves that can be drawn on the surface without separating it. HECC is a type of public-key cryptography which is based on point multiplication and point addition such as ECC.

Considering k to be a field, the general equation of hyperelliptic curve C of genus g over k is depicted in Eq. (4) where $h(x)$ is a polynomial of degree $\leq g$ over **F** and $f(x)$ is a monic polynomial of degree $2g + 1$ over **F**

$$C : y^2 + h(x)y = f(x) \quad (4)$$

Security Arrangement using HECC

Two organizations P and Q which use their own data and software are assumed to act like public clouds. A situation is considered such that organization P wants to send some data on demand of Q. After getting the request from Q, P will retrieve the data from its database. After that P will sign the document with its private key using the encryption scheme explained below. Thus P will send the encrypted message along with the signature. After receiving it, Q will verify the sign using P's public key. After verification Q will decrypt the ciphertext and get the original message.

Digital Signature Algorithm using HECC

This section includes three main categories of cryptographic schemes based on hyperelliptic curve cryptography. They are key agreement, encryption and digital signature schemes.

A. *Key Agreement*

The Diffie-Hellman key agreement protocol uses the multiplicative group of numbers with a prime modulo, but it may be formulated using general groups also. Let G be a group whose elements may be represented efficiently in a proper way, and in which the group operations are well explained as well. The group is Jacobians of hyperelliptic curves [15, 16].

Now, we consider the publicly known system parameters as follows:

- The group G.
- An element $R \in G$ of large prime order r.

The steps that Organization P performs are the following:

1. Choose a random integer $a \in [1, r - 1]$.
2. Compute $A = aR$ in the group G, and send it to Q.
3. Receive the element $B \in G$ from Q.
4. Compute $S = aB$ as common secret.

The steps that Organization Q performs are:

1. Choose a random integer $b \in [1, r - 1]$.
2. Compute $B =$ bR in the group G, and send it to P.
3. Receive the element $A \in G$ from P.
4. Compute $S = bA$ as common secret.

Note that both P and Q have computed the same values S, as

$$S = a(bR) = (ab)R = b(aR)$$

B. *Encryption/decryption scheme*

This scheme is based on the ElGamal encryption process which uses the complexity of discrete logarithm both for encryption and decryption. The scheme is again discussed for a general group G [16]. The following publicly known system parameters are considered:

- The group G.
- An element $R \in G$ of large prime order.

Organization P wants to send Q a message M, which is assumed to be encoded as an element of the group G. Organization P wants to encrypt M using Q's public key B, such that only Q can decrypt the message again, using his secret key b.

To encrypt M, P goes through the following steps:

- Obtain Q's public key i.
- Choose a secret number a $[1, r - 1]$.
- Compute C1 $= aR$.
- Compute C2 $= M + aB$.

- Send (C1, C2) to Q.

Q can decrypt the encrypted message by obeying the following steps:

- Obtain the encrypted message (C1, C2) from P.
- Compute $M = C2 - bC1$.

C. *Signature generation and verification*

The digital signature algorithm can be used for any group G where it is difficult to execute the Discrete Logarithmic problems [17]. To create a key pair, P chooses a secret integer $p \in Z$, and computes $A = pR$. The number p is P's secret key, and A is it is the public key. Consider the following system parameters are publicly known:

- A group G
- An element $R \in G$ with large prime order,
- A hash function H that maps messages m to 160-bit integers.

If P wants to sign a message m, it has to do the following:

- Choose a random integer k $[1, r - 1]$, and compute $B = kR$.
- Compute $s \approx k^{-1}$ $(H(m) + a\varphi\ (Q))$ mod r.
- The signature is (m, Q, s).

To verify this signature, a verifier Q has to do the following:

- Compute $v_1 \approx s^{-1}H\ (m)$ mod r and v_2 s^{-1} (Q) mod r.
- Compute $V = v_1R + v_2A$.
- Accept the signature if $V = Q$. Otherwise, reject it.

4 Authentication and Digital Signature Algorithms

Authentication is a procedure to confirm that the messages are coming from the alleged source and have not been tampered. If required, message authentication may also be used to verify sequencing and timeliness. An authentication technique that also takes measures to counter denial of service by either source or destination is known as Digital Signature. Message authentication or digital signature mechanism uses some sort of functions for producing an authenticator which is a value to be used to authenticate a message.

There are three types of functions that are used to produce an authenticator:

A. Message encryption

- Here ciphertext itself serves as authenticator.

B. Message authentication code (MAC)

- It is a fixed-length value generated from a function of the message and a secret key that serves as the authenticator.

C. Hash function

- Here the authenticator is a fixed-length hash value produced from the message of any length being mapped by a public function.

4.1 Message Encryption

Ciphertext generated by a conventional encryption can serve as authenticator. It provides authentication as well as confidentiality. This concept is explained via the diagram in Fig. 6.

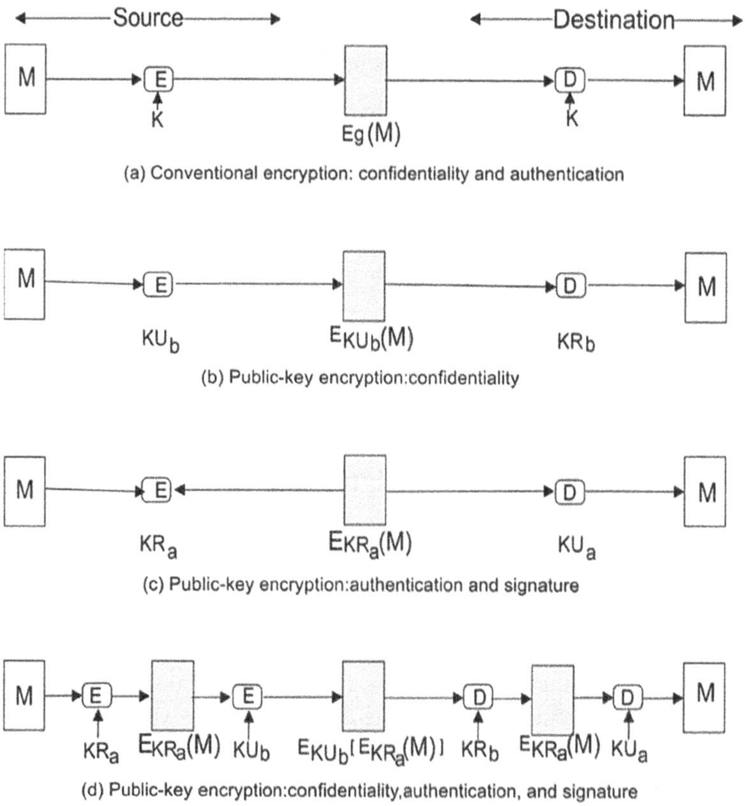

Fig. 6 Public key encryption providing confidentiality and authentication

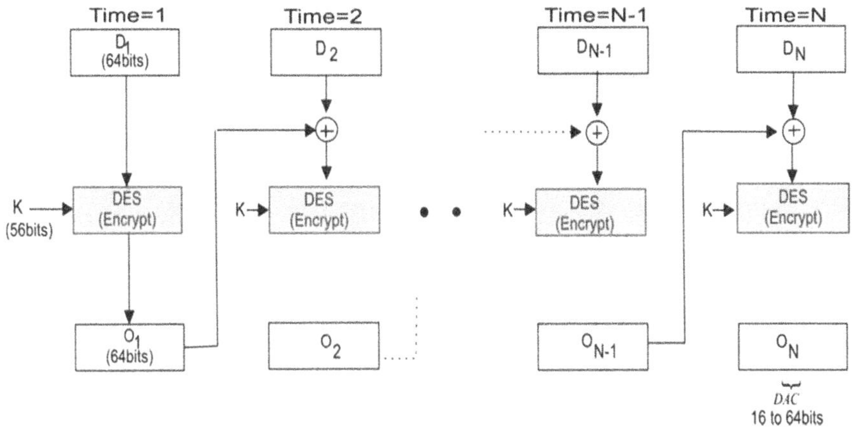

Fig. 7 MAC using conventional encryption (DES)

4.2 *Message Authentication Code*

Message Authentication Code (MAC) employs a shared secret key to generate a fixed-length block of data (also known as cryptographic checksum) which is used to append to the message before transmission. At the receiver side, the MAC is again generated with the help of the shared secret key and compared to the received MAC. If a match is found then the message is considered intact otherwise it is considered fraudulent or hacked. Sometimes the MAC is encrypted before appending to the message and sometimes the whole message plus MAC is encrypted after amalgamation to increase the security level. The MAC assurances that the message has not been altered and it is from the alleged sender, as well as the sequence of the message, are unaltered. The MAC algorithms are similar to encryption but need not be reversible.

MAC uses a shared secret key to generate a fixed-size block of data also known as a cryptographic checksum. It is represented as MAC $= C_K(M)$, where k is the secret key. An example of a DES-based MAC generator [18] is shown in Fig. 7.

4.3 *Hash Function*

Among the several algorithms generating hash functions the most popular one is the message digest (MD) algorithm. The advanced version of this is known as a secure hash algorithm (SHA). Another popular hash function is hash-based message authentication code (HMAC) which is a combination of Hash and MAC algorithm.

Message Digest Algorithm is a cryptographic Hash function, first developed by Ronald Rivest in 1989. The first version was named message digest algorithm 2 or

MD2. Although the advanced version of this algorithm has been already proposed since, such as version 4(MD4), version 5(MD5) and SHA.

Message Digest Algorithm Version 5

Message Digest 5(MD5) is a widely used cryptographic hash function with a 128-bit hash value. MD5 has been exploited in a wide variety of security applications and also used to check the integrity of files. An MD5 hash is typically a 32 character hexadecimal number. MD5 was designed by Ronald Rivest in 1992 [19] to replace MD4 which was broken by that time. MD5 is also not completely flawless.

Whatever be the length of the input message, they are processed in 512-bit blocks. The number of such blocks is considered as K. If the input message length is not a multiple of 512, padding is done as per requirement from 1 to 512 bits. First, a single bit 1 is appended to the end of the message. This is followed by as many zeros as are required to bring the length of the message up to 64 bits less than a multiple of 512. The remaining bits are filled up with a 64-bit integer representing the length of the original message. The message length is considered as k mod 2^{64}. The length of the output is 128 bits and is called the message digest. The algorithm is explained in Fig. 8.

There are four rounds with sixteen mathematical operations in each round.

g is a nonlinear function: one function is used in each round.

$X[k]$ denotes a 32-bit block of the message input.

$T[i]$ denotes a 32-bit constant, different for each operation.

All $\oplus$ symbols denote modulo 2^{32} additions.

The main MD5 algorithm operates on a 128-bit state, divided into four 32-bit words, denoted by A, B, C and D. These are initialized to certain fixed constants.

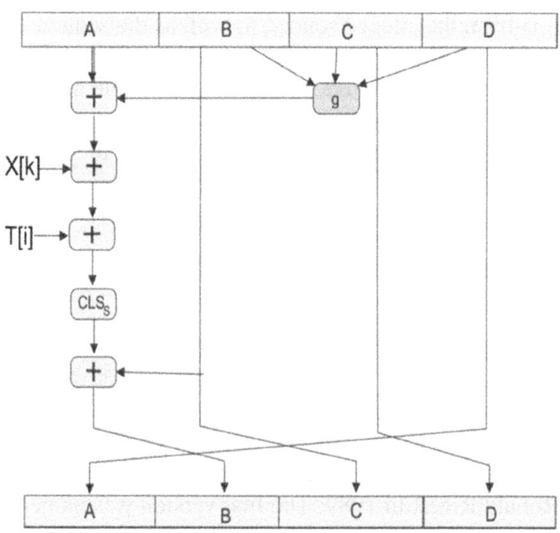

Fig. 8 MD5 algorithm

Secured Hash Algorithm (SHA)

SHA-1 was proposed by NIST as a message digest algorithm [20]. Actually, the SHA hash functions refer to five approved algorithms for computing a message digest that is, to a high degree of probability, unique for a given input data sequence. As the five algorithms are denoted by SHA-1, SHA-224, SHA-256, SHA-384, SHA512, the latter four variants are collectively referred to as SHA-2. These algorithms are called secured because for a given algorithm, it is computationally infeasible to find a message that corresponds to a given message digest. It is also difficult to find two different messages that produce the same message digest. If the message is changed then the digest will change accordingly. It is also very fast to compute.

Here the first step is to prepare the message. Given m, create 1000…000mmm…mxxxxx….x and append it to the length of m ($<=2^{64}$, so can be written in 64 bits). We append a 1 and then enough zeros to make the total a multiple of 512 bits. The whole message is broken into L, 512-bit chunks. Each will be used to compress into a 160-bit total message digest. The operations involved in each SHA round are explained in Fig. 9.

There are 4 rounds of 20 iterations each, that is, altogether 80 rounds.

Each iteration of SHA-1 includes compression function A, B, C, D and E which are 32-bit words of the state. A nonlinear function is used that varies in each round.

$W[i]$ denotes a 32-bit block of the message input.

$K[i]$ denotes a 32-bit constant, different for each iteration.

All $\oplus$ symbols denote modulo 2^{32} additions.

Hash Massage Authentication Code

Hash Massage Authentication Code (HMAC) resulted from an effort to find a MAC algorithm that could be proven to be secure if there is underlying MD's compression function f [21]. They are defined as secure for having the Collision Resistance property. An attacker, who does not know the key, would not be able to compute the proper digest. The HMAC would be represented as,

$$\text{HMAC}_K = \text{Hash}\ [(K^+ \oplus \text{opad})||\text{Hash}\ [(K^+ \oplus \text{ipad})||M)]]$$

where K^+ is the key padded out to input block size of the hash function and opad, ipad are specified padding constants.

Key size: $L/2 < K < L$

MAC size: at least $L/2$, where L is the hash output.

Figure 10 illustrates the overall operation of HMAC algorithm.

The terms of the figure are defined as follows:

H = Embedded Hash function (e.g., SHA-1)

M = Message input to HMAC

Y_i = ith block of M, $0 \leq i \leq (L-1)$

L = Number of blocks in the message

b = Number of bits in a block

n = length of Hash code produced by Embedded Hash function

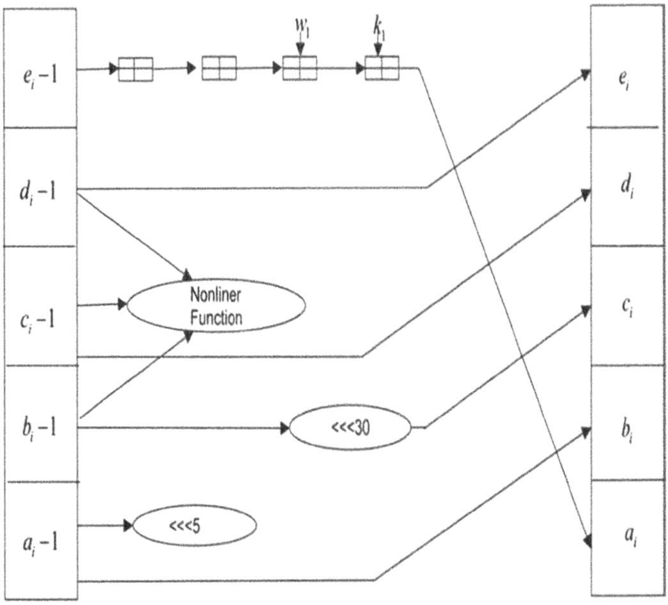

$$f_t(B,C,D) = \begin{cases} (B\wedge C)\vee((\neg B)\wedge D) & if\ 0 \le t \le 19 \\ B\oplus C\oplus D & if\ 20 \le t \le 39 \\ (B\wedge C)\vee(B\wedge D)\vee(C\wedge D) & if\ 40 \le t \le 59 \\ B\oplus C\oplus D & if\ 60 \le t \le 79 \end{cases}$$

Fig. 9 Each round of SHA-1

K = Secret key; if the key length is greater than b, the key is input to the Hash function to produce a b bit key; the recommended length of b is $\geq n$.

K^+= K padded with zeroes on the left so that the result is b bits in length.

ipad = 00110110 (36 in Hexadecimal) repeated $b/8$ times

Opad = 01011100 (5C in Hexadecimal) repeated $b/8$ times

Security of HMAC relates to that of the underlying Hash algorithm. There is no known practical attack against HMAC if used with a Secure Hash Functions (e.g., SHA-1) according to the proper specifications. In general, HMAC may be attacked either by brute force on the key space or attacks on the Hash function itself. Birthday attack is also possible, although the use of key makes this attack more difficult. There may be attacks against the compression function also.

Digital Signature

Digital signatures are analogous to the handwritten signature. However, it must verify the author and the date and time of the signature. It should also be able to authenticate

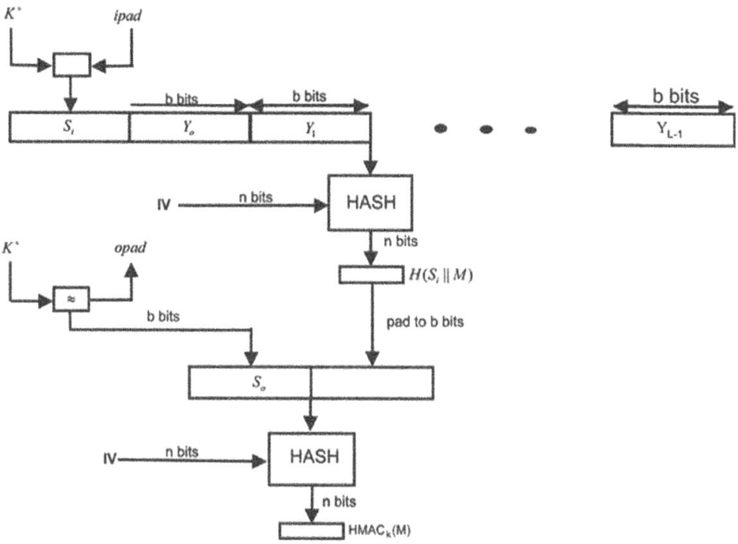

Fig. 10 HMAC algorithm

the contents at the time of the signature and must be verifiable by third parties to resolve the dispute. Generally, the signature is a bit pattern that is generated as a function of the message being signed. It uses some information unique to the sender, to prevent both forgery and denial. A digital signature must be relatively easy to produce and should be relatively easy to recognize and verify. It must be computationally infeasible to generate a false digital signature. Hackers may try to generate false signatures either by constructing a new message for an existing digital signature or by constructing a fraudulent digital signature for a given message. There are two types of digital signatures: direct signature and signature via arbitrated digital signature.

The direct digital signature involves only the sender X and the receiver Y. In the case of an arbitrated digital signature, every signed message from the sender to the receiver goes at first to an arbiter A, who investigates the origin and content of the message and its signature through a number of tests. After the verification satisfies the arbiter, the message is then dated and sent to Y. The strength of the direct digital signature scheme completely depends on the security of the sender's private key. The sender can claim that the private key was stolen or lost and that someone has forged the signature with it. Again the private key may actually be stolen from X at time T. This may also happen that the opponent can send a message signed with X's signature and stamped with a different time which is before or equal to T. The presence of A solves these problems faced by the direct signature schemes. The arbitrated digital signature is also of two types: (1) the arbitrator may see the message and (2) the arbitrator would not be able to see the message. The scheme is presented in a nutshell in Table 4.

Table 4 Different digital signature algorithms

(a) Conventional encryption, arbiter sees message
(1) X → A: M ‖ E_{Kxa} [ID_X ‖ H(M)] (2) A → Y: E_{Kay} [ID_X ‖ M ‖ E_{Kxa} [ID_X ‖ H(M) ‖ T]
(b) Conventional Encryption, Arbiter does not see Message
(1) X → A: ID_X ‖ E_{Kxy} [M] ‖ E_{Kxa} [ID_X ‖ H(E_{Kxy}[M])] (2) A → Y: E_{Kay} [ID_X ‖ E_{Kxy} [M] ‖ E_{Kxa} [ID_X ‖ H(E_{Kxy}[M])] ‖ T]
(c) Public key Encryption, Arbiter does not see Message
(1) X → A: ID_X ‖ E_{KRx} [ID_X ‖ E_{KUy}(E_{KRx} [M])] (2) A → Y: E_{KRa} [ID_X ‖ E_{KUy}(E_{KRx} [M]) ‖ T]

Elliptic Curve Digital Signature Algorithm

Elliptic Curve Digital Signature Algorithm (ECDSA) is a digital signature scheme which utilizes the complexity of Elliptic Curves to generate computationally infeasible digital signatures. The scheme was proposed by Don Johnson, Alfred Menezes and Scott Vanstone in 1999 and was accepted as an ANSI standard ANSI X 9.6 2ECDSA [22]. It was also accepted as IEEE and NIST standards in the next year. Here, the entity A has domain parameters $D = (q, a, b, G, n, h)$, public key Q_A and private key d_A. The entity Y has authentic copies of D and Q_A. To sign a message m, X does the following:

- Select a random integer k from $[1, n-1]$.
- Compute $kG = (x_1, y_1)$ and $r = x_1 \bmod n$. If $r = 0$ then go to step 1.
- Compute $\mathrm{k}^{-1} \bmod n$. Compute $e = \text{SHA-1(m)}$.
- Compute $s = \mathrm{k}^{-1}\{e + d_A.\, r\} \bmod n$.
- If $s = 0$ then go to step 1.
- A's signature for the message m is (r, s).

To verify A's signature (r, s) on m, B performs the following steps:

- Verify that r and s are integers in $[1, n-1]$.
- Compute $e = \text{SHA-1(m)}$.
- Compute $w = \mathrm{s}^{-1} \bmod n$.
- Compute $u_1 = ew \bmod n$ and $u_2 = rw \bmod n$.
- Compute $(x_1, y_1) = u_1 G + u_2 Q_A$
- Compute $v = x_1 \bmod n$.
- Accept the signature if and only if $v = r$.

Here, SHA-1 denotes the 160-bit hash function.

5 Quantum Cryptography

Moore's law reveals the fact that the requirement of transistors on Integrated Circuit (IC) chips is becoming approximately double in every two years as the availability of transistors with smaller size and faster processing speed has increased enormously. This is inviting huge energy consumption as well as heat production in the chip which is increasing day by day with the number of transistors. Moore's law is still working, however, gradually it would be obsolete. Quantum theory offers a more powerful computational approach to solve this problem. In quantum theory, the operations are reversible, which implies that the inputs can be assumed from the outputs. This process prohibits the heat production and energy dissipation as well. This allows the circuit to save much more energy than the current technology. Moreover, the quantum theory algorithms are able to enhance the computational speed for mathematical problems beyond the capability of normal computers. For example, it could easily carry out complex mathematics such as prime factor decomposition, which is the core part of the RSA algorithm.

In quantum theory, a particle is called a qubit and it is assumed that each qubit is spinning in its own direction. This spinning direction of the particle is called its state. A qubit's state can be changed by performing quantum operations. For digital information, binary numbers are assumed depending on the spinning direction of the qubit. The qubits that spin up represent the "0", and conversely the qubits that spin down represent "1". Other qubits with different spinning directions will change their directions to either up or down through measurement. The term *measurement* means measuring the probability of spinning up or down. However, before measuring we cannot be sure about their directions. This is an incredible feature because it implies that a qubit may be "0" and "1" simultaneously before measurement. This phenomenon is called superposition. This special feature uses parallel computing to speed up the computation time. Another interesting feature is the entanglement. The spinning directions of two or more qubits are correlated to each other. There are two kinds of entanglement states. The first is due to qubits spinning in the same direction. The second is spinning in opposite directions. It is assumed that there are two qubits, namely qubit1 and qubit2 which are in entanglement with each other. In the case of the first type of entanglement, qubit1 is measured at first. If it spins up, then it can be immediately assumed that qubit2 spins up as well. The second type of entanglement is where if qubit1 spins up, it is predicted that qubit2 spins down. It is like some kind of super relationship that exist between them. It does not matter how far they are from each other, we can always correctly predict one's direction from the other. The quantum mechanics features seem to be so strange to us that we may doubt their reality. But many experiments are conducted so far and all results show that they are real and that this is how the universe works. If the advantages of these features are utilized intelligently that would define a new set of ways to solve our difficulties. Shor et al. published a quantum algorithm in 1994 which dramatically reduced the time for integer factorization [23] as well as is able to carry out prime factor decomposition which is the heart of the RSA algorithm. It uses qubit superposition to represent all possible solutions and to find the cycle for all solutions divided by the product of two prime numbers in polynomial time. So in the quantum world, encryption techniques such as DES and RSA would become unsafe in the future due to this high-speed computation regardless of the key length. Quantum computers try all solutions at the same time, which would make it very useful for the brute force attack.

The two widely accepted security protocols of quantum computing need to be explained before discussing the application of quantum cryptography in network security. These two protocols are Quantum Key Distribution (QKD) protocol and Quantum Secret Direct Communication (QSDC) [24].

5.1 Quantum Key Distribution

All the encryption techniques discussed so far use fixed-length keys. But a limited key does not always guarantee the absolute security as they are vulnerable to Brute force attack. On the other hand, unlimited keys which are also known as one-time pad are useful in this context. QKD is one type of one-time pad which perfectly solves the cryptographic key transmission problem in quantum theory. It is a modification of BB84 [25] which is a renowned protocol for creating a one-time pad. It can create the shared key and is also able to detect intruders. Let us first discuss the BB84 process. In this process, the sender and the receiver share a random key with each other before communication. The first step of doing that is to define spin type: up and down as well as left to right to represent "0" and "1," respectively. The spin up and down type is defined as key1 and spin left and right type as key2. The sender at first arranges a series of qubits with a random spin type. He then sends the sequence to the receiver. If the receiver chooses the same position type as prepared by the sender, they can exactly get bit "0" or "1." The positions having exactly bit 0 or 1 form their random shared key and the rest of the qubits would be discarded. This shared key would be used for both encoding and decoding. Now, eavesdroppers would try to steal the sequence and measure them. But in this case error rate would be higher than the secure communication. So, by checking the error rate we may determine whether the information is stolen or not.

5.2 Quantum Secret Direct Communication

Quantum Secret Direct Communication (QSDC) combines the BB84 concept with direct communications. This is a kind of protocol for direct communications with quantum mechanics that guarantees a high level of security [26]. Like BB84 the sender at first generates a sequence of qubits with different spine types. These qubits contain test qubits and message qubits. After receiving these bits the test qubits are checked to find if any eavesdropper has interrupted. If no tampering is found then the sender tells the spin type of each message qubit to the recipient, Bob. When Bob measures the message qubits with the correct spin type, he can decode the information correctly. If no eavesdropper is detected, the test qubits are thrown away and communication proceeds.

5.3 Quantum Secret Sharing

In spite of these two protocols, there is another well-known secret sharing technique of quantum theory, known as Quantum Secret Sharing (QSS). If a sender divides a secret message into m components and sends them to m agents, it means that the m agents will have to cooperate with each other to decode the secret message. Now, let us discuss different steps of the QSS protocol to explain the structure of the protocol [27, 28]. Let us consider there are four agents A, B, C, D. It requires four steps to complete the whole process. For simplification, we assume that A is sending a message containing eight classical bits and does not require any channel checking.

Step 1: A prepares three pairs of two entangled qubits ordered by (1, 2), (3, 4), and (5, 6). A then sends one qubit of each pair to each receiver.

Step 2: A at first measures all the qubits he have and then the relations of the entangled qubits are changed. There is a special feature of quantum theory known as entanglement swapping. Here, if there are a number of pairs and first qubits from each of the pairs are entangled with each other, then other qubits would be mutually entangled with each other. According to this theory, three of the qubits, which A holds now are entangled with each other as well as other qubits are also mutually entangled.

Step 3: After this swapping process, A encrypts his message. After encryption, he measures these three qubits and announces the outcome publicly. Three additional bits of information are also announced to the receivers for decoding. The three bits come from the computational result of A's secret message.

Step 4: B, C and D cooperate to decode A's message using the qubits they have and the classical bits that are published by *A*.

Most QSS protocols require a code table to decode the message. This QSS protocol does not need a code table. The required decoding information is obtained using calculations and the entangled qubits stored in the receivers. Few QSS schemes are also available which require a code table for decoding.

5.4 Applications of Quantum Computing for Network Security

Generally, most of the security protocols are based on mathematical complexity. Quantum computers are able to eliminate this mathematical complexity. Luckily, data scientists have already developed various cryptographic algorithms and security protocols based on quantum mechanics whose security is governed by physical laws. Physical laws completely assure the communication security as well as computational security. In this section, we shall discuss various such applications for the Internet.

Secure computation provides the final decryption result without revealing the required details. There is such a secure computation problem [29] which is called the dating problem. Suppose A and B decide to date each other. But they do not want to expose their intention to the public except the ultimate result. The formal way to solve this dating problem is to use a prime factor decomposition concept. A quantum machine may help us to achieve the desired result. Here we consider the quantum machine as a black box. Imagine a black box exists between A and B. Inside the box, there are four entangled qubits. A and B, each has two qubits in their possession. A and B each at first measure one of their qubits. If the outcome is in accordance with their intentions, the process continues; otherwise, they restart the process. If the process goes on, they measure the remaining qubits and publish their second outcome. Now the final result may be obtained by computing the published outcomes. The dating problem is solved by using the quantum machine. But the truth is that, the quantum machine does not exist. The whole process is actually accomplished using four entangled qubits. Only the entanglement property can help us solve the dating problem. This kind of computation concept may be used for cloud computing as well to enhance the security. We may assume that this black box is a cloud data center. A and B encode their data into qubits and then send the encoded qubits, to the cloud data center. This center performs the computation and sends the outcomes back to them. In the end, they obtain the wanted outcomes without revealing their own data. Moreover, it is not possible for the cloud data center to know anything about the encoded data because any measurement would certainly disturb the qubit states. In this way, quantum mechanics guarantees the transmission and computation security.

There is another application of e-commerce that uses the concept of QKD and QSDC [30]. Suppose A and B being customer and seller, respectively, have committed a deal through an online shopping mall. There are some constraints as follows. First, the business platform is provided by an online shopping mall. Second, the process by which the customers have to find and buy goods must be through the online shopping mall. Third, sellers provide their services through the online shopping mall. We have achieved two advanced improvements in this application. The online shopping mall will control the information transmission. A and B transfer their messages simultaneously. We ignore the test qubits in this protocol for simplification, but the test method is similar to BB84.

There is yet another quantum protocol developed mainly for on-line transactions over the Internet. It is known as One Time Pad or OTP, the unbreakable encryption method. In this protocol, random key pairs which are as long as the plain text is used to encrypt the plain text before sending it to the receiver. The keys are known to each other before the encryption takes place. The key is mixed by (XOR-ing) bit by bit. A bit of the key is blended with a bit of the plain text to create a bit of ciphertext. Later, the encrypted message is mixed (XOR-ed) with the duplicate copy of the One Time Key and the plain text is recovered [31].

Another quantum cryptographic protocol namely Three Stage Quantum Protocol was first proposed in 2006 and implemented in 2012. It is asymmetric cryptography-based protocol, where each applicant utilizes one's own private key to encode the messages sent over the Internet. Here the message remains quantum in each stage in comparison to the BB84 protocol, where the stream of quantum bits are transferred in a single direction [32] and classic data are switched consequently. There is a probability of the Multi-qubit Quantum Cryptography, as this protocol, proceeds multiple qubit to interchange the information between the sender and receiver which improves the transmission of data.

6 Homomorphic Cryptography

In the last decade, cloud computing technology has been expanded very rapidly with the evolution of IoT and WSN. It is providing a new facility to store and work with large amounts of data. Cloud computing has many characteristics, for example, multi-user, virtualization, scalability and so on. Due to all of these characteristics, traditional security technologies cannot make Cloud computing fully safe. So data security remains a vital issue particularly when the public cloud service provider is used. The new scheme of homomorphic cryptography would ensure the data privacy of the users on the cloud increasing the prospective market for cloud computing. Cloud computing provides many services to its users, which includes the option of storing and computation of huge amounts of data to the cloud. However, to take the benefit of it, users must trust and share their data with the cloud service providers. To ensure the privacy of the data it is encrypted before uploading it to the cloud. Now if users desire to make some computation on the data, the data must be downloaded and decrypted which may lead to the loss of confidentiality. One solution to this problem is to use the homomorphic encryption. This type of encryption scheme allows computations on encrypted data, without fully decrypting the data on the cloud. Partially homomorphic encryption schemes have been used for many years, providing the facility to carry out a few particular operations like addition or multiplication on the ciphertexts without decryption. Examples of such schemes are the additively homomorphic Paillier [33] or the multiplicatively homomorphic ElGamal [17] cryptosystems. However, this partially homomorphic schemes are not sufficient except few applications. The idea of a Fully Homomorphic Encryption (FHE) scheme was first proposed by Rivest, Adleman and Dertouzos in 1978 [34]. At that time they referred it as privacy homomorphism. After a long time, the same concept was revisited by Craig Gentry in 2009 in his Ph.D. thesis [35], in which he introduced the first plausible fully homomorphic encryption scheme. A FHE scheme is an encryption scheme that follows an efficient estimate of an arbitrary depth circuit (composed of additions and multiplications) to be assessed directly on encrypted data. Gentry introduced the concept of constructing an FHE from a so-called Somewhat Homomorphic Encryption (SHE) scheme. SHE is an encryption scheme that allows the evaluation of a limited depth circuit. To date, three main branches of homomorphic encryption schemes have been developed: Lattice-based, Integer-based and Learning-With-Errors (LWE) or Ring-Learning-With-Errors (RLWE)-based encryption.

The literature survey reveals that the homomorphic cryptography has been grabbing the maximum research attention over the last few years; yet the implementations of such encryption schemes remain unsuitable for real-time applications. For example, key generation in Gentry and Halevi's lattice-based scheme [36] takes from

2.5 s to 2.2 h. Moreover, a recent implementation required 36 h for a homomorphic evaluation of AES using FHE by Gentry et al. [37]. This limitation is overcome by using the Graphics Processing Units (GPUs) and Field Programmable Gate Arrays (FPGAs) [38, 39]. Another significant limitation of FHE schemes is memory usage. Generally, very large ciphertext and public key sizes are required to strengthen the security to prevent possible lattice-based attacks. Gentry and Halevi's FHE scheme uses public key sizes ranging from 17 MB to 2.25 GB [40]. Also in the above-mentioned AES scheme, memory is found to be the main limiting factor during implementation [41]. A further limitation to FHE schemes is the costly bootstrapping operation which is required to reduce and manage the noise during the homomorphic encryption process. Recent research has therefore focused on optimizations to improve the efficiency of the FHE schemes, such as minimizing the need for the expensive bootstrapping operation [42] or employing batching techniques for multiple bit encryptions [43]. Although there has been a lot of recent research in the field of homomorphic cryptography, many open problems are still there. One such problem is parameter selection as per theoretical research. Parameter selection is a very complicated process as each scheme has specifically selected parameters that are interlinked with each other and are generally selected on the basis of current possible lattice-based attacks and their existing limits. More research is needed in this area to ensure the selection of the most suitable parameters to guarantee both security and efficiency.

7 Light Weight Cryptographic Techniques for Wireless Sensor Network and IoT

The WSNs consist of several tiny sensor nodes. These sensor nodes have capabilities of sensing and processing various real-world data collected from the environment. When these sensor nodes come in the range of communication, they form an ad hoc network and are able to communicate among themselves. For communication purpose, they use to have in-built transceiver. The data collected by the sensor nodes reach the server through a base station which in turn is connected to the Internet. Moreover, as they are ad hoc in nature, hence they may be deployed anywhere for military applications, environment monitoring, medical applications, etc. With the advancement of IoT there are many applications such as continuous health monitoring or monitoring home appliances from the remote area where these sensor nodes are used. Because of its broad usage in varied applications and in different environments, security becomes the main issue in WSN. When we provide security to sensor nodes, it is more complicated than that of other networks because of the limitations of the sensor nodes. Sensor nodes have resource constraints such as limited energy and power supplies, limited memory, low computation and communication capabilities. This is the main reason why conventional cryptographic algorithms cannot be applied to sensor networks; hence require different cryptographic approach with high security and low computational complexity. So, lightweight cryptographic schemes are preferred for sensor nodes which require limited resource but provide high security. Here we have discussed few lightweight techniques some of which are the lightweight versions of known cryptographic models.

7.1 Scalable Encryption Algorithm

Scalable Encryption Algorithm (SEA) was designed for low-priced embedded systems with limited resources (memory size and processor capability) in 2006 by François-Xavie Standaert et al. [44]. The design of SEA is based on a symmetric block cipher approach which utilizes small memory size, small code size, and limited instruction set. To meet the above-mentioned design criteria, SEA uses basic bit operations such as XOR, bit or word rotations, modular addition, and s-box substitution. SEA, which is defined as SEA (n, a), has a very flexible structure. It can operate on different plaintext and key sizes. In addition, SEA has a Feistel structure with a variable number of rounds. It is defined by the following parameters (Standaert et al. 2006):

(i) n: plaintext and key size

(ii) a: word size

(iii) $n_a = n/2a$: number of word per Feistel branch

(iv) n_r: number of rounds

Step 1: Bitwise XOR $\oplus$ operation: The bitwise XOR on $n/2$-bit vectors is presented in Eq. (5)

$$\begin{aligned} \oplus : & z_2^{n/2} \times z_2^{n/2} \rightarrow z_2^{n/2} : x, y \rightarrow z : x \oplus y \Leftrightarrow z(i) \\ & = x(i) + y(i), 0 \leq i \leq n/2 - 1 \end{aligned} \tag{5}$$

Step 2: Substitution box *S*: SEA (*n*, *a*) uses the following 3-bit substitution table:

ST: $= \{0, 5, 6, 7, 4, 3, 1, 2\}$, in C-like notation. For efficiency purposes, it is applied bitwise to any set of three words of data using the following recursive definition:

$$\begin{aligned} S : & z_{2a}^{na} \times z_{2a}^{na} : x \rightarrow x = S(x) \Leftrightarrow \\ & x_{3i} = (x_{3i+2} \wedge x_{3i+1}) \oplus x_{3i}, \\ & x_{3i+1} = (x_{3i+2} \wedge x_{3i}) \oplus x_{3i+1}, \\ & x_{3i+2} = (x_{3i} \vee x_{3i+1}) \oplus x_{3i+2}, 0 \leq i \leq n_a/3-1 \end{aligned} \tag{6}$$

where $\bigwedge$ and $\bigvee$, respectively, represent the bitwise AND and OR.

Step 3: Word rotation *R*: The word rotation is defined on n_a word vectors as shown in Eq. (7)

$$R : z_{2a}^{na} \times z_{2a}^{na} : x \rightarrow y = R(x) \Leftrightarrow y_{i+1} = x_i, 0 \leq i \leq n_a - 2, y_0 = x_{n_a - 1} \quad (7)$$

Step 4: Bit rotation *r*: The bit rotation on n_a word vectors is defined using Eq. (8)

$$r : z_{2a}^{na} \rightarrow z_{2a}^{na} : x \rightarrow y = r(x) \Leftrightarrow y_{3i} = x_{3i} \gg 1, y_{3i+1} = x_{3i+1}, y_{3i+2} = x_{3i+2}, 0 \leq i \leq n_a/3 - 1 \quad (8)$$

where $\gg$ and $\ll$ represent the cyclic right and left shifts inside a word.

Step 5: Addition mod2^a: The mod 2^a addition on n_a-word vectors is defined in Eq. (9)

$$z_{2a}^{na} \times z_{2a}^{na} \rightarrow z_{2a}^{na} : x, y \rightarrow z = x \bmod 2^a y,\ 0 \leq i \leq n_a \quad (9)$$

For the implementation of SEA, n and parameters can be configured in respect of target processor attributes. However, the bit size of the key and plaintext must be in multiples of six such as 48, 96… 192 and so on. Another crucial point is that we have to meet the following conditions (Standaert et al. 2006) to maintain an acceptable security level:

$a \geq 8$ and

$n_r = 3n/4 + 2\ (n_a + [a/2])$.

7.2 Chaotic S-Box for Wireless Sensor Network

In the Internet of Things (IoT) and WSN, many constrained devices are used to interact with each other via the public network. Therefore the security of these constrained nodes is becoming an issue, especially when they exchange private information. However, it is very difficult to implement standard cryptographic models such as DES, AES, etc., on constrained devices due to their limited resources. Lightweight cryptography is a better alternative which is more efficient in the constrained environment inside the hardware used in IoT that have limited computational power, battery power, and memory.

The function of an S-box is to provide confusion in the system. But there are no strict design criteria for S-box. The main property of a common lightweight S-box the block cipher is of smaller size (4 ×4-bit S-box) and usually uses the easiest form of look-up tables. Those S-boxes are static, and substitution tables are fully defined. Here, in this section, we are discussing a new method to obtain a good chaotic (4 × 4)-bit S-box which is suited for hardware implementation on 8-bit processors. This type of S-boxes has Bijective property as well as non-linear property [45]. The chaotic map chosen for this type of S-boxes is known as discretized skew tent map is defined in Eq. (10)

$$F_k^1(X) = \text{ceil}\left(M \times \frac{X}{K}\right) \text{if } 1 \le X \le K$$
$$\text{floor}\left(M \times \frac{M - X}{M - K}\right) + 1 \text{ if } K < X \le M \quad (10)$$

The inverse function of the discretized skew—tent map is used to create the inverse S-box in the decryption. The inverse of F_k is given as

$$F_k^1(Y) = X_1, \text{if } m(Y) = Y \text{ and } \frac{X_1}{K} > \frac{M - X_2}{M - K}$$
$$X_2, \text{if } m(Y) = Y\text{and } \frac{X_1}{K} \le \frac{M - X_2}{M - K}$$
$$X_1, \text{if } m(Y) = Y + 1 \quad (11)$$

where,

$$\text{X1} = \lfloor M^{-1}KY \rfloor; X2 = [(M^{-1}K - 1) + M]$$

$$(Y) = Y + \lfloor M^{-1}KY \rfloor - \lceil M^{-1}KY \rceil + 1$$

It is clear that F^1K (X) is a one-to-one map as M is a positive integer $M \geq 2$. In general, this kind of S-box takes n input bits and produces n output bits, the input value can be any possible value in a closed interval $X = (0, 1, 2, \ldots, 2n - 1)$ with $M = 2n$. X is represented in bits as $X = ({}_1, {}_2, {}_3, {}_4)$, where $x_i \in \{0, 1\}$. For given = 4, it is obvious that = 16 is obtained.

The S-box is one of the major components in block cipher. Many methods generating the (4 × 4)-bit S-box for the lightweight block cipher have been suggested in recent years. A common feature of these methods is to use the Boolean minimization tool or simple nonlinear functions to obtain Look-Up Tables of the static S-box. In spite of the existing algorithm, the structure of dynamic (4 × 4)-bit S-box based on chaos has been proposed. The (4 × 4)-bit S-boxes in this study have fulfilled the cryptographic properties of the "good" ones.

7.3 Cognitive Radio Encryption Standard Algorithm

Cognitive Radio Encryption Standard (CREnS) Algorithm is a symmetric key encryption algorithm designed for tiny embedded cognitive radio sensor nodes [46]. The CREnS is a modified version of Blowfish algorithm to minimize the requirement of hardware resources. The Blowfish uses a static predefined S-box whose value is pre-computed and stored in the memory before encryption operation. The performance of CREnS is improved by designing a dynamic S box utilizing a convolutional coder and P box using Pseudo Noise (PN).

The standard S-box properties are fulfilled by the convolutional coder that in turn, provides an acceptable level of avalanche effect. A (2, 1, 2) serial bit convolutional coder is used to make a substitution of plaintext bit which also expands the input bitstream into two times in numbers. Figure 11 represents the functional block diagram of the S-box. The inherent ability of a convolutional coder to produce parity bit is utilized here to design S-box.

The P-box value is generated by using the PN sequence as depicted in Fig. 12. The nonlinear PN sequence is generated to enhance the security level. The matrix is

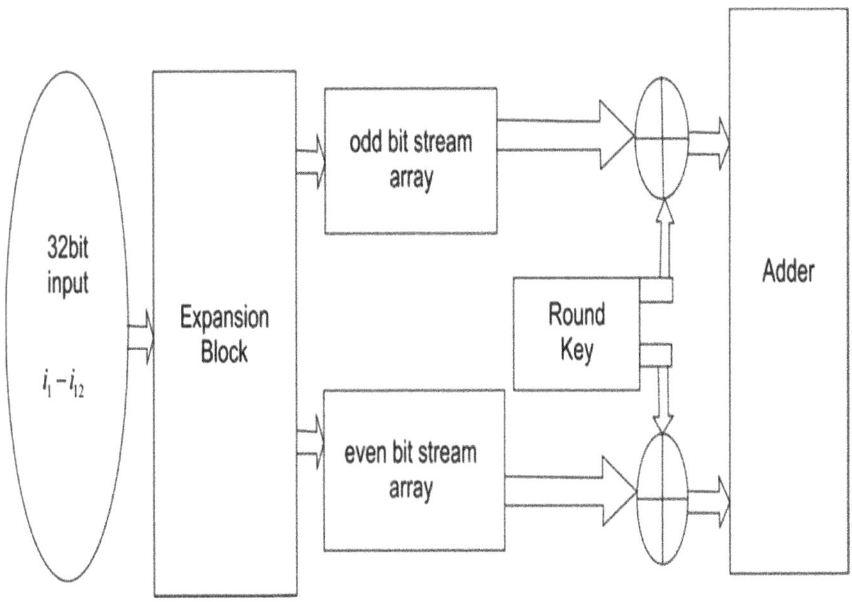

Fig. 11 Functional block diagram of the S-box

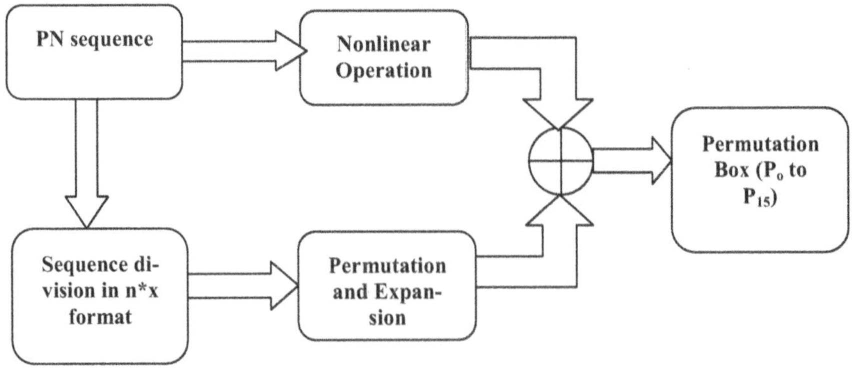

Fig. 12 Functional block diagram of the P-box

generated from the PN sequence whose number of "row and column" is decided by the number of iteration of the algorithm.

8 Network-on-Chip Security

In recent years, Network-on-Chips [47] have evolved as an emerging area of academic and research interest. The main function of NoC is to handle the data transmission among different intellectual property blocks embedded in a complex System-on-Chip. It contains multiple diverse processor cores, memory cores and application-specific IP cores incorporated in a communication network between SoC modules which are used to manage a large number of applications including security. However, such a complex infrastructure may introduce several flaws in the system that should be carefully addressed. Security is one of such flaws of NoC which is not much explored so far. As discussed in [48], security attacks to an embedded system may be categorized in different ways. If the attacks are classified in terms of the attacking agents only, they may be classified as software attacks, physical attacks and side-channel attacks.

- *Software Attacks*—These include attacks through software agents viz. worms, viruses and Trojan horses. Buffer overflow attacks are examples of this category.
- *Physical Attacks*—These kinds of attacks involve physical interference at some level. They involve the use of micro-probing techniques, de-packaging as *Side-channel attacks* as well as reconstruction of the layout, etc., to import granularity to the architecture.
- *Side-Channel Attacks*—These attacks exploit information collected from the physical infrastructure of the system, such as timing information, power consumption or electromagnetic leakage.

If categorized in terms of structure-specific attacks, three types of attacks can be identified [49].

- *Denial of Service* (DoS) *attack*: This kind of attack aims to reduce the system performance in several ways. One example of such an attack is band-width reduction attack which aims to increase the latency of the on-chip networks to the saturation level by sending several useless packets frequently in the network. This in turn reduces the communication bandwidth. Another type of DoS attack is Draining or Sleep Deprivation attack. This type of attack directly hampers the operation life of a battery used in the embedded systems (for battery operated systems). It is executed by sending continuous requests to the victim to make it perform power-hungry tasks.
- *Extraction of secret information*: This type of attack tries to read secret data, critical instructions or information stored in the secure areas of the memory or

registers in particular targets. It can be performed by exploiting software weaknesses of the system. For example, exploitation of buffer overflow may be a way to make such an attack.

- *Hijacking attack*: This kind of attack targets the configuration or execution of the system and tries to make some changes in it in order to make it bound to perform some extra tasks created by the attacker. It makes use of buffer overflow and reconfiguration of internal registers.

Several security frameworks are designed to prevent each and every attack subjected to NoC. NoC-based security schemes generally follow two basic approaches. One type is based on lightweight cryptographic algorithms and the other type exploits several error control strategies. Few of them are mentioned here to focus on the future research direction on this subject.

The authors of [50] proposed a general security framework on NoC at both the network level (or transport layer) and the core level (or application layer). In this scheme, each IP core has a security wrapper and a key-keeper core that are used to protect encrypted private and public keys. Here, unencrypted keys are prohibited from leaving the cores and NoC which would prevent any unreliable software (on or off the NoC) to get the access of the keys. This security framework is illustrated at the core level (application layer) with modified software to resist power attacks with extremely low overheads in energy.

An Authenticated Encryption (AE)-based security framework for NoC-based systems is developed in [51] which resides in the Network Interface (NI) of every secure IP core allowing secure communication among such IP cores. This proposed scheme also verifies the authentication at the Network Interface (NI) of each secure IP cores. In this framework, NI of each secure IP core will give an authentication status on receiving any packet from other IP cores. This scheme results in tolerable area overhead and does not affect the network performance except some initial latency.

Another security scheme proposed in [52] exploits transient and permanent error control methods to address Hardware Trojan (HT) issues in NoC links. The hardware-efficient error control methods, when applied to NoC links have the advantage of having much reduced overall hardware cost than that of cryptography-based security algorithms. In this scheme, an error control coding method for transient errors is used to detect the HT induced link errors. For faulty links, it is proposed to reshuffle the links and isolate the HT-controlled link wires. Here, the utilization of partially failed links is resumed to improve the bandwidth and the average latency of NoCs in spite of rerouting packets via alternative paths.

9 Conclusion

This chapter has focused on various cryptographic techniques in the light of network security. There are several techniques of quantum cryptography, homomorphic cryptography and lightweight cryptography which are used to enhance the security level of networks. Current research works on these techniques are discussed systematically. The innovation of quantum computers has enriched the computing ability which in turn has strengthened network security, however, at the same time, it increases vulnerability in the system. To combat these vulnerabilities will become the challenge for post-quantum cryptography which would snatch the research attention. The homomorphic cryptography has the advantage of providing a high level of security without disturbing the business process or application functionality. It allows computations on encrypted data, without fully decrypting the data on the cloud. This facility provides the opportunity to a user to execute any computation on other user's private data without knowing its secrets. This is possible only if the first user performs this computation in a faster and cheaper way. The homomorphic encryption also has the advantage to decrypt the data less often, which reduces the overhead of the security system. On the other hand, this scheme has the disadvantage of being extremely slow and expensive. This inferiority defers it from being a realistic security solution in the cloud. This is the challenge for the next generation researchers to overcome its shortfalls and establish it as a superior security scheme. Future research direction on homomorphic cryptography includes efficient parameter selection, suitable hardware designs and optimizations of existing schemes to enhance the speed. The requirement of lightweight cryptographic techniques would also increase day by day with the advancement of IoT-based smart technologies. So, strengthening the lightweight cryptographic models would be another challenge for researchers. The main constraint of the lightweight cryptographic models is low battery resources. To overcome these constraints researchers have to focus on designing lightweight ciphers with fast confusion and diffusion techniques in less number of rounds. The strong security algorithms such as AES, RSA, ECC, etc., are based on modular arithmetic which requires a large number of multiplication operations. To reduce the multiplication overhead Vedic multiplier may be an option. Security solution for On-chip networks is an emerging research area. Lots of open problems are present in designing of NoC security system. Individual schemes are required to counterfeit different security attacks. Hardware-based designs using physical unclonable functions would draw future research interest.

References

1. Smith JL The design of lucifer, a cryptographic device for data communication, RC 3326. IBM Research, White Plains
2. National Bureau of Standards (1977) Data encryption standard, U.S. Department of Commerce, FIPS Publication 46, Jan 1977
3. American National Standards Institute (1998) New York. ANSI X9.52–1998, Triple Data Encryption Algorithm—Modes of operation
4. Daemen J, Rijmen V (2002) The design of Rijndael, AES—the advanced encryption standard. Springer, p 238
5. The Blowfish Encryption Algorithm (1994) Dobb's J 19(4):38–40
6. Rivest RL (1994) The RC5 encryption algorithm. Fast Software Encryption. FSE 1994. Lecture Notes in Computer Science, vol 1008. Springer, Berlin, Heidelberg

7. Juels A, Ristenpart T (2014) Honey encryption: security beyond the brute-force bound. EUROCRYPT 2014, LNCS 8441, pp 293–310
8. Cormen TH, Leiserson CE, Rivest RL, Stein C (2009) Introduction to algorithms, 3rd edn. MIT Press, pp 428–436
9. Berry N (2012) PIN analysis. DataGenetics blog
10. Rivest RL, Shamir A, Adleman L A method for obtaining digital signature and public key cryptosystems. Communications of the ACM
11. Diffie W, Hellman ME (1976) New directions in cryptography. IEEE Trans Inform Theory 22(6)
12. Koblitz N (1987) Elliptic curve cryptosystems. Math Comput 48:203–209
13. Miller VS (1986) Use of elliptic curves in cryptography. Advances in Cryptology CRYPTO'85, Lecture Notes in Computer Science, vol. 218. Springer, pp 417–426
14. Koblitz N (1989) Hyperelliptic cryptosystems. J Cryptol 1:139–150
15. Mukhopadhyay D, Shirwadkar A, Gaikar P, Agrawal T (2014) Securing the data in clouds with Hyperelliptic curve cryptography. In: Proceedings of international conference on information technology. IEEE
16. Scholten J, Vercauteren F (2015) An introduction to elliptic and hyperellipticcurve cryptography and the NTRUCryptosystem. IEEE (2015)
17. ElGamal T (1985) A public key cryptosystem and a signature scheme based on discrete logarithms. IEEE Trans Inf Theory 31(4):469–472
18. National Institute of Standards and Technology (1985) Computer data authentication. U.S. Department of Commerce, FIPS Publication 113
19. Rivest R (1992) Step 4. Process message in 16-word blocks. The MD5 message-digest algorithm. IETF, p. 5. sec. 3.4
20. National Institute of Standards and Technology (2002) Secure Hash Standard (SHA-1). U.S. Department of Commerce, FIPS Publication 180-1, April 1995 superseded by FIPS 180-2
21. National Institute of Standards and Technology (2008) The keyed-hash message authentication code (HMAC). U.S. Department of Commerce, FIPS Publication, pp 198–1
22. ANSI X9.62 (1999) Public key cryptography for the financial services industry: the elliptic curve digital signature algorithm (ECDSA)
23. Shor PW (2014) Algorithms for quantum computation: discrete logarithms and factoring. In: Proceedings of the 35nd annual symposium foundations of computer science, pp 124–134
24. Chen C-Y, Zeng G-J, Lin F-J, Chou Y-H, Chao H-C (2015) Quantum cryptography and its applications over the internet. In: Proceedings of the IEEE network
25. Bennett CH, Brassard G (1984) Quantum cryptography: public key distribution and coin tossing. In: Proceedings of the IEEE international conference on computers, systems & signal processing, Bangalore, India, pp 175–179
26. Chou YH et al (2013) Using GHZ-state for multiparty quantum secret sharing without code table. Comput J 56(10):1167–75
27. Chou YH et al (2012) Enhanced multiparty quantum secret sharing of classical messages by using entanglement swapping. IET Info Secur 6(2):84–92
28. Chou YH et al (2011) Quantum entanglement and non-locality based secure computation for future communication. IET Info Secur 5(1):69–79
29. Chou YH, Lin FJ, Zeng GJ (2015) An efficient novel online shopping mechanism based on quantum communication. Quant Inf Process 14:2211–2225
30. Mils electronic (1947) Mils electronic, 1947. [Online]. Available: https://www.mils.com/
31. Kak S (2006) A three-stage quantum cryptography protocol. Foundations of physics letters, vol 19
32. Krawec WO (2016) Asymptotic analysis of a three state quantum cryptographic protocol. In: Proceedings of the IEEE ISIT. Springer, Barcelona, pp 2489–2493
33. Paillier P (1999) Public-key cryptosystems based on composite degree residuosity classes. In: International conference on advances in cryptology "EUROCRYPT 1999". Lecture Notes in Computer Science, vol 1592. Springer, Berlin, Heidelberg, pp 223–238

34. Rivest RL, Adleman L, Dertouzos ML (1978) On data banks and privacy homomorphisms. Foundations of secure computation, pp 169–180
35. Gentry C (2009) A fully homomorphic encryption scheme. Ph.D. dissertation, Stanford University
36. Gentry C, Halevi S (2011) Implementing Gentry's fully-homomorphic encryption scheme. In: EUROCRYPT, pp 129–148
37. Gentry C, Halevi S, Smart NP (2012) Homomorphic evaluation of the AES circuit. IACR Cryptology ePrint Archive 2012:99
38. Wang W, Hu Y, Chen L, Huang X, Sunar B (2012) Accelerating fully homomorphic encryption using GPU. In: HPEC, pp 1–5
39. Moore C, Hanley N, McAllister J, O'Neill M, O'Sullivan E, Cao X (2013) Targeting FPGA DSP slices for a large integer multiplier for integer based FHE. Workshop on applied homomorphic cryptography, vol 7862
40. Brakerski Z, Gentry C, Vaikuntanathan V (2011) Fully homomorphic encryption without bootstrapping. Electronic Colloquium on Computational Complexity (ECCC), vol 18, p 111
41. Smart NP, Vercauteren F (2011) Fully homomorphic SIMD operations. IACR Cryptology ePrint Archive 2011:133
42. Brakerski Z, Gentry C, Halevi S (2012) Packed ciphertexts in LWE based homomorphic encryption. IACR Cryptology ePrint Archive 2012:565
43. Coron J-S, Lepoint T, Tibouchi M (2013) Batch fully homomorphic encryption over the integers. IACR Cryptol ePrint Archive 2013:36
44. Standaert F-X, Piret G, Gershenfeld N, Quisquater J-J (2006) SEA, a scalable encryption algorithm for small embedded applications. In: International conference on smart card research and advanced applications "CARDIS 2006", Lecture Notes in Computer Science, vol 3928. Springer, Berlin, Heidelberg, pp 222–236
45. Hue TTK, Hoang TM, Tran D (2014) Chaos-based S-box for lightweight block cipher. IEEE
46. Roy Chatterjee S, Mukherjee S, Chowdhury J, Chakraborty M (2018) CREnS: a convolutional coder based encryption algorithm for tiny embedded cognitive radio sensor node. In: Proceedings of international ethical hacking conference 2018. Advances in Intelligent Systems and Computing, vol 811. Springer, Singapore (2018)
47. Benini L, De Micheli G (2002) Networks on chips: a new SoC paradigm. IEEE Computer
48. Kocher P, Lee R, McGraw G, Raghunathan A, Ravi S (2004) Security as a new dimension in embedded system design. In: Proceedings of DAC 2004, pp 7–11
49. Evain S, Diguet J (2005) From NoC security analysis to design solutions. In: IEEE workshop on signal processing systems design and implementation, pp 166–171
50. Gebotys CH, Gebotys RJ (2003) A framework for security on NoC technologies. In: Proceedings of the IEEE computer society annual symposium on VLSI (ISVLSI'03)
51. Sajeesh K, Kapoor HK (2011) An authenticated encryption based security framework for NoC architectures. In: IEEE international symposium on electronic system design
52. Yu Q, Frey J (2013) Exploiting error control approaches for hardware trojans on network-on-chip links. In: IEEE international symposium on defect and fault tolerance in VLSI and nanotechnology systems (DFTS)

Part IV
Machine Learning and Artificial Intelligence in Network Security

Cyber Security with AI—Part I

Bhanu Chander and Gopalakrishnan Kumaravelan

Abstract As a result of continuous and extreme inclusion of the Internet, computer networks, and social life, there has been a complete transformation of how people learn and work. With the expansion of the Internet and its application to our lives, it opens an abysmal for cyber security attacks. The continuous increase in cyberattacks has given rise to Artificial Intelligence (AI) and Machine Learning (ML)-based techniques that have a vital measurement in detecting security risks, security breaches and alerts, progress triage events, and malware detection to defense issues. {ML, AI} is the set of statistical and mathematical forms to clarify higher non-linearity troubles of dissimilar themes such as data organization, prediction, and classification. Moreover, it is an undeniable fact that information is an attractive reasonable presence for each corporation and big business. For that reason, protecting security models driven by the real data sets logically turns out to be important. Hence, this chapter presents the role of ML and AI in cyber security, describes a variety of active ML techniques, how and where to add ML and AI models for network security, cyber security threats classification. This chapter presents commonly used ML techniques and network data sets. Finally, challenges and future works are discussed.

Keywords Internet of Things · Machine learning · Artificial intelligence · Cyberspace · Network datasets

1 Introduction

The exploitation of Artificial Intelligence (AI), as well as Machine Learning (ML) in the field of cyber security, has turned to be a common business in recent times. With the continuous innovations on the Internet, cyber threats are varying quickly moreover cyber security circumstances are not optimistic also [1–3]. Almost the major part of cyber security, researchers employ ML techniques to transform a reactive to a predictive tactic for threat exposure. Mostly, cyber threats show signs of unique

B. Chander (✉) · G. Kumaravelan
Computer Science and Engineering, Pondicherry University, Pondicherry 609605, India
e-mail: gujurothubhanu@gmail.com

M. Chakraborty et al. (eds.), *The "Essence" of Network Security: An End-to-End Panorama*, Lecture Notes in Networks and Systems 163,
https://doi.org/10.1007/978-981-15-9317-8_6

activity prototypes, allowing practitioners to use ML to perfectly recognize the attack. However, the issue in detecting attacks with ML is that the outcomes are very rarely arranged in real-world solutions. The trend of implementing ML in cyber security is influenced because of the increasing amount of data instances. The potential risks of networks relying on ML techniques are not new, in the recent period it has got attention from various research fields [2, 3]. From the recent reports realized by Internet Security Threat Report (ISTR), roughly 430 million new malware varieties, 362 crypto ransomwares, as well as additional Internet threats were originated in 2015. Both cyber threats and cybercrimes turn to be essential parts of our daily life; on the other hand, the consequence of network safety measures has continuously come into view as a key distress. WannaCry the latest ransomware strikes more than 10,000 companies in greater than 250 countries which proves the rising severity of the promising cyber threats [3–6].

AI is developed based on three rules: learning, reasoning, and problem-solving. Here learning, works on information gathered with principles to use data instances, reasoning finds the suitable information from learning, and problem-solving selects instant response on the bases of learning plus reasoning. ML stands on the inspiration of mechanized find-outs from exemplars and skill, exclusive of being programmed [3–5]. In the recent decade, most of ML techniques have been transferred from laboratory toward the front position of outfitted methods such as Google, Facebook, Flipcart, Virgin, Space-X and Amazon, which are some of the biggest techno giants that use ML in every appliance to improvise customer experience, easy personal associations, suggestions while purchasing products, innovations in space research, etc. Because of the powerful ability of ML, it is also applied to cyber security and recent research showed that it notably reforms the cyber security view. ML in cyber security is used to enhance malware detection, triage actions, identify infringes and alert organizations to security issues. Additionally, ML can recognize advanced targets, threats such as organizational reports, transportation vulnerabilities and probable inter-reliant weakness. It was very much known that malware by itself could represent more than 3.5 million novel shapes in an hour. Hottest molests plus complicated malware intelligence to evade set of connections furthermore end-point exposure to dispense cyberattacks has increased by shocking rates. ML should be powered to deal with emergent malware troubles [3–12].

The intention of this chapter is to present a suitable review and introduction on the role of the ML and Deep Learning (DL) approaches in cyber security. It is designed to give the readers a profound background on various ML and DL and the latest progressions in this part. The rest of the chapter arranged as follows: firstly we describe cyber security then start a summary of various ML, DL and AI approaches in cyber security, then after contemporary appliances, error evaluation strategies, cyber network-based datasets and finally future research challenge and conclusion are described.

2 Background of Cyber Security

In the year 2015, Global Information Security Workforce Study (GISWS) for a healthier worldwide security workforce had been done. The study accomplished an analysis of almost 14,000 information security experts through International Security Consortium (ISC). The labor force originated the space for safety persons, and it has been predicted to reach above 2 million by 2020. In 2015, Forbes anticipated that the cyber security marketplace would increase twofold by an amount from $75 billion to $170 billion by 2020. In related studies, Microsoft estimated that above 35% of associations were not capable to pack major cyber security posts; moreover, the bulk of these businesses are likely to face a cyberattack within the next 12 months [11–14].

ML procedures are a group of mathematical models, which are designed to explain high non-linearity troubles of unrelated areas such as data conceptualization, prediction, classification, and association, etc. [3–7]. Internet, the word changes the entire working style of people around the world and becomes an important resource for the populace. The North Atlantic Treaty Organization (NATO) delineates the Internet as *a critical national resource for governments, a vital part of national infrastructures, and a key driver of socio-economic growth and development*. In 2015, nearly 40% of the world's residents utilize the Internet; this percentage rises toward 85% in urbanized nations. Most up-to-date network surveys acknowledged that the Internet usage will reach 65% in the entire world population, coming to the urbanized population it will reach 98%. Besides the raising of Internet handling, malevolent software codes come into sight to demolish computer structures and wipe out the data records they control, much like profile passwords, abuser information and credit card details, etc. [1–7]. At present malware is the main issue and it is expressed as software capable of damage, interrupting data, and get unofficial control over systems. It is the biggest risk for persons as well as organizations, companies, and governments, which are the possibility of drop precious information and their names. There are numerous issues that have happened in recent times such as gathering information on individuals' credit details from web payment systems, pinching power grids details, and human tracking and many more. Based on the above discussions the idiom "Cyber security" has been formally defined as "the preservation of confidentiality, integrity and availability of information in the Cyberspace" by the ISO/IEC, is also known as Confidentiality, Integrity and Availability (CIA) rules. For better understanding, we described some fundamental security concepts below [8–10].

Integrity: It ensures that the information can be changed, modified, or altered by the legitimated entities only. Protects data in a constant, accurate, and trustworthy manner.

Authentication: It guarantees both receiver and senders are authenticated with each other before transmitting information to each other.

Availability: The services which are mentioned for security must be present when the system needs. Various systems have dissimilar availability requirements; they must be accessed easily whenever they are needed.

Confidentiality: It makes sure that the message is received by the authenticated entity. Restricts the illegitimate entities and grant access to the legitimate ones.

Accountability: Accountability itself cannot stop networks to be attacked but supportive in making sure other security mechanisms are working appropriately or not. In addition, it adds liability and confident procedures to improvise security.

Information Security: It completely depends on the defense of secrecy, reliability, and accessibility of data records and to provide the correct information about the abuser.

No-repudiation: The intention of non-repudiation states that it delivers certain confirmations in some cases where the abuser cannot deny an act.

Network Security: This deals with the design, execution, and action of a set of connections to attain the use of information security on networks within, among and between associations and abusers.

Internet Security: It relates to defending Internet-related services as well as linked ICT schemes along with the system as an expansion of system safety in associations and at home, to accomplish the intention of protection.

Critical information infrastructure protection: It guarantees to facilitate the methods, networks are privileged and tough not in favor of information safety threats, network protection threats, Internet protection threats along with Cyber security hazards.

Auditing: It is a well-organized progression of measuring security how well it is established for a particular entity. It concludes any vulnerable faults that place data in danger.

Cyber Crime: It describes the illegal action where works or functions in the Cyberspace utilized on behalf of or else the objective of a fault, or where the Cyberspace is the cause, tool, objective, or consign of a fault.

Cyber Safety: It exploits the circumstance of being confined not in favor of physical, communal, spiritual, economic, political, poignant, work-related, emotional, educational or other kinds or consequences of malfunction, harm, mistake, disaster, damage or any other incident in the Cyberspace which might be measured non-desirable.

Hence, the intention of Cyber security has to guard individual, legislative along with industry data from mishandling or else mismanagement by new groups, in detail focal point on three most important farm duties: first—taking procedures to shield tools, software along with the information they have, second—promising the shape or eminence of being sheltered from numerous intimidations and finally—execute as well as humanizing mentioned dealings. In current years, countless associations plus assignments to facilitate with the intention of opposite above-mentioned intimidations.

3 Summary of Artificial Intelligence, Machine Learning and Deep Learning in Cyber security

From the time of innovation, ML has been tremendously applied in a mixture of domains such as pattern recognition and computational learning, etc. ML makes the systems to learn the ability to perform exclusive of candidly planned, make replicas that can be trained from a dataset, and formulate the data-driven conclusions or estimations. From the period of past decades, ML has to fetch huge control over our daily life with examples including smart-homes, proficient web search, and smart healthcare management, self-driving systems, image processing, military surveillances, optical character, speech and voice recognition [1–6]. However, every capability of ML is to be considered for improving the model, regarded as numerous factors such as computation instance, actualization ability as well as complication. Based on employment, the preference might diverge [10–12].

3.1 Support Vector Machine

Support Vector Machine (SVM) which is well-known as Support Vector Classification (SVC) or Support Vector Regression (SVR) defined by an eminent researcher Vapnik. SVM working theory is optimizing separating the hyper-plane of a dataset in a classification problem, which means the design of the decision boundary. A decision borderline separates a group of data records encloses dissimilar class importance among paired groups. SVM is one of the most perfect and dynamic models in ML. It contains the closest point to the partition hyper-plane that includes the most favorable separation hyper-plane. For example, if the data points are not linearly divisible, then kernel operations applied to SVM which push or map them into upper dimensional spaces where they detachable in those spaces. However, sorting out hyper-planes has two central limitations: the requirement of linear separability of the sample and its linear temperament and training algorithm. To utilize SVMs in a well-organized, effectual way, it was compulsory to stand in mind two main characteristics: The preparation model and Hyper-parameter selection methods [1–5, 8–10].

Chandrasekhar and Raghuveer [12] suggest a filtering model based on the SVM classifier for first-rate multiple intrusion taxonomy responsibilities on the NSL-KDD standard dataset. The model accomplishes 91% taxonomy correctness with three input features and 99% classification precision by 36 input features. Yan and Liu [13] proposed a hybrid model based on ANN and SVM for intrusion detection with fuzzy C-means practice. Fuzzy C-means collects the mixed training data that decreases complexity; ANN trained is on resulted data from the clusters and finally implements linear SVM to execute the classification. Simulation is done on KDD CUP 1999 dataset and outcomes show great exactitude. Kokila et al. [14] developed a transductive model for SVM. The experiment proceeds with a subpart of the DARPA 1998 dataset. Especially for DoS-type attack, 200 ordinary samples, as well as 200, attack samples are picked; moreover, samples are arbitrarily separated into a training set and test set according to a percentage of 6:4. The investigational outcomes illustrate that the correctness and False Positive Rate (FPR) are 80.1% and 0.47%, respectively. Peng and Jiang [15] spotlights on DDoS attacks based on the Software Defined Networking (SDN) regulator with a similar dataset of Kokila et al. [14]. The authors implemented a mixture of ML models to compare and evaluate. The SVM classifier has an inferior False Acceptance Rate (FAR) and a superior classification precision of 0.8 and 95.11%.

3.2 Decision Tree

A decision tree is a predictive model in ML; it tries to make a mapping among entity attribute and entity values. It follows like a tree structure, every node in the tree symbolizes an entity, every deviation trail stands for potential element importance and every leaf node communes to the importance of the entity represented by trail from root to leaf node. Most probably decision tree has solitary productivity; if you wish for multifaceted productivity, you can set up an autonomous decision tree to hold dissimilar outputs. ID3, C4.5, and CART are some of the frequently developed decision tree representations [1–5]. The iterative Dichotomiser 3 (ID3) technique generates a multiway tree, discovers every node categorical element that will give the largest information for categorical goals. The C4.5 model is frequently utilized in Data mining and Statistical models as a Decision tree for classification problems by creating a decision in both univariate and multivariate interpreters. It translates the trained trees to if-then policy. The precision of every policy is evaluated for the order in which they should be applied. It advances ID3 approach by dealing with continuous as well as discrete characteristics and omits values after building. Classification and Regression Trees (CART) is related to C4.5, however, it diverges numerical object variables and moreover does not calculate rule sets. CART builds binary trees with the element and threshold that yields the largest information increase at every node.

Rai et al. [16] proposed a Correlation Feature Selection (CFS) model with the basis of a decision tree for IDS, simulated with the NSL-KDD dataset. The model reported a precision detection rate of 88.5% and FAR 5.45%. Azad and Jha [17] came with a pair of C4.5 decision tree practices with pruning on behalf of feature selection. Here both trainings, as well as test taxonomies, utilize KDD Cup 99, NSL-KDD datasets, respectively. The analysis results confirm that C4.5 with pruning has superior accuracy along with lesser FAR of 98.45 and 1.55% than a C4.5 decision tree. Additional works of Puthran and Shah [18] make use of C4.5 for interference exposure on the NSL-KDD dataset. The proposed model solves feature assortment as well as segmentation concerns. The information helps to decide the appropriate elements and segmentation principles ensure that taxonomy has no bias on most recurring standards. A total of 16 characteristics are elected as features on the NSL-KDD dataset. But, the accuracy of the technique is just 79.52%.

3.3 k-Nearest Neighbor

In classification, distance measurement has its reputation and k-Nearest Neighbor (k-NN) is the foremost distance-based classifier in ML. Calculation of differences or similarities among two data instances x, y is done with standard Euclidean distance d(x, y) defined as:

$$d(x, y) = \sqrt{\sum_{k=1}^{n}(xk, yk)^{2}}$$

Here, xk is the kth featured factor of instance x, yk is the kth featured element of the instance y and n is the overall quantity of features in a dataset. Presume that the design set for k-NN taxonomy is U. Let us make the following assumptions:

S = complete set of samples in a dataset.

X = {X1, X2...XL} = L different class labels that are accessible in S

x = an input vector for which the class label must be predicted.

Let yk put the kth vector in the dataset set S. The k-NN model is to locate the k neighboring vectors in dataset set S for input vector x. Then the input vector x is classified to class Xj if the superior part of the k adjoining vectors encloses their class as Xj.

Vishwakarma et al. [19] designed a novel technique called farthest neighbor (k-FN) and k-NN for the classification of data instances and evaluate model work on the NSL-KDD dataset. Investigational results illustrate that these techniques are enhanced with respect to precision, exposure rate and decrease of FAR. However, the work did not recognize an explicit kind of attacks in irregular states. Ingre et al. [20] planned a k-NN-based Ant Colony Optimization (ACO) algorithm for intrusion exposure and model pre-train by KDD Cup 99 dataset. The accuracy report for the projected technique is 94.17% and FAR is 5.82%. Another study [21] makes use of the same KDD Cup 99 dataset and improved the end results of methods designed in Ingre et al. [22] and Relan and Patil [22]. Recommended, watermark motivated statistical feature mining approach and 3 supervised ML models k means Nearest Neighbor (k-NN) at DE using ADFA-LD as DS are recognized. The suggested HADS is competent to accomplish sufficient results while defending the computational cost at modest, Detection Rate (DR) is 78% as well as FAR is 21%.

3.4 Clustering

The major objective of cluster analysis is to divide data instances and grouping them into a standardized group based on the association flanked by them. And the designed groups must be determined with no earlier information concerning the classes; moreover, they presume fully from data records. Cluster analysis is mostly applied to investigate datasets, prepare statistical premises that can verify afterward, make the data simpler and their corresponding variables as well as the forecast from other groups. Researchers are continuously developing various cluster methods that are classified into hierarchical-based and partition-based. Hierarchical methods establish hierarchies between the clusters; means they present a sequence of successive partitions where every separation is found as unified or separated clusters [1–4]. Partition methods rely on the success of the most favorable separation of an anonymous sharing in the input space by shaping a specific quantity of clusters based on a definite comparison appraise. Thus the substances that fit into every meticulous cluster can be symbolized as the center of the cluster. The k-means models are one of the recognized models that come under this category [4–6].

3.5 Genetic Algorithm (GA) and Genetic Programming (GP)

Genetic Algorithm (GA) as well as Genetic Programming (GP)are the two most well-liked computational processes based on the theory of continued existence of robustness. Both processes operate on the count of chromosomes to facilitate progress based on definite operations such as selection, crossover, and mutation. At the initial stage, the model starts with an arbitrarily produced populace; a robustness assessment is calculated for every entity. This indicates that individuals are able to crack difficulties with the highest possibility of choosing for the next level and two competent individuals will execute the next pace crossover and at last, each one goes through mutation. Finally, the highest robust chromosome will enter into the subsequent production [3–8]. Authors of [23] implemented a simple GP representation to explore taxonomy for cyberattacks. Authors utilized Linear Genetic Programming (LGP), Gene Expression Programming (GEP) and Multi Expression Programming (MEP) in analysis work. The DARPA 1998 data set applied as a major dataset to certify the shaped replica and simulation shows great outcomes.

3.6 Hidden Markov Models (HMM)

Hidden Markov Model (HMM) is a statistical practice with a collection of inter-related circumstances with transition likelihood that finds out the structure of the proposed system. It consists of the unseen parameters, forward-backward correlation utilized for the identification of these hidden parameters. Because of likelihood, every circumstance is special, the system modifies circumstances eventually and additionally, it is proficient on behalf of modifying successions [3–8].

Joshi and Phoha [24] prepared a model to utilize HMM to build-up an intrusion exposure method. Authors implemented five specific states, which are exercised on six observation representations for each state. Moreover, the inter-links among states are executed on a special condition where any state can transition into any dissimilar state. For justification of representation the KDD 1999 dataset is utilized where it holds extensive 41 features but 5 were preferred for scrutiny work. The positive rate of the replica boosted to 79% with an FPR of 21%.

3.7 Inductive Learning

The undertone, inference and presumption from any standard datasets were recognized as a deduction. The procedure affecting particular inspection to enlarge theories along with prototypes is acknowledged as inductive learning. The inductive analysis extends several broad-spectrum prototypes that employ several enlarged imaginary executions [2–6].

Fan et al. [25] proposed an artificial outlier producer to create random actions as well as anomalous works. In this proposed work authors utilized circulation-based abnormality creation along with filtered artificial abnormality. This produced data was randomly merged with the DARPA 1998 dataset and data were exploited to learn the presentation of the proposed inductive knowledge structure. The replica illustrates a victorious exposure rate of 94 and a little FAR of 2%.

3.8 Random Forest Tree

Random forest or Random Decision tree is a procedure that functions by building numerous decision trees at the preparation period for classification and further specific tasks. Every branch of the tree symbolizes a feasible result, incident, or act. The decision volume of the tree is preferred with the random forest as the ultimate decision. It has some best advantages as compared to other ML techniques: *No over-fitting*—utilization of multiple trees reduce the over-fitting risk as well as timeless training, *High accuracy*—for huge datasets it gives the best outcomes with accurate predictions, *Estimates missing data*—if the random forest is trained with a large part of datasets it tries to maintain accuracy if some part of data were missing [2–6].

3.9 Self-organizing Map

The design of Self Organizing Maps (SOMs) is motivated by the general structure of neural networks, which builds the features maps during matrix functions of synthetic neurons. SOM tries to build a typological plot for optimal results by placing a predetermined quantity of vectors in an input space of the superior aspect, in consequence, make data easier towards recognition. It is a well-liked non-linear as well as unsupervised ANN model which is also utilized for the solution of dimensionality shrinking troubles [3–5]. The basic learning procedure of SMOs is as follows: at first, it starts with an arbitrary selection of the input prototypes after that handovers it to the current network; secondly, it measures spaces of neurons that are mapped primarily. Moreover, it includes the nearest neuron to the input vector. Thirdly, the winner neuron and its neighbor's weights are updated and finally, preceding steps are repeated until the predefined conditions of the technique are fulfilled [6–8].

3.10 Artificial Neural Networks

A neural set of connections have been inspired by human biological neurons, which consist of input, hidden, and output layers. Here the number of hidden layers may vary depending on the abuser as well as the application. At this juncture, the works of two well-known persons Psychiatrist McCulloch and Mathematician Pitts are considered as the foundation for the development of Artificial Neural Networks (ANN). Some major elements must be considered while implementing any neural network model. (a) Model constructed with neurons that are prearranged in layers. (b) Each neuron has a numeral value of input associations mutually with a set of weights to ease adjust the energy to each input signal. (c) An activation function takes action upon input signals [2–6, 21–28].

3.11 Deep Learning

ML and DL are very close to each other although there are some variations. DL is a fresh field in ML study and its inspiration lies in the concern of NN to imitate the human-brain for methodical knowledge. Geoffrey Hinton a well-known computer scientist and researcher at Google called the father of deep learning, shaped back-propagation restricted Boltzmann machines and AlexNets. ML is a learning process where computational schemes assume to progress machinery performance by noticing the solidity, variation in test and training datasets. DL discovers from a huge number of representations throughout a chain of commands of manifold layers and it begins to realize and react functionally. DL gives additional flexibility, enormous authority by learning as compared to ML. ML integrates an extensive mixture of methods but as a rule, none of them shows taxonomy performance as of DL.

Auto-Encoder

Auto-encoder, also known as representation learning is used mostly to reduce the dimension as well as feature learning. The Auto-Encoder is poised of a hidden layer (h) that refers to input along with two major components: encoder operation $h = f(x)$ plus decoder operation also acknowledged such as reconstruction error: $r = g(h)$. The most important point of auto-encoder is that both the encoders, as well as the decoder are trained mutually, and the divergence among the novel data and its renovation can be diminishing. Auto-encoder is classified into two taxonomies: (i) Under-complete auto-encoder—hidden code measurement is less than input data, and tries to find out more salient features from input, (ii) Over-complete auto-encoder—hidden code measurement is alike or the same as the input data. De-noising auto-encoder, Contractive auto-encoder, and sparse auto-encoders fall under this taxonomy.

Recurrent Neural Networks

ANNs produce great results in various applications, but the issue was every data instance is dependent. Most of the recent applications are real-time-based, so if the data instances fall inside the time or are space linked, then there is the option for dropping the state of network location. Recurrent NN (RNN) is sequence-based and hence it can represent an input or output comprising of autonomous sequenced rudiments. The motive of RNN is that the existing production of a series is associated with production before it. In short, the solid expression of a system can memorize the records of preceding instant and affect it to the computation of contemporary output; this means nodes linking hidden layers turn into associated layers and input to the hidden layer comprises both the outputs of the input layer along with the last instant of the hidden layer efficiently. Hypothetically, RNN can process any series of data. However, in practice, to decrease the complication, it is often unspecified that the existing state is only connected to preceding states. The RNN could exploit either unsupervised or supervised. If it is an unsupervised learning approach, the anticipation of the data sequence from the proceeding data samples is achievable, however, this model occasionally is seldom used due to the training complexity [1–6, 26–28].

Staudemeyer [26] designed RNN-based Intrusion Detection System (RNN-IDS). It symbolizes an RNN. The model is analyzed with the NSL-KDD dataset and the performance is calculated in binary as well as multi-class classification. The training precision and the test precision attained in binary classifications correspondingly are 99.91 and 84.58%. The training, as well as test precision with multi-classifications, are 99.63 and 83.39% [27]. Executed an LSTM-RNN classifier in support of intrusion detection. The consequences explained that LSTM classifiers have definite compensations above other well-built fixed classifiers with a 10% KDD Cup 99 dataset having DR and FAR value of 93.85% and 1.62%, respectively. Kim and Le [28] used LSTM as their model using 10% KDD Cup 99 dataset. The authors used 80 hidden layer and 0.01 learning rate. The model reports 96.83% DR, 98.9% standard precision and 10% normal FAR.

Deep Belief Networks

Deep Belief Network (DBN) is a probability-based generative replica consisting of numerous layers of theoretical as well as hidden variables. There is some close relationship between Restricted Boltzmann Machine (RBM) and DBN because the former encourages hidden layers to train the data instances resourcefully by activating it in favor of remaining training phases. The Bernoulli–Bernoulli RBM is designed for twofold variables, although the Gaussian-Bernoulli RBM (GBRBM) is used for constant data.

Zhao et al. [29] designed a joint neural network, through semi-supervised learning to attain the best precision values with a little number of labeled samples. For evolution authors utilized KDD Cup 99 dataset, and then used DBN to categorize the data instances, with a calculated precision value is 99.18%. Alrawashdeh and Purdy [30] evaluated dissimilar DBN arrangements, to use many layers and hidden units in the system. In addition, it is also used to find out a four-layer DBN replica. The KDD Cup 99 dataset is utilized for testing. Precision, accuracy, and FAR of representation were 92.89%, 93.44%, and 0.82%, respectively [31]. Applied DBN with Logistic-regression soft-max on behalf of deep set-up. It uses multi-class Logistic Regression level expert with 10 periods on improved pre-trained records to develop the performance of the set-up. The technique reaches a DR of 97.9% on a total 10% KDD Cup 99 test dataset furthermore shaped a small FNR of 2.47% after being trained by 40% NSL-KDD datasets. Yin et al. [32] used exploration of intrusion exposure potentiality of DBN throughout a sequence of trials. The skilled DBN system is capable of efficiently classifying unidentified attacks; however, the projected system realizes 97.5% precision after 50 iterations.

Convolution Neural Networks

Convolution Neural Networks (CNN) is a brand of ANN and DL that has turned out to be a hot topic in speech analysis as well as image recognition. It is one of the successful learning models for training multi-layer neurons and the purpose is to reduce the data preprocessing prerequisites. Local receptivity, weight sharing, and pooling are three major resources for CNN to trim down network-training. The influential component of CNN learns feature hierarchies commencing a huge quantity of unlabeled records. Hence, CNN is reasonably talented for network intrusion detection appliances.

Kolosnjaji et al. [33] planned a neural network consisting of convolution plus feed-forward neural formations. This structural design represents a hierarchical attribute extraction that mingles the features with the easy vectorized convolution. Saxe and Berlin [34] suggest an ExposedNN (ENN), which utilizes the DL method to extract and learn for feature extraction and classifications concurrently with character-level insertion and CNN. In automating the feature extraction ENN outperformed the physical feature mining for each intrusion detection. The disclosure rate and reduction in FAR are 92% and 0.1%, respectively. Wang et al. [35] worked for a malware traffic taxonomy technique with a CNN by considering traffic statistics such as images. The technique required no hand-made features however directly took uncooked interchanges as input records for a classifier. Wang et al. [36] planned a one-dimensional CNN that was continuously encrypted by interchanging taxonomy technique. The technique amalgamates feature extraction, assortment as well as classifier into a combined uninterrupted frame, which involuntarily learns the nonlinear affiliation among the unique input as well as the probable output. This technique applies an open ISCX VPN-non-VPN traffic dataset for proof.

4 Difference between Deep Learning and Machine Learning

There are numerous studies based on ML, DL, and AI. AI is the latest technology that studies and develops theories, processes, procedures, and appliances that mimic human brainpower. It is a branch of computer science, which tries to construct a category of intelligent machines that react in a mode similar to human intelligence. Research in this area contains expert systems, robotics and computer vision [1–5, 26–29]. ML is a branch of AI and is closely related to computational statistics and has well-built connections to mathematical optimization, which transports procedures, theories and appliances to the domain. The definition of ML is as follows. "*Field of study that gives computers the ability to learn without being explicitly programmed.*" ML can also be unsupervised and used to gain knowledge of behavioral outlines for a variety of entities and then employ to discover significant abnormalities. DL is the latest field in ML, its enthusiasm lies in the association of an NN that simulate the human intelligence for methodical learning. DL imitates the human brain structure to understand data such as texts, voice, video frames, and images with the implementation of an unsupervised layer-by-layer algorithm. Some of the dissimilarities between ML and DL are described below.

Data dependencies

ML models show the best results when the amount of data is small, on the other hand, DL models do not perform sound if the number of datasets is small since DL algorithms need a large dataset to identify the features of the data.

Hardware dependencies

Most of the ML models can easily operate with a single computer system. DL models need lots of matrix functions. GPUs are mainly utilized to optimize matrix functions resourcefully. For that reason, the GPU hardware is essential for the DL to work accurately.

Feature processing

The accumulated real-time data from various regions is increasing at a fast rate and there is a chance that they may be populated with the unnecessary matter. Feature processing is the procedure of analyzing datasets towards a feature extractor to lessen the complication of data as well as to produce patterns that formulate learning algorithms to work in a healthier manner. In ML, most of the data instances are encoded by an expert and then ML models are applied on them. The accuracy of ML models depends upon the precision of the features extorted. Gaining advanced features instantly from datasets is a foremost discrepancy among DL and ML models.

5 Artificial Intelligence and Machine Learning Applied to Cyber Security

This section covers a few applications of AI and ML in cyber security. Authors developed numerous methods and techniques on behalf of the recognition, exposure of threats in cyberspace. A mixture of fields such as Spam detection in software, industrial power schemes, intrusion exposure in Supervisory Control and Data Acquisition (SCADA) method, power system security, intrusion exposure in favor of vehicular ad hoc network (VANET), malware scrutiny, etc., have use of ML techniques in modern days. In the context of malware, the antivirus can exploit high-quality consequences but their schemes may be immobilized by bugs whose progression and enhancement can be typically quicker than the expansion of malware exposure program. This express advancement in attacks leads towards the discoveries of new methods for recognition of unidentified, strange malware plus a mixture of ML practices. In this modest analysis, we tried a little illustration towards the current readers with an experience of applicability of ML into cyber security [1–4, 6–9].

Power Security

Power Grids are essential for any national financial system as well as for their safety. Blackout is one of the rigorous appearances of power failure to a quite broad region that redistribution far-fetched communal costs, cloud effect from contradictory reasons such as an error at a power diffusion line or anticipated molests. Some of the blackouts turn out to be extended through the primary failures transmits into dissimilar as well as complicated odd events. If any power system is troubled, stupid line connections might intensify the rigorousness of the results, may disturb the geographical transmission otherwise may show the way to a cascade blackout. Hence, the power system is within a worried status, a partial cyber molest could provide evidence to be catastrophic via tumbling breakdowns. In the literature, numerous researchers propose countless innovative digital skills to protect broadcast communications in power systems.

Sortomme et al. [37] planned an adaptive protection technique called Phasor Measurement Units (PMUs) to avoid cyberattacks. Bernabeu et al. [38] developed an ML-based Decision tree algorithm to classify the power system status whether it was in stress, worried condition or not. Authors incorporated probable interpreters, such as, current and voltage scales, mega volt-amps (MVAr), angle variations, etc., taken by PMU that are used for teaching classifiers. Moreover, ML models have been applied to optimize the number of PMU's to be employed along with positions. Here, the decision tree assists in the partition of attributes, which decide the position. Regardless of the rising conditions of complication as well as improbability in power structures, AI & ML assists power-system engineers to take action on problems and use upcoming power schemes with a satisfactory intensity of protection.

Spam Detection

Spam detection is normally associated with incoming e-mails, however, search engines, tweets and blogs are the well-known goals. Most of the spam detection techniques that are developed are based on filters that evaluate the content and make a judgment whether it is a spam or valid message, blog, or website. The early-stage filters are user-friendly and easy to use, which were effortlessly moved by spammers. In recent days, ML models are applied to improve the detection of spam, and numerous models have been developed. There are two strategies in spam detection: Textual analysis—exploits a text categorization problem, Image-based analysis—images are occupied with implanted texts, it was essential to execute image-based study. While developing any spam detection the following characteristics must be verified before it is applied: complex text patterns, changing class distributions, changing target, message-misclassification costs, and intelligent adaptive adversaries. As mentioned in the above discussion filters consist of a regular structure. They mine the major terms of the communication by trimming them down to their source structures, frequent, inappropriate terms are removed and the resultant words are taken as input.

Zhao et al. [39] proposed a spam filter that shows the outcomes in electronic mail taxonomy to decide the terms in the mail. Almeida et al. [40] showed the manufacture of Naive Bayes classifiers and discussed dimensionality reduction. They also found the preference of crucial features in the training phase. Naive Bayes' simple execution, momentum, and spirited works are the characteristics that make this method most liked. Image-based spam was the successive weapon of spammers that formulates text-analysis. Biggio et al. [41] showed a collection of computer visualization, plus pattern identification systems useful to the exposure of spam letters implanted within images. Spammers may create additional locations apart from e-mails to circulation their spam substances.

Spam blogs are formed to attract the flow of interchange initial blog search engines to promote categorically. Many ML procedures similar to SVM, Decision trees, Naive Bayes as well as neural networks were cynical. Nonetheless, SVM spot spam blogs in numerous modern works.

Intrusion Detection System for VANET

Vehicular Adhoc Networks (VANETs) are a promising research area, which takes a huge part in contemporary transport structures for producing well-being and helpful information. VANETs are frequently exposed to numerous mallicious activities as well as passive attacks such as Man-in-middle, packet loss attacks.

Zhang and Zhu [42] planned an ML-based shared IDS called Collaborative Intrusion Detection Systems (CIDS) architecture. The proposed model teaches a classifier to notice impositions in VANET. CIDS assists the vehicle to utilize the facts of the label training data of other mediums, moreover, vehicles distribute data records without exchange the training data. However, analysis of these collected data at each vehicle increases the computation work, to reduce the workload; this work is circulated to every motor vehicle inside the network. Kumar et al. [43] designed a CIDS to detect threats such as control breaches and unofficial interruptions by detecting uncharacteristic, malevolent behaviors. Distributed ML is a tactic for drawing a collaborative detection method on VANET. However, privacy and security play a major part while nodes exchange sensitive information. A malevolent knot can block in addition to achieve responsive records regarding new contributing knots. Boyd [44] projected a Distributed ML-based Alternating Direction Method for Multipliers (ADMM), the main interest is confidentiality as every molest interloper can monitor their taxonomy as these authors initiated a privacy-preserving process in their model. As a result, this approach promises road security and makes the driver's information safe.

Malware Detection

Malware is a special kind of software that is specifically designed to interrupt damage or gain unofficial admission to any computer system. The suitable definition defined by Ranveer and Hiray [45] as "malware refers to a program that inserted into a system, usually covertly, with the intent of compromising the confidentiality, integrity, or availability of the victim's data, applications, or operating system or of otherwise annoying or disrupting the victim." In the past, malware was recognized with the help of signature-based procedures; however, these procedures tolerate complexity in noticing highly developed malware. Hence, researchers included ML models, which not only identify famous malware but also bring facts for the discovery of new-fangled malware. The exposure of malicious code in its dissimilar varieties is approached by numerous ML techniques to evaluate code patterns and comparisons. Some of them are an unsupervised ML structure to observe, access to workstation furthermore Markov representations to learn the explanation of malware. Clustering models valuable to categorize software as malware in addition to SVM have been tested in computer methods.

In recent times, great diligence has been set in malware recognition; SVM, as well as Naive Bayes schemes, have been broadly exploited with satisfactory results. Other than Naive Bayes and SVM, other ML processes such as decision trees, efficiently functional to malware recognition. Clustering was utilized to sense and sort malware. Enhanced Bayesian networks have the utmost precision and poor false positive rate in contrast to decision trees, K-NN, Naive Bayes, regression, and random forest.

Exposure of Cyberattacks to Industrial Control Systems

In recent times, cyberattacks have turned out to be a serious hazard even for the control system. To solve these, ML models are applied to have an involuntary intrusion detection system that can secure control systems as well as Critical Infrastructure (CI).

Wihersaari [46] proved that CI depends on interconnected communication technology where most of them are connected and vulnerable. Farwell and Rohozinski [47] worked on STUXNET, a CPU worm, that mostly attack industrialized controller and Programmable Logic Controller (PLC). Here, PLC is a small system with an integral operating system, its PLC compromised means; it results in a massive loss. Therefore, there is a vital call for ICS to enlarge consistent protection and security features. Figueiredo and da Costa [48] proposed a SCADA scheme and operators which can be utilized explicitly for all. Principal Component Analysis (PCA) exploits to sense every anomaly. PCA assists in decision prototypes to decrease the measurement of the dataset with nominal loss data records. Huang et al. [49] introduced an adversarial attack on the above-mentioned ML. In some way, an intruder on set-up can monitor and notice the prototype of training data after that he can modify the consequences of the ML model. Suppose if some trespasses into the system by altering and insert forged records to deceive the controller, the model will not detect the divergence and might offer false outcomes. This adversarial consequence can be sensed by restraining the interruption to a region or sector. This decreases the possibility of a trespasser to handle both the regions; hence, it assists the security system to spot the intrusion archetype.

Intrusion Detection in SCADA Systems

SCADA methods are important in supervising and monitoring of CI such as electric power production, transmit and allotment of water/sewage behavior plant. However, at this moment safety measures of SCADA are difficult since they have been associated with the IT arrangement. It has amplified the intimidation and threats posed by attackers, hence, there is an advice to propose and build up little watchful schemes that are capable of the present CI operatives with a mechanism that is capable to maintain them in perceive constant impositions. Maglaras and Jiang [50] have planned a pair of scattered IDS, talented to sense attack which takes place in SCADA classification. Both models are executed as well as estimated for the Cockpit CI scheme [51]. It is based on concurrent perimeter IDS, which endows with central part cyber analysis to assist in reviewing along with protecting the security boundaries of every CI. The main principle of this is to achieve abnormality exposure in a time capable mode as well as high precision with little transparency.

Phishing Detection

A phishing attack is an effort to steal sensitive data records through masking oneself as an honest approach in electronic communication. Fette et al. [52] employed ML base taxonomy advance entitles as PILFER, works such as random decision trees, with high exactness and little false-negative rates. Santhana Lakshmi and Vijaya [53] evaluated a few ML schemes specifically logistic-regression, random forests, neural networks, and SVM with uncertain consequences. Random forests shown low error-rate, although logistic-regression offers poor FPR along with a one-sided error-rate. SVM is applied on many occasions: phishing webs, e-mails, and URLs. Decision tree-based models show high-quality results within various appliances. Almomani et al. [54] used decision trees that show better results. Gu et al. [55] surveyed various techniques for phishing e-mail filtering along with network-level security and verification practices. Moreover, they made a discussion on the advantage and disadvantages of dissimilar filters plus classifiers. Feature selection along with it manipulates in the presentation of classifier have been considered as healthy because it is a key point for phishing exposure and sorting.

6 Error Decisive Factors

The evaluation of a model is an essential part of any ML assignment. Different ML assignments contain a variety of evaluation signs, while the parallel nature of ML assignments too has dissimilar valuation signs, each by a dissimilar prominence such as clustering, classification, SVM and regression, etc. Numerous researchers express the importance of analysis because the error rate is not the major aspect. It is compulsory to think of additional factors such as computation time, difficulty. The main concern could be different in each machine. The confusion matrix is a table that explains the taxonomy consequences in detail, whether they are properly or improperly classified. When it comes to classifying inward e-mails, for illustration, a novel error decisive factor arises; FPR and FNR. FPR is important, since the outcomes and losses of information suitable to the misclassification of an e-mail. As a result, cyber security schemes are recurrently charged according to the below-mentioned definitions [1–10, 16–24]:

True positive (TP): number of positive samples or several harmful applications are properly classified by the proposed model.

True negative (TN): number of negative samples or number of benevolent applications are correctly classified by the proposed model.

False-positive (FP): number negative samples or several benevolent applications are misclassified as harmful by the proposed model.

False-negative (FN): number of positive sample or number of unsafe applications are misclassified as benevolent.

Precision: calculates one-hundredth of all accurate positives detected over the whole number of detected positives recognized.

$$\text{Precision: TP}/(\text{TP} + \text{FP})$$

Accuracy: shows the entitlement of the right predictions, which means the ratio of appropriately categorized illustrations to the entire number of illustrations for a specified examination dataset.

$$\text{Accuracy: }(\text{TP} + \text{TN})/(\text{TP} + \text{TN} + \text{FP} + \text{FN})$$

Sensitivity or Recall: fraction of optimistic items properly recognized as dangerous over the total and estimates the relation of every properly detected item to every item that must be detected.

$$\text{Sensitivity or Recall: TP}/(\text{TP} + \text{FN})$$

False Negative Rate (FNR): shows false negatives over the total number of negatives identified or ratio of misclassified positive examples to the number of positive examples.

$$\mathrm{FNR} = \mathrm{FN}/(\mathrm{TP} + \mathrm{FN})$$

False Positive Rate (FPR): determines false positives over the entire amount of positives identified or ratio of the amount of misclassified pessimistic sample to the whole quantity of pessimistic samples.

$$\mathrm{FPR} = \mathrm{FP}/(\mathrm{FP} + \mathrm{TN})$$

True Negative Rate (TNR): percentage of the amount of appropriately classified pessimistic trials to the amount of negative sample.

$$\mathrm{TNR} = \mathrm{TN}/(\mathrm{TN} + \mathrm{FN})$$

F1-score: estimates the harmonic mean of the precision as well as the recall.

$$\text{FI-Score} = 2 * \mathrm{TP}/\,(2 * \mathrm{TP} + \mathrm{FN} + \mathrm{FP})$$

Error rate: imperfectly categorized instances over the total.

$$\mathrm{FP} + \mathrm{FN}\ \text{Error rate} = \mathrm{TP} + \mathrm{TN} + \mathrm{FP} + \mathrm{FN}$$

Total cost ratio: price of miss-classified data occurrence considered, where the cost is relative to both faults.

$$\mathrm{TCR} = \mathrm{FN} + \mathrm{TP}\ (\mathrm{FP} + \mathrm{FN})$$

Weighted Error: designed through a specific load or weight.

$$\mathrm{WErr} = \mathrm{TN} + \mathrm{TP}\,\mathrm{FP}$$

Receiver Operating Curve (ROC): illustrates the exchange of the classifier among TP with FP.

Area under the Curve (AUC): volume of the region below the ROC arc.

In cyber security, metrics frequently extracted in assessment form are precision, F1-score, and recall. Improvement is done by precision along with evoking of replica analysis in cyber security; however, these two in some cases are ambiguous and forcefully impartial depending on assignment requests. The F1-score is a vocal set of exactness along with reminding, given their consequences. In broad-spectrum, more the superior F1-score, more the superior the replica will be.

7 Cyber Security Datasets

Cyber security is an art of adding security preventions to provide CIA rules concerning information. Numerous authors suggested various cyber security descriptions in the literature. Cyber security is an active research area since most of the functions operated in government, industrial, military, hospital management, commercial and financial sections gather, stores and process tremendous information on computers. These companies must follow some set of rules if they want to be on the protective side of cyber security [7–10]. However, with the exponential enhancement of computer appliances and computer systems usage, securities turn into more significant. Adversaries are potentially implementing various attack paths to do chaos to your business. Because of this huge increase in cyberattacks AI and ML tactics are an essential part to detect security threats [7]. To afford the most excellent security appliances and proper level of security, security associated benchmark and standard datasets are astonishingly essential. In the literature there are numerous studies in the domain of cyber security with particular datasets, on the other hand, there are very few studies on other datasets. At present well-known research groups put collectively several types of datasets for their study intention as well as for private community repositories. Some of them are discussed below [4–7, 22–34].

KDD Dataset: In the year 1998, Defense Advanced Research Projects Agency (DARPA) assembled the first benchmark dataset to estimate computer network intrusion detection systems in the MIT laboratory. KDD cup 1999 is a part of the DARPA project includes tcpdump and BSM files. It is considered as a point of reference data for the evaluation of intrusion exposure systems. Dataset holds four major kinds of attacks that are Denial-of-service (DoS), Probing attack, user-to-root attack, and Remote-to-local-attack. Moreover, it contains 38 numerical features and 3 contented features which consist of individual TCP links and connections recommended by domain information. To date, KDD is one of the most widely; popular datasets utilized for anomaly or intrusion detection systems.

ECML-PKDD Dataset: The ECML-PKDD dataset was generated for the European Conference on Machine Learning and Knowledge Discovery which held in 2007, it was purely confronted with data mining contests. The interesting thing is entire data features described in XML, with three main parts that are context—it explains features of OS running on the webserver, HTTP server targeted by request and SQL database. Class—a type of data and query—method, query, headers, and body.

HTTP CSIC Dataset: This dataset is developed by the Spanish Research National Council; it encloses thousands of involuntarily engendered web requests. The main objective behind creating HTTP dataset is for testing web attack defense methods. Dataset holds 6 thousand normal web requests and higher than 25,000 anomalous requests. For user handiness dataset divided into testing, anomalous and training parts. Furthermore, dataset inbuilt holds 3 kinds of attacks those are: Static attacks—session Id, default files, request for the hidden resource, etc. Dynamic attacks: buffer overflow, SL injection, cross-site scripting, etc. Unintentional illegal requests—illegal requests with malicious intention, not follow the mentioned principles of web appliances.

UNSW-NB15 Dataset: It was formed by the Australian Center for Cyber security by the Cyber Range Lab. It contains the nine most important cyberattacks along with 49 features. To extract these features Argus tool or any equally qualified mechanism utilized. Some of the features are categorized into content features, flow features, time features, basic features, and additional engendered features. Compared with an existing dataset, this one holds numerous attacks that eventually reproduce the majority of contemporary attacks.

Czech Technical University (CTU-13) Dataset: The goal of this dataset, is to detect, notice, or discover valid mixed botnet traffic. It was the leading, well-liked dataset engendered by CTU Prague in the Czech Republic in the year of 2013. Dataset consists of thirteen dissimilar malware or set-ups in a non-efficient network atmosphere, each of these set-ups examined by particular malware which implements various protocols and actions. Here, a major benefit of CTU-dataset has a warily labeled dataset, detaining procedure carried out in the prescribed atmosphere.

ADFA Dataset: In the year 2013, the Australian Defense Force Academy (ADFA) at the University of New South Wales (NSW) released a Linux base Dataset entitle ADFA, which is principally related to Host-based abnormality detection. Most of the existing benchmark datasets for Host-based imposition or abnormality detection schemes are old moreover they are not efficient to detect fresh attacks. It contains normal and Linux attack system call traces. Whenever a sampling stage, the host that is configured to symbolize a recent Linux server detain the system call traces wherever genuine programs are functioning as normal. At the same time, cyberattacks instigated against the host.

Information Security and Object Technology (ISOT): The ISOT dataset was gathered from the French chapter of a honey-net project which consists of various botnets especially Storm, Waledac botnets for malicious traffic in ISOT. Its traffic dataset, joins with another dataset formed by Lawrence Berkeley National Lab (LBNL). Non-malicious datasets were collected from Traffic Lab Ericson Research in Hungary. It is an important traffic-based dataset for the Ericson industry. LBNL further have intermediate size arrangements as well as 5 massive datasets.

NSL-KDD Dataset: It is a fresh edition from KDD Cup 99 dataset. It develops a few boundaries of KDD Cup 99 dataset and useful at Intrusion Detection. It planned the 3rd International Knowledge Discovery and Data Mining Tools competition. It classifies attributes among interfering plus standard relations for constructing a network intrusion detector. In this dataset, every illustration has the individuality kind system data. It encloses KDDTrain+ as a training set as well as KDDTest+, KDDTest−21 as a testing set, which has dissimilar standard records along with four special styles of attack proceedings. The KDDTest−21 dataset is a detachment of the KDDTest+ moreover is not easy toward classify.

8 Future Research Directions and Challenges

1. The accessible benchmark datasets are few, contains old, unbalanced classes of data categories and redundant information. In some works, the same dataset is used for development shows the same outcomes. In some of the methods, data preprocessing is applied but there is an issue of inadequate data. Thus, implementing anomaly, attack detection datasets with large data samples, a balanced number of attack categories become the main concern in the field of cyber security.
2. The simulation metrics are not identical, numerous studies simply evaluate the accuracy of tests, and so the end result is unfair. But another study makes use of multi-criteria estimations frequently approve special metrics like that the research consequences cannot be weighed against one another.
3. Hybrid detection techniques generally unite ML and DL models, like a combination of deep learning along with machine-learning models for cyberattacks are less studied.
4. The way in which the cyber security is growing day by day and there is exciting work to shape the fresh data with the trained model. Nowadays, researchers are applying transfer learning to adjust the model with an undersized quantity of label data, however, results are not in expected way more research must be done here.
5. No study has been given for employment effectiveness, and the majority of the research resides inside the lab irrespective of time complication and competence of detection in the authentic set of connections.
6. Employing decreasing detection time and improving the speed of models depends on hardware features. Hardware can exploit numerous computers for comparable computing that can easily handle ML and DL, merger models.
7. The learning of crossbreed representations has been noticeable in current years, in addition to enhanced data metrics that are achieved by reasonably merging special algorithms.
8. The introduction of DL has made an end-to-end learning, including managing a large quantity of data with no human participation. However, the enhancement needs numerous trials and practices; interpretability is not up to the mark.

9 Conclusion

ML, as well as DL, provides an individual or combinational explanation to a mixture of cyberattack dilemmas such as malware and intrusion detections. Notably, security concerns associated with cyberattacks are power system safety, industrial control systems and intrusion recognition for VANET, etc. These troubles absorb disciplined and successful training and taxonomy of data in enormous size. This chapter has described a few regions of cyber security where AI, ML, and DL are being utilized.

In addition, various literatures help us to understand ML and DL purposes and drawbacks in cyber security. Moreover, various techniques, network security datasets, and future research directions are also described.

References

1. Torres M, Comesaña JI, Carla G-N, Paulino J (2019) Review: machine learning techniques applied to cybersecurity. Int J Mach Learn Cybern 10(10):2823–2836
2. Handa AS, Shukla A, Sandeep K (2019) Machine learning in cybersecurity: a review. Wiley Interdiscip Rev Data Mining Knowl Discov 9(4):1–7
3. Das RM, Thomas H (2018) Machine learning and cyber security. In: 2017 international conference on computer, electrical and communication engineering, ICCECE 2017, pp 1–7
4. Fraley JB, Cannady J (2017) The promise of machine learning in cybersecurity. In: Conference proceedings—IEEE southeastcon
5. Xin Y, Kong L, Liu Z, Chen Y (2018) Machine learning and deep learning methods for cybersecurity. IEEE Access 35365–35381
6. Vljqlilfdqw S, Iru S, Frpsdqlhv DOO (2017) Cyber security data sets. In: 2017 IEEE international conference on big data (BIGDATA)
7. Liu W, Wang Z, Liu X (2017) A survey of deep neural network architectures and their applications. Neurocomputing 234(10):11–26
8. Kwon D, Kim H. A survey of deep learning-based network anomaly detection. Clust Comput 6(22):949–961
9. Vapnik V (1982) Estimation of dependences based on empirical data. Springer, Berlin
10. Drucker H, Burges C, Kaufman L, Smola A, Vapnik V (1997) Support vector regression machines. MIT Press, Cambridge
11. Osuna E, Freund R, Girosi F (1997) An improved training algorithm for support vector machines, In: Proceedings of the 1997 IEEE signal processing society workshop, Amelia Island, Florida, USA, pp 1–10
12. Chandrasekhar AM, Raghuveer K (2014) Confederation of FCM clustering, ANN and SVM techniques to implement hybrid NIDS using corrected KDD cup 99 dataset. In: International conference on communications and signal processing, pp 672–676
13. Yan M, Liu Z (2017) A new method of transductive SVM-based network intrusion detection, in computer and computing technologies in agriculture IV. In: IFIPTC 12 conference, CCTA 2010, Nanchang, China, October 22–25, 2010, Selected Papers, pp 87–95
14. Kokila RT, Selvi ST, Govindarajan K (2015) DDoS detection and analysis in SDN-based environment using support vector machine classifier. In: Sixth international conference on advanced computing, pp 205–210
15. Peng XU, Jiang F (2014) Network intrusion detection model based on particle swarm optimization and k-nearest neighbor. Comput Eng Appl
16. Rai K, Syamala M, Devi, Guleria A (2016) Decision tree based algorithm for intrusion detection. 07(4):2828–2834
17. Azad C, Jha VK (2015) Genetic algorithm to solve the problem of small disjunction the decision tree based intrusion detection system. 7(8):56–71
18. Puthran S, Shah K (2016) Intrusion detection using improved decision tree algorithm with binary and quad split. In: International symposium on security in computing and communication, pp 427–438
19. Vishwakarma S, Sharma V, Tiwari A (2017) An intrusion detection system using KNN-ACO algorithm. Int J Comput Appl 171(10):18–23
20. Ingre B, Yadav A, Soni AK (2017) Decision tree based intrusion detection system for NSL-KDD dataset. In: International conference on information and communication technology for intelligent systems, pp 207–218

21. Malik AJ, Khan FA (2017) A hybrid technique using binary particle swarm optimization and decision tree pruning for network intrusion detection. Clust Comput 3:1–14
22. Relan NG, Patil DR (2015) Implementation of network intrusion detection system using variant of decision tree algorithm. In: International conference on nascent technologies in the engineering field, pp 1–8
23. Abraham A, Grosan C, Martin-Vide C (2007) Evolutionary design of intrusion detection programs. Int J Netw Secur 4(3):328–339
24. Joshi SS, Phoha VV (2005) Investigating hidden Markov models capabilities in anomaly detection. In: Proceedings of the 43rd annual southeast regional conference, vol 1. ACM, pp 98–103
25. Fan W, Miller M, Stolfo S, Lee W, Chan P (2004) Using artificial anomalies to detect unknown and known network intrusions. Knowl Inf Syst 6(5):507–527
26. Staudemeyer RC (2015) Applying long short-term memory recurrent neural networks to intrusion detection. 56(1):136–154
27. Kim G, Yi H, Lee J, Paek Y, Yoon Y (2016) LSTM-based system-call language modeling and robust ensemble method for designing host-based intrusion detection systems. arXiv:1611.01726
28. Le TTH, Kim J, Kim H (2017) An effective intrusion detection classifier using long short-term memory with gradient descent optimization. In: International conference on platform technology and service, pp 1–6
29. Zhao G, Zhang C, Zheng L (2017) Intrusion detection using deep belief network and probabilistic neural network. In: IEEE international conference on computational science and engineering, vol 1, pp 639–642
30. Alrawashdeh K, Purdy C (2017) Toward an online anomaly intrusion detection system based on deep learning. In: IEEE international conference on machine learning and applications, pp 195–200
31. Tan Q, Huang W, Li Q (2016) An intrusion detection method based on DBN in ad hoc networks. In: International conference on wireless communication and sensor network, pp 477–485
32. Yin Q, Zhu YF, Fei JL, He XZ (2017) A deep learning approach for intrusion detection using recurrent neural networks. IEEE Access 5:21954–21961
33. Kolosnjaji B, Zarras A, Webster G, Eckert C (2016) Deep learning for classification of malware system call sequences. In: AI 2016: Advances in artificial intelligence, pp 137–149
34. Saxe J, Berlin K (2017) eXpose: a character-level convolutional neural network with embeddings for detecting malicious URLs, file paths and registry keys. arXiv:1702.08568
35. Wang W, Zhu M, Wang J, Zeng X, Yang Z (2017) End-to-end encrypted traffic classification with one-dimensional convolution neural networks. In: 2017 IEEE international conference on intelligence and security informatics (ISI), pp 43–48
36. Wang W, Zhu M, Zeng X, Ye X, Sheng Y (2017) Malware traffic classification using convolutional neural network for representation learning. In: International conference on information networking, pp 712–717
37. Sortomme E, Venkata S, Mitra J (2010) Microgrid protection using communication-assisted digital relays. IEEE Trans Power Deliv 25(4):2789–2796
38. Bernabeu EE, Thorp JS, Centeno V (2012) Methodology for a security/dependability adaptive protection scheme based on data mining. IEEE Trans Power Deliv 27(1):104–111
39. Wang ZJ, Liu Y, Wang ZJ: E-mail filtration and classification based on variable weights of the Bayesian algorithm. Appl Mech Mater 513–517
40. Almeida J, Almeida T, Yamakami A (2011) Spam filtering: how the dimensionality reduction affects the accuracy of Naive Bayes classifiers. J Internet Serv Appl 1(3):183–200
41. Biggio B, Fumera G, Pillai I, Roli F (2011) A survey and experi- mental evaluation of image spam filtering techniques. Pattern Recognit Lett 32(10):1436–1446
42. Zhang T, Zhu Q (2018) Distributed privacy-preserving collaborative intrusion detection systems for vanets. IEEE Trans Signal Inf Process Netw 4(1):148–161
43. Kumar V, Srivastava J, Lazarevic A (2006) Managing cyber threats: issues, approaches, and challenges, vol 5. Springer Science & Business Media, New York, NY

44. Boyd S (2011) Alternating direction method of multipliers. In: Talk at nips workshop on optimization and machine learning. Now Publishers, Boston
45. Ranveer S, Hiray S (2015) Comparative analysis of feature extraction methods of malware detection. Int J Comput Appl 120(5):1–7
46. Wihersaari K (2015) Intelligence acquisition methods in cyber domain: examining the circumstantial applicability of cyber intelligence acquisition methods using a hierarchical model
47. Farwell JP, Rohozinski R (2011) Stuxnet and the future of cyber war. Survival 53(1):23–40
48. Figueiredo J, da Costa JS (2012) A SCADA system for energy management in intelligent buildings. Energy Build 49:85–98
49. Huang L, Joseph AD, Nelson B, Rubinstein BI, Tygar J (2011) Adversarial machine learning. In Proceedings of the 4th ACM workshop on security and artificial intelligence. ACM, New York, NY, pp 43–58
50. Cruz T, Barrigas J, Proenca J, Graziano A, Panzieri S, Lev L, Simões P (2015) Improving network security monitoring for industrial control systems. In: IFIP/IEEE international symposium on integrated network management (IM) IM2015, pp 878–881
51. Maglaras LA, Jiang J (2014) Intrusion detection in SCADA systems using machine learning techniques. In: Science and information conference (SAI), pp 626–631
52. Fette I, Sadeh N, Tomasic A (2007) Learning to detect phishing emails. In: Proceedings of the 16th international conference on World Wide Web (WWW '07). ACM, New York (US), pp 649–656
53. Santhana Lakshmi V, Vijaya MS (2012) Efficient prediction of phishing websites using supervised learning algorithms. Procedia Eng 30:798–805
54. Almomani A, Gupta BB, Atawneh S, Meulenberg A, Almomani E (2013) A survey of phishing email filtering techniques. IEEE Commun Surv Tutor 15(4):2070–2090
55. Gu X, Wang H, Ni T (2013) An efficient approach to detecting phishing web. J Comput Inf Syst 9(14):5553–5560

Cyber Security with AI–Part II

Moutushi Singh, Hrithik Lall, and Shaurya

Abstract Internet practice has become an essential part of human life. Without Internet, even a miniscule work may not be finished. On the contrary various forms of cybercrimes or network, outbreaks are also increasing with the same rhythm and swiftness. In this crucial time protection of information on the cyber world has become an enormous task. Therefore, everyone should have knowledge of different strong security dealings to protect their information. The Artificial Intelligence approaches are applied on existing data to constantly improve its functionalities and tactics over time. It studies and comprehends standard user behavior with the ability to detect the smallest amount of dissimilarity from the given outline. Nonetheless, besides collecting information to perceive and classify threats, an artificial intelligence system also uses gathered information to advance its own functionalities. It is difficult to create software with standard algorithms or hardware for successful alarms against the network attacks which are powerfully designed. It is obvious that several cyber security issues are additionally coming day by day. So, it is into consideration that artificial intelligence applications may be incorporated with cyber security computing issues to analyze the improvements against cyberattacks.

Keywords Cyber security · Network attacks · IoT · DoS · DDoS · IDS · Artificial intelligence

1 Introduction

The digital world in this era is facing the biggest threat known as "Unethical Hacking" or "Cracking". Cyber security is one of the genres that could be used to protect the cyber world from these threats. With the development of technology, the concept of Artificial Intelligence (AI) came into the scene which helps the security analyst to stay ahead of threats to protect their system. An outsized variety of methodologies

M. Singh (✉) · H. Lall · Shaurya
Institute of Engineering & Management, Kolkata, India
e-mail: moutushi.singh@iemcal.com

M. Chakraborty et al. (eds.), *The "Essence" of Network Security: An End-to-End Panorama*, Lecture Notes in Networks and Systems 163,
https://doi.org/10.1007/978-981-15-9317-8_7

are developed within the AI field to find painstaking issues that require intelligence from the human perspective [1].

In the recent past, billions of US dollars were invested for cyber security startups especially those that specialize in AI and machine learning. Those were the top investment areas. Technology development has brought in the concept of AI, which is causing works to be done more easily, more precisely and more efficiently. In the case of the Humanoid "Sophia" developed by Hanson Robotics, it has the capacity to do the work of ten persons single-handedly. As security features, it has a self-bug detection system and intrusion detection system (IDS). Firewalls and IDS which are available in the market nowadays are configured in such a way that they detect any type of malicious activity on their own and inform the user [2]. Therefore, the concept of AI is required in the field of cyber security.

Cyber security has the following genres such as *Network*, *Web*, *Android* and most importantly *Internet of Things (IOT)*. All these genres when configured or tested, security issues keep on arising time to time. If a Firewall or IDS can be placed in these systems, the malicious activities can be blocked automatically, and a log of those activities gets created so that the defaulter can be identified. Precisely an AI-based application layer can be configured on the system to improve the security features.

Therefore, the collaboration of AI and cyber security is important to save our systems as well as AI-based projects in the digital world from upcoming cyber threats automatically [3].

With the prolonged research, an AI-based IOT device can be developed which will have an automatic intrusion and bug detection system, and will block malicious activities, create logs and will inform the user about the activities on a regular basis.

Organization: In Sect. 2 the broader cyber security concepts are introduced whereas network security concepts are introduced in Sect. 3. Review of types of network security is in Sect. 4. Section 5 discusses the role of AI. Section 6 discussed how AI can improve the security of the cyber world. Finally, the chapter is concluded in Sect. 7.

2 Cyber Security

Cyber security has become one of the most important fields amongst the vast genre of existing fields of research and technologies. Some cyber security statistics to understand its importance are:

- 94% of malware is distributed via emails.
- More than 80% of recorded incidents include phishing attacks.
- 60% of breaches constitute vulnerabilities for which patch was available but not used.
- In the 1st half of 2019, attacks on IoT computers tripled.
- Large-scale DDos attacks are rising by 500%.

- Approximately 6 trillion is projected to be spent on cyber security worldwide by 2021.
- 95% of information security violations are due to human errors.

From the above data it is clear that Unethical Hacking or "*Cracking*" is posing a threat on the whole cyber space and IoT devices [4]. The IoT devices which are all responsible for controlling a fridge for playing music are becoming new targets of unethical hackers. Users unpack their products and use them without modifying the default password setup by the manufacturer leaving them highly vulnerable. The industrial uses of IoT are at a greater risk of attacks that may lead to damage of property and lives. The smart TV, smart speakers, connected printers, smart fridge, smart air conditioner, smart coffee machines, have been reported with security issues that may result in breach of information on a mass scale for malicious purposes.

Apart from firms, the nuclear facilities and critical command and control systems are not immune to cyberattacks [5]. Such an attack could facilitate cyber terrorism or a catastrophic act of sabotage. Cyber terrorism is not only limited to breaching the nuclear facilities but also includes the use of malwares in order to achieve political or ideological gains. To prevent all the threats and risks it is very crucial to empower the field of cyber security with Artificial Intelligence.

The attackers can be divided into 4 categories on basis of sources:

Nation States: They are the ones who launch cyberattack on other countries.

Cybercriminals: These include the crackers, scammers, defacers, carders, and are the people who disrupt the systems or perform malware attacks to steal information in order to gain profits as well as defaming an organization [6].

Cyberterrorists: They are ones who pose a threat to life and property through malwares and breaches in order to harm life resources or for political and ideological gains.

Hacktivists: These threat actors are motivated by politics and societal grounds. For example, *Anonymous* is a perfect example of hacktivists [7].

3 Network Security

Network security is any operation planned to safeguard network and data accessibility and integrity. It includes both hardware equipment and software. It addresses several threats and prevents malware from infecting the network and users.

3.1 Importance of Network Security

The importance of network security varies for different users. For a company, the network security is important for the sake of the information regarding the ongoing projects to the transaction-related details. Hence, to highlight the importance of Network Security, here are some facts:

- Over half of the world's Internet traffic is now done on mobile devices. This brings new patterns and new network challenges not only to manage the mobile traffic smoothly but also to secure it.
- 56% of the Internet traffic is initiated by an automated source such as bots, hacking operations, spammers, bitcoin miners, impersonators, etc.
- Approximately 4,000 ransomware attacks happen every day. Out of which 73% are done for monetary gains.
- Nearly 46% of consumers say that if they have a poor or suspicious digital experience, or worse, they may not buy products or services from that business again.

According to security analysts, a careless or ignorant employee can cause major security risks. This reflects directly on customers who are dependent on the company to keep them safe. Even repeated hacks in a company can badly ruin its reputation [8].

3.2 Disruptions Due to Lack of Network Security

Interrupted Business: Even minor cyberattacks can disrupt business and risk everything from financial information and interrupted inventory to a complete digital shutdown [9].

Data Loss: The loss is the least of uncertainties during a cyberattack. Customer privacy and pertinent business data as well as sensitive information are compromised.

Fines and Legal Complications: Apart from the depth and breadth of a cyberattack, a business could face precise government-instructed mishandling fines and lose compliance or standard certifications.

Overall Loss of Business: Few consumers may trust a company whose resume comes discolored with digital mismanagement. This directly affects our bottom line and the ability to stay open [9].

3.3 Types of Network Security Attacks

Networks are subject to attacks from malicious sources. Attacks can be from two categories:

Passive: A network intruder intercepts data traveling through the network and.

Active: An intruder initiates commands to disrupt the network's normal operation or to conduct reconnaissance and lateral movements to find and gain access to assets available via the network [10].

Whereas the data breaches of a firm can take place in two ways:

Social Attacks: Hackers or configured bots can spam the employees, usually via emails or as concealed traffic on a website. They will try to persuade top employees to enter the passwords in fake logins through emails (spear phishing) or while opening a file to access materials that pretend to be something secure and vital to the company. If a hacker can get a direct entry point into the network, then the vital part of security has already been compromised [11].

Network Attacks: A skilled cracker can research network defenses, such as weak passwords, outdated operating systems or even ineffective anti-virus software. The tunnel through such weak spots can be used to implant a backdoor or worm in the system. The worst network attacks can pose indefinite data extraction risks that will take a complete system overhaul to patch [12].

On basis of methods there are five major types of network security attacks:

1. ***Malware***: It is a malicious version of the software that is planted into a network that can wreak all sorts of chaos [13]. It does this either by using a social or network attack. Malware is commonly planted through human mistakes, such as clicking a media link with a preinstalled malware bug or downloading an unauthorized file. Such malware involves ransomware, backdoors, botnets, etc. Threats of malware include risks such as transmitting data from the server, monitoring system activity, history, screen behavior and even keystrokes, accessing system camera and microphone, taking maximum hardware control and locking all devices via ransomware.
2. ***Phishing***: Phishing attacks are executed in a multitude of ways but their main purpose is the same as having an actual user to share confidential information. Business-critical details such as network passwords or personal data such as emails, passwords, credit card numbers, bank accounts and medical history may be the information at risk. Often phishing operations are more effective than other forms of data-stealing methods as they use a human input to get what they need. Phishing hackers also send messages and frame data as if it had originated from a trusted source. They can also look like emails from established individuals or services such as a co-worker or bank including links that redirect to a non-intrusive website.
   By gaining access to the credentials obtained in phishing, the hacker can implant malware and backdoors onto the server while still pretending to be authentic. The phishing attacks are also combined with email spoofing attacks through which the email pretends to be sent from the actual email address of a known person. This is known as Spear Phishing.
   In contrast to bulk phishing, spear phishing attackers often gather and use personal information about their target to increase their probability of success. Apart from Spear Phishing there are also different varieties of phishing such as Clone Phishing, Phone Phishing, DNS-Based Phishing, etc. [14] (Fig. 1).
3. ***Denial of Service (DoS) Attack***: Denial-of-service (DoS) attacks are malicious attempts to make an online service unavailable to clients by flooding the servers with very large traffic. This form of flooding of the network is dangerous because it threatens something that companies and customers prefer to take for granted. It can also be launched from multiple points of attack, synchronizing hundreds of computers or computer programs or authentic webservers with backdoors/botnets

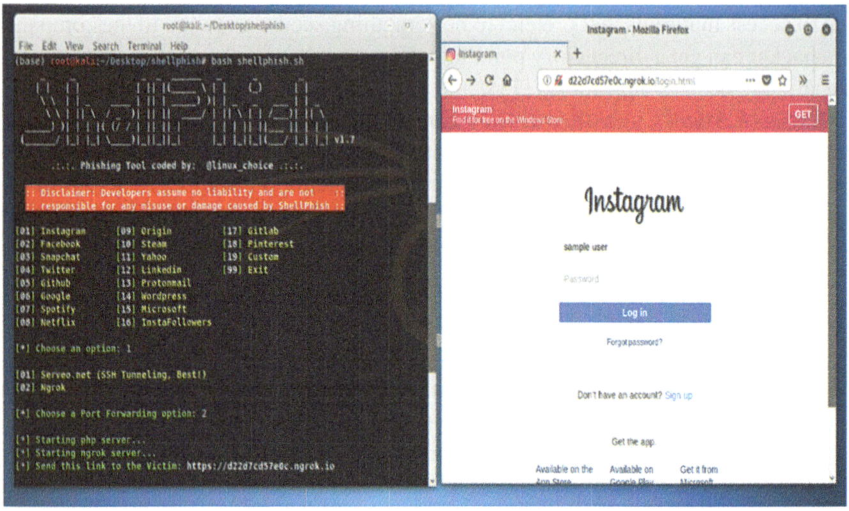

Fig. 1 Phishing attack

to attack a business network and block network traffic. Denial-of-service attacks do not generally intend to extract data, as it is expensive and destructive.
Panix, the third-oldest ISP in the world, was the target of what is thought to be the first *DoS* attack. On September 6, 1996, *Panix* was subject to a *SYN flood* attack that brought down its services for several days while hardware vendors, notably Cisco, figured out a proper defense [15].
A Distributed DoS (DDoS) attack is a collective effort from a secluded location over the Internet to prevent computer systems from working as usually. Here numerous machines work together for a single targeted machine [16, 17] (Fig. 2).

DoS and DDoS attacks are basically classified into three categories:

(I) ***Volume-Based Attacks***: It includes *UDP flooding*, *ICMP flooding* and other floods of spoofed packets. The aim of the attack is to saturate the attacked site's bandwidth. It is measured in *Bits per second (Bps)*.
(II) ***Application Layer Attacks or Layer 7 (L7) Attacks***: It includes *low-and-slow attacks*, floods of spoofed *GET and POST* requests, resulting in more damage with less bandwidth. The purpose of these attacks is to crash a webserver by using floods of seemingly legitimate requests. The magnitude of Application layer attacks is measured in *Requests per second (Rps)*.
(III) ***Protocol Attacks***: It includes *SYN floods*, Multiple *SYN-ACK Spoofed session flood, RST/FIN Floods*, fragmented packet attacks, *Smurf DDoS*, etc. This form of attack consumes server resources or intermediary communication equipment resources such as firewalls and load balancers. It is measured in *Packets per second (Pps)* (Table 1).

A short description of some *DoS* and *DDoS* attacks:

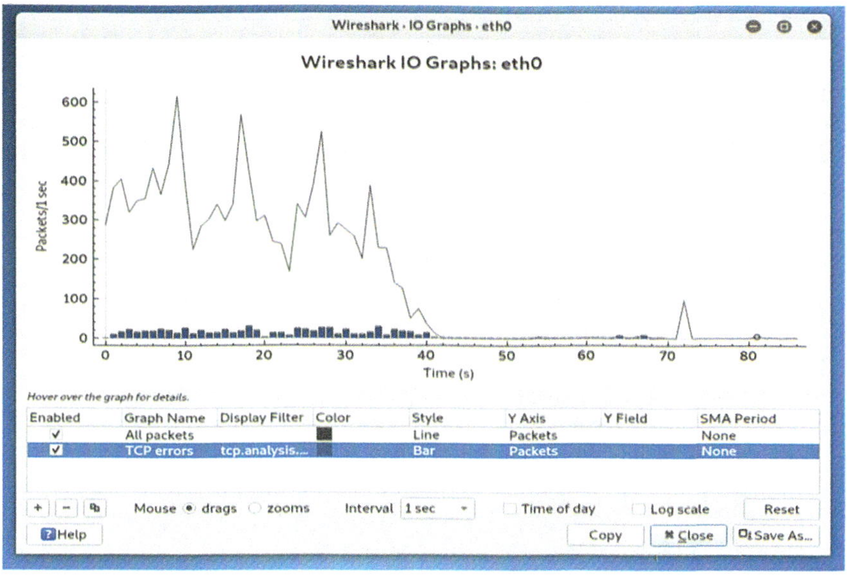

Fig. 2 DoS Attack

Table 1 Comparison of DoS and DDoS attack

DoS: Denial of Service	DDoS: Distributed Denial of Service
In DoS single system attacks the target server	In DDos multiple systems attack the target server
Can be blocked easily as only one system is used	It is difficult to block this attack as multiple devices are sending packets and attacking from multiple locations
Easy to trace	Difficult to trace
Volume of traffic in DoS attack is less as compared to DDoS	DDoS attacks allow the attacker to send massive volumes of traffic to the victim network
Types of DoS attacks are: • Buffer Overflow attack • Ping of Death or ICMP Flood • Teardrop attack	Types of DDoS attacks are: • Volumetric attack • Fragmentation attack • Application layer attack

UDP Flood: By definition, a User Datagram Protocol (UDP) flood is any DDoS attack that floods a target with UDP packets. The attack's purpose is to flood random ports to a remote host. Hence, it causes the host to search repeatedly for the application listening at port and respond, *Destination Unreachable*, when no application is found [18, 19].

ICMP (Ping) Flood: In concept such as the UDP flood attack, the target resource is overwhelmed by an ICMP flood of ICMP Echo Request (ping) packets, generally sending packets as rapidly as possible without waiting for replies. This form of attack

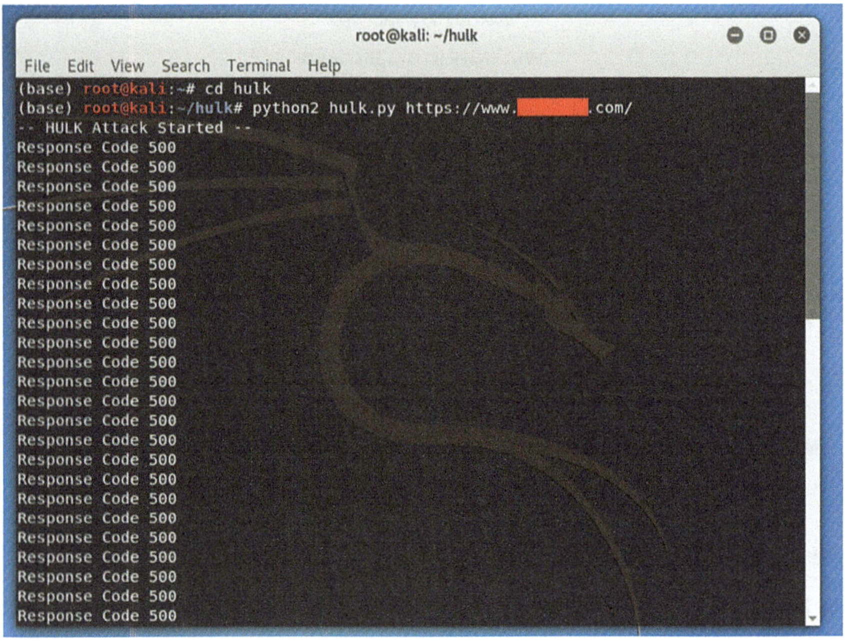

Fig. 3 ICMP Flood attack

will consume both outgoing and incoming bandwidth, as the victim's servers will often attempt to reply with ICMP Echo Response packets, leading to a slowdown in the overall network. This overloads the victim computer and can even make it unusable during such attack [20] (Fig. 3).

SYN Flood: A SYN flood DDoS attack exploits the known flaw in TCP communication sequence ("three-way handshake"), in which a SYN request to establish a TCP link with a host must be answered by a SYN-ACK response from that host, and then validated by the requester's ACK response. The requester sends numerous SYN requests in a SYN flood scenario but either does not respond to the SYN-ACK replies from the host or sends SYN requests from an anonymous IP address. Each way, with any of the requests the host network continues to wait with acceptance, binding resources until no further connections can be made and eventually result in DoS [21] (Fig. 4).

Ping of Death: Ping of Death is a DDoS attack in which the intruder tries to disrupt a targeted server by sending the data packets larger than the allowed size [22]. Hence, it results in a server crash. The IPv4 packet's total packet length (including header) is 65,535 bytes. But the Data Link Layer typically presents limitations to the overall frame size to 1500 bytes on an Ethernet network. In this case, a large IP packet will be fragmented into several IP packets, and the receiver host must reassemble the IP fragments into the corresponding whole packet. In a Ping of Death scenario, the receiver ends up with an IP packet that is greater than 65,535 bytes when reassembled

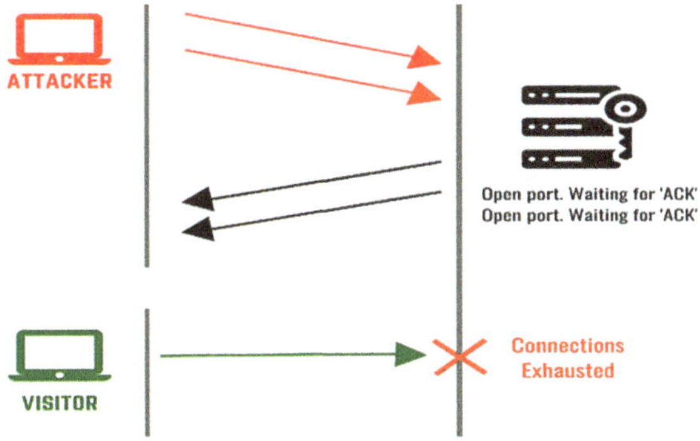

Fig. 4 SYN-ACK Response

after malicious abuse of the fragment content. It can overload the allocated memory buffer size for the packet and trigger the denial of legitimate packets.

Slowloris: These attacks enable one webserver to access another server without disrupting the target network's any of the services or ports. It does this by maintaining as many links as possible, available to the target webserver. A connection is created to the target server but only by sending a partial request. Slowloris sends more and more HTTP headers continuously but never completes a request. Then the targeted server is flooded by these fake connections. This inevitably overflows the permissible numbers of connection and leads to rejection of connections from legitimate clients [23] (Fig. 5).

DNS Amplification attack: In DNS amplification attacks, the hacker exploits the vulnerabilities present in DNS servers to convert small initial requests sent by botnets into massive payloads [24]. Reducing the number of open DNS servers can be a good measure of prevention. It is a type of reflection attack.

HTTP Flood: The attack exploits apparently legitimate HTTP GET or POST requests to assault a webserver or application in a DDoS assault by HTTP flood. HTTP floods do not use malformed packets, spoofing or reflective techniques and need less bandwidth to put down the targeted site or server in comparison with other types of DoS attacks. The attack is most successful when it causes the server or application to assign to every single request the maximum resources possible. There are several DDoS security mechanisms available including separate firewalls and traffic monitoring programs that can be used to resolve the threat of attacks [25].

4. ***Session Hijacking***: The Session Hijacking attack exploits the control mechanism of a web session, which is controlled for a session token in a general case [26]. Since the HTTP communication uses several different TCP connections, a method is required for the webserver to recognize the connections of every user. The Session Hijacking attack exploits the session token by forging or predicting a legitimate session token in order to obtain unauthorized access to the webserver. The session token may be exploited in various ways; the most common is a repetitive session token, session sniffing, attacks on the client side (XSS, malicious JavaScript codes, trojans, etc.), man-in-the-middle attack, man-in-the-browser

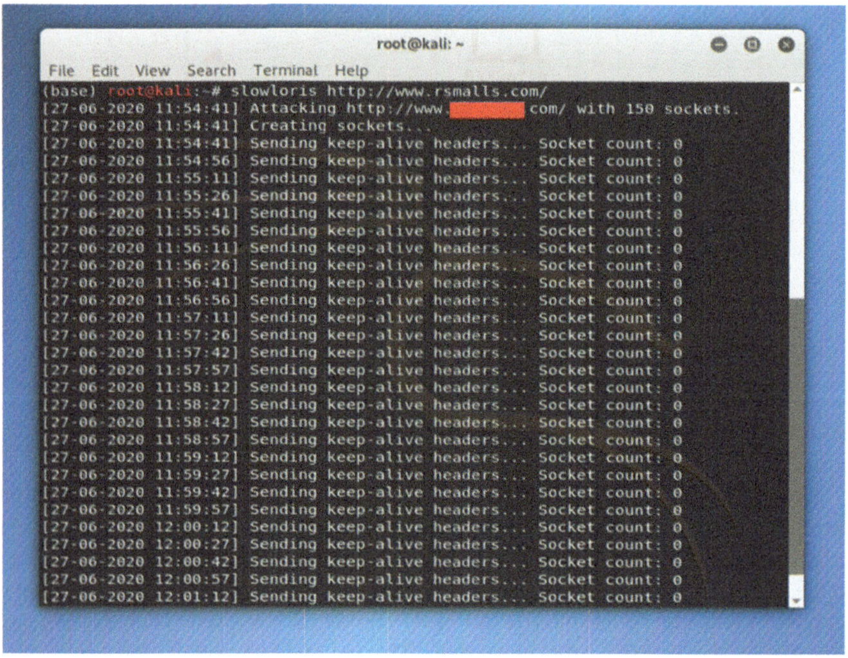

Fig. 5 Slow HTTP attack

attack, etc. There are many tools available in Kali Linux such as Wireshark, Browser Exploitation Framework (*BeEF*), Burp Suite, etc., that allow session hijacking attacks [27].

5. ***Brute Force Attacks***: Whilst other computer threats aim to deceive users or get through device defenses, brute-force hackers only charge a network. Cyberattackers use a brute-force technique usually to saturate a network with multiple attempts with trial-and-error passwords. Most of the applications would use specialized apps that may attempt hundreds of variations of passwords in a minute. By this hacking is accelerated and allowance is obtained through a single point of data to an entire network. Brute force attacks affect the systems with simple and poor passwords the most. Though Brute force attacks can be prevented by proper control systems and strong passwords [28].

4 Types of Network Security

Network security is the progression of planning a self-justifying method to secure data and resources over the computer network set-up in contradiction of any potential threat or unauthorized access. It uses both software and hardware technologies to achieve the optimum solution for network defense [29].

There are different methodologies available by which network security can be achieved.

- ***Firewalls***: Firewalls create a wall between the secured internal network and untrusted external networks, including the Internet [30]. They use a series of rules specified to permit or block the traffic data. Both hardware and software Firewalls are available.
- ***Email security***: The principal vulnerability vector for a data breach is email gateways. Attackers use personal information and social media techniques for creating complex phishing strategies to confuse the users and transfer them to malware-serving sites. An email protection program blocks incoming threats and monitors outgoing messages to prevent confidential data from leaking out.
- ***Antivirus and antimalware software***: Malware is the short form of "**Mal**icious Soft**ware**" which includes viruses, worms, trojans, ransomware, spyware, etc. After infecting a system, a Malware remains inactive for days or even weeks. In addition to scan for Malware upon entry, the better antimalware applications often actively monitor files afterward to detect irregularities, delete them and repair damages.
- ***Network segmentation***: Software-defined segmentation puts network traffic into various classifications and makes it easy to implement security policies. Ideally, the classifications are based on the identification of endpoints, not pure IP addresses. The admin grants access rights based on role, location and more so the right individuals and suspicious devices are controlled and rectified with the right level of access.
- ***Access control***: All the users do not have access to the network. To eliminate the possible threats, a network administrator needs to recognize and allot roles to each user and device. After that, the security policies are enforced by the admin. To achieve better security, an administrator can also block noncompliant endpoint devices. So, limited access is given to the devices or users. This process is known as Network Access Control (NAC). It is an approach that attempts to amalgamate endpoint security technologies (e.g., antivirus, host intrusion prevention, and vulnerability assessment), the user or system authentication and network security enforcement.
- ***Application security***: Every application software intended to manage customer needs, must be secured, whether it is developed by in-house IT personnel or purchased from outside. Unfortunately, there remain flaws and vulnerabilities in those programs that attackers may use to penetrate the network. Application security requires time to time hardware, applications, and procedure to patches to the vulnerabilities.
- ***Behavior Analytics***: To detect unusual behaviors in the network, knowledge about normal behavior must be there. Behavioral analytics systems automatically detect those behaviors that deviate from the standard. The compliance staff will then help to identify vulnerability signs that pose a possible risk and address threats very quickly.

- ***Intrusion prevention systems***: To effectively block threats, an Intrusion Prevention System (IPS) always checks the network traffic. The program achieves so by correlating vast volumes of global vulnerability intelligence not only to combat malicious activity but also to monitor the movement of suspicious files and ransomware through the network to avoid outbreaks and spreading of malwares.
- ***Data loss prevention***: Every organization must ensure that its employees are not transferring confidential information outside the network. For Data Loss Prevention (DLP), technology may halt copy, distribution or even printing of sensitive information in an inappropriate manner.
- ***Wireless security***: Wireless networks are not as secured as wired networks. Therefore, to ensure the security of wireless networks, different solutions are specially designed.
- ***Virtual Private Network (VPN)***: A VPN encrypts the connection by routing the traffic from users to a private server instead of ISP. Hence, the traffic seems to be originating from the VPN server rather than the user's device. Applications running across a VPN may therefore benefit from the functionality, security and management of the private networks [31].

5 Artificial Intelligence and Its Types

Artificial Intelligence (AI) is a broad range of computer science engaged in the development of smart machines capable of carrying out tasks that physically involve human intelligence [32]. AI is a multidisciplinary science with several approaches. Nowadays it is used to implement cyber security. It can be classified based on both capability and functionality.

On basis of capability AI is of three types:

(i) ***Weak/Narrow AI***: Narrow AI enables intelligence to perform a dedicated task. These are the most common and currently available AI. It has limitations since it can operate for one task. Therefore, often it is called a Poor AI. Even if a Narrow AI goes beyond its limits, it may fail miserably in many unpredictable ways. In 2010, weak AI trading algorithms led to a "flash crash" causing a temporary but significant dip in the market [33].

(ii) ***General AI***: General AI is expected to perform human-like intellectual tasks efficiently. It is aimed to create a machine that would be intelligent and possess decision making abilities such as any human. Currently, no such AI framework has been developed that can make decisions and perform intellectual tasks such as humans. The worldwide researchers are now concentrating on designing machines with General AI. As this is still under research, so the production of these systems would take a lot of effort and time.

(iii) ***Super AI***: Super AI is defined as an intelligence level of a system where the device will exceed human intelligence and perform any function with cognitive properties better than human beings. Some main features of Strong or Super AI include the ability to think, reason, solve the puzzle, make decisions, prepare,

Table 2 Categories of AI

Capability	Functionality
Weak or Narrow AI	Reactive machine
General AI	Limited memory machine
Super AI	Theory of mind
	Self-awareness

learn and interact autonomously. Super AI is still a hypothetical concept of Artificial Intelligence and the development of such systems in real-world is a challenging task.

On basis of functionality AI is of four types:

(i) ***Reactive Machines***: The simpler forms of AI are reactive devices. These AI systems do not store memories for potential acts. These machines only concentrate on current scenarios and respond to them according to the best possible way. The *Deep Blue* program from *IBM* is one example of reactive systems. *Google's AlphaGo* is also a good example for reactive machines.

(ii) ***Limited Memory***: Limited memory machines can store previously processed data for a short period of time. These computers can only use the stored information for a limited period. One of the latest examples of Reduced Memory systems is the *self-driving vehicles*. These cars can store nearby cars' recent speed, distance from other cars, speed limit, and other information required to navigate the route.

(iii) ***Theory of Mind***: Mind AI should consider human feelings, personalities, value and should be able to communicate socially such as humans. The development of this type of AI is not getting any great success, but researchers are making a lot of effort for developing these AI.

(iv) ***Self-Awareness***: Self-Awareness AI is the future of AI. These devices empowered with self-awareness will be extremely intelligent and will possess their own consciousness, emotions and self-awareness. Those systems are going to be even smarter than human beings. AI Self-Awareness does not exist and is still a hypothetical concept [34] (Table 2).

6 Role of AI in the Improvement of Cyber Security

AI is now considered very crucial for cyber security. For detecting suspicious behavior of the network, AI techniques can be used to learn how to eliminate interference or unwanted data and to empower security experts to understand the cyber environment. It can very quickly identify newer types of threats. AI is advantageous in preventing a full-scale breach being released.

The AI also enhances capabilities of cyber security with automated responses to identify, inform and generate patches against the cyber threats that are detected. AI can assess massive amounts of data and enable the development of existing systems and software to reduce different sorts of cyberattacks in an appropriate manner. AI can be deployed to strengthen the currently available solutions such as firewall, network segmentation, intrusion detection, bug detection, email security, behavior analytics, etc. [35].

Intrusion Detection Systems (IDS) are designed specifically for detecting any malicious attempts to destroy, steal, alter, resource or gain unauthorized access. Any form of intrusion activity or violation of the rule is typically reported either to an administrator or collected centrally using a Security Information and Event Management (SIEM) system. A SIEM system combines outputs from multiple sources and uses alarm filtering techniques to distinguish malicious activity from false alarms [36].

To block attacks, an Intrusion Prevention System (IPS) scans network traffic to identify the malformed data. The IPS often sits directly behind the firewall and provides a complementary layer of analysis that negatively selects for dangerous content [37]. The system does this by analyzing vast amounts of available data on possible threats to monitor and block the progression of suspected files and malware throughout the network to prevent outbreaks in the devices connected to the network. Hence, Intrusion Detection and Prevention System (IDPS) along with AI is capable of handling emerging cyber threats.

7 Conclusion

Offering security measures to prevent cyberattacks has also appeared as a huge profit business. The government and many organizations are investing plenty of money to secure their personal, financial and other official information. To mitigate the effect of cyber threats, AI came into the market. Despite plentiful uses, AI can also be used for abolishing the human life that is the reason human interference is required to monitor its actions. Cybercrime has become a regular phenomenon and arrives as a daily news item. It is a global problem. Without strong security trials, AI is valueless as very easily it may go to a wrong hand. As a result, government, banks, healthcare system and multinational companies face a big threat from online hackers. Lots of personal and commercial information may be leaked and exploited by hackers. So, in the recent times, a rigorous research in the area of cyber security with AI has grown importance. And hopefully, it will last in the coming future because of its dynamic and sensitive issues associated with human life.

References

1. Patil P (2016) Artificial intelligence in cyber security. Int J Res Comput Appl Robot 4(5):1–5. ISSN 2320–7345
2. Ashok Kumar D, Venugopalan SR (2017) Intrusion detection system: a review. Int J Adv Res Comput Sci 8(8). ISSN No. 0976-5697
3. https://def.camp/artificial-intelligence-cybersecurity
4. Abomhara M, Køien GM (2015) Cyber security and the Internet of Things: vulnerabilities, threats, intruders and attacks. J Cyber Secur Mobility 4(1):65–88. https://doi.org/10.13052/jcsm2245-1439.414
5. Cherdantseva Y et al (2016) A review of cyber security risk assessment methods for SCADA systems. Comput Secur 56:1–27. https://doi.org/10.1016/j.cose.2015.09.009
6. Vadza K (2013) Cyber crime & its categories. Indian J Appl Res 3(5):130–133. https://doi.org/10.15373/2249555X/MAY2013/39
7. Romagna M (2019) Hacktivism: conceptualization, techniquesand historical view. In: The Palgrave handbook of international cybercrime and cyber deviance, pp 1–27
8. https://www.abusix.com/blog/the-importance-of-network-security-in-any-organization
9. https://consoltech.com/blog/why-network-security-is-more-important-than-ever/
10. Wright J, Harmening J (2009) 15 Computer and information security handbook. Elsevier Inc., Morgan Kaufmann Publications, p 257
11. Salahdine F, Kaabouch N (2019) Social engineering attacks: a survey. Future Internet. https://doi.org/10.3390/fi11040089
12. Pawar M, Anuradha J (2015) Network security & types of attacks in network. Procedia Comput Sci 48. https://doi.org/10.1016/j.procs.2015.04.126
13. Denning DE (2012) Stuxnet: what has changed? Future Internet. https://doi.org/10.3390/fi4030672
14. Nazreen Banu M, Munawara Banu S (2013) A comprehensive study of phishing attacks. Int J Comput Sci Inf Technol 4(6):783–786
15. Distributed denial of service attacks. Internet Protocol J 7(4). Cisco
16. Sahu SS, Pandey M (2014) Distributed denial of service attack: a review. I. J. Mod Educ Comput Sci 1:65–71. https://doi.org/10.5815/ijmecs.2014.01.07
17. https://www.us-cert.gov/ncas/tips/ST04-015
18. Singh A, Junefa D (2010) Agent based preventive measure for UDP flood attack in DDoS attacks. IJEST 2(8):3405–3411
19. https://www.cloudflare.com/learning/ddos/udp-flood-ddos-attack/
20. Bogdanoski M, Risteski A (2011) Wireless network behavior under ICMP ping flood DoS attack and mitigation techniques. Int J Commun Netw Inf Secur (IJCNIS) 3(1):17–24
21. Vishwakarma R, Jain AK (2020) A survey of DDoS attacking techniques and defense mechanisms in the IoT network. Telecommun Syst 73:3–25. https://doi.org/10.1007/s11235-019-00599-z
22. Selvara V (2018) Distributed denial of service attack detection, prevention & mitigation service on cloud environment. J Comput Eng Inf Technol 7:2. https://doi.org/10.4172/2324-9307.1000204
23. Suroto (2017) A review of defense against slow HTTP attack. Int J Inform Vis 1(4). ISSN: 2549-9904
24. Dragoni N et al (2018) DDoS-capable IoT malwares: comparative analysis and Mirai investigation. Secur Commun Netw. https://doi.org/10.1155/2018/7178164.
25. Gupta BB, Badve OP (2017) Taxonomy of DoS and DDoS attacks and desirable defense mechanism in a Cloud computing environment. Neural Comput Appl 28:3655–3682. https://doi.org/10.1007/s00521-016-2317-5
26. Baitha AK, Vinod S (2018) Session hijacking & technique. Int J Eng Technol 7:193–198
27. https://owasp.org/www-community/attacks/Session_hijacking_attack
28. Raza M, Iqbal M et al (2012) A survey of password attacks and comparative analysis on methods forsecure authentication. World Appl Sci J 19(4):439–444

29. https://www.csoonline.com/article/3285651/what-is-network-security-definition-methods-jobs-and-salaries.html
30. Ingham KL (1994) Network firewalls. In IEEE Commun Mag 32. https://doi.org/10.1201/978 0849330452
31. Mason AG (2002) Cisco secure virtual private network. Cisco Press, 7 pp
32. Singh G et al (2013) An overview of artificialintelligence. SBIT J Sci Technol. ISSN 2(1):2277–8764
33. https://cyberlaw.stanford.edu/blog2011/08/sorcerers-apprentice-or-why-weak-ai-interesting-enough
34. https://hackr.io/blog/types-of-ai
35. Shamiulla AM (2019) Role of artificial intelligence in cyber security. Int J Innov Technol Explor Eng (IJITEE) 9(1). ISSN: 2278-3075
36. Martellini M, Malizia A (2017) Cyber and chemical, biological, radiological, nuclear, explosives, challenges: threats and counter efforts Springer. ISBN:9783319621081.
37. https://www.paloaltonetworks.com/cyberpedia/what-is-an-intrusion-prevention-system-ips

Detection of Malicious URLs Using Deep Learning Approach

Kumar Sankar Ray and Rajni Kusshwaha

Abstract A phishing attack is one of the cyberbullying activities over the Internet. Most of the phishing websites try to imitate the legitimate websites. Web contents of phishing websites and URL features mimic the legitimate website and URL. Existing techniques for detecting and analyzing these malicious URLs are very much time consuming and costly due to their complexities. Traditional black and white listing is used for the detection of malicious websites. But this technique is not effective for real-time application. Various other techniques for the detection of malicious URL have been designed from machine learning approaches such as Support Vector Machines, Random Forest, etc. In such approaches, lexical properties of the URL are considered using the Bag-of-words feature. But it does not provide information about semantics and sequential information. It requires a considerable amount of manual feature engineering. Further, it cannot handle new features. To deal with the above limitations we focus on a real-time and language-independent phishing detection model by analyzing the anatomy of the URLs. We first detect the static and dynamic features manually. After getting the feature values, we find the lexical features of URL using Convolutional Neural Network (CNN) which can capture characters and words of the URL. After that, we merge features that are manually extracted and features obtained from CNN. Subsequently, we apply the bidirectional Long Short Term Memory (LSTM) model to generate the sequence of information of the URL. Thus, a hybrid model of CNN and Bidirectional LSTM are used for the detection of malicious URLs. We analyze the URL without accessing the web content of the corresponding website. It eliminates the time latency. We compare the performance of our network with some other existing ones and try to obtain better results.

Keywords Malicious URL · Feature extraction · Lexical analysis · CNN · LSTM · Deep learning neural network

---

K. S. Ray (✉)
Electronics and Communication Sciences Unit, Natural Computing Lab, Indian Statistical Institute, Kolkata, India
e-mail: kumarsankarray@gmail.com

R. Kusshwaha
R Systems International Ltd., Greater Noida, Uttar Pradesh, India

M. Chakraborty et al. (eds.), *The "Essence" of Network Security: An End-to-End Panorama*, Lecture Notes in Networks and Systems 163,
https://doi.org/10.1007/978-981-15-9317-8_8

1 Introduction

1.1 Web Phishing and Malicious URL Detection

Phishing is a fraudulent act to extract the secret and personal information of an Internet user; for example, bank details, pan card number, login credentials, etc. Generally, phishing attack is done using techniques of sending spoofed emails and spoofing of websites, as shown in Fig. 1 [1–4].

In the recent past, lots of research had been developed to prevent malicious attack in order to prevent Internet crimes. Malicious URL is a serious threat to cyber security. Malicious websites broadcast unsolicited content over the Internet and Internet users may visit such websites; thus become a victim of phishing attacks. As a result secret and personal information of the user are misused for cybercrime [5].

The inexpensive availability of the Internet and the escalating use of digital payment processes demand web security and protection against Internet fraud. By web phishing, the users are provoked to visit fake and harmful websites, which tries to hack sensitive information of the user such as, credit card details, bank account details, token for payment, etc.

The first recorded incident of a phishing attack occurred in early 1995. In this incident, a phisher successfully hacked the personal information of AOL (America Online) users. In early 2016 the phishing activity was recorded to be highest (1,220,523 recorded attacks) since its first occurrence and a sharp 65% increase as compared to 2015. There were 1,609 phishing attacks per month in the last quarter of the year 2004, which has increased by 5,753% over 12 years; that is, on an average 92,564 phishing attacks were recorded in the last quarter of 2016. In February 2014, the 3rd Microsoft Computing Safer Index Report revealed that the annual worldwide impact of phishing could be around $5 billion.

With the predominance of the use of Internet, phishing has become one of the most serious security threats. Thus it is an essential research domain to detect the phishing activities in order to defend the user; both private users as well as enterprises.

In Feb 2019, Google statistics shows 1,300,000 malicious URL block per day (as shown in Fig. 2) [6]. The existence of these phishing URL leads to great threats to the security of web application. Many researchers and practitioners have been working

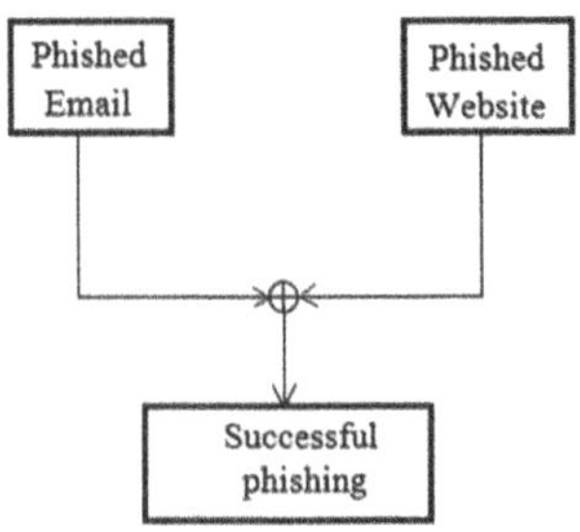

Fig. 1 A combination of phished email and websites

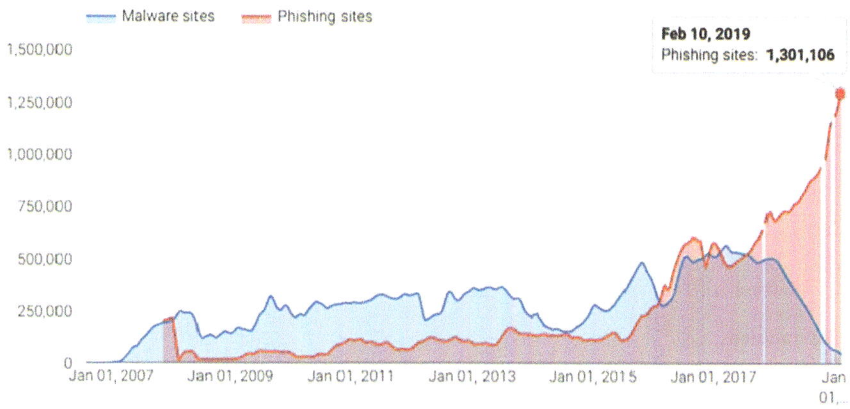

Fig. 2 Select dataset: number of sites deemed dangerous by safe browsing

on designing an effective and real-time system to detect malicious URLs. The most common method to detect such URLs is using a black and white list method. But, this method is inefficient for real-time detection, because the exhaustive list of malicious URLs cannot be provided real-time due to the generation of new URLs. But attackers try to use various approaches to evade the blacklist and modify the malicious URL which looks similar to legitimate via obfuscation. Also, the stated technique has several limitations, and bypassing them seems almost trivial, particularly because blacklists are ineffective to predict the fresh URLs. Therefore, an immediate issue is how to solve this problem. Identification of kinds of attacks is helpful for the understanding of the nature of a prospective threat.

1.2 Existing Phishing Scenario

Figure 3 represents a practical phishing scenario which is based on spoofing. The typical phishing attempts proceed as follows:

- Over a webserver, a spoofed version of the targeted website is run by the phisher and then spoofed emails are generated and sent to the users [7].
- Typically the email bears a false message about some emergency situation that demands an immediate action and obtain confidential information.
- The email contains a spurious link that redirects the user to the fake webserver, where the phisher has created a similar login page, which deceives the user.
- The user unknowingly provides his/her secret credentials on the fake website, which the phisher gets immediately.
- This phisher then uses the confidential information of the user to accomplish the fraudulent acts.

The various types of web phishing techniques are listed below:

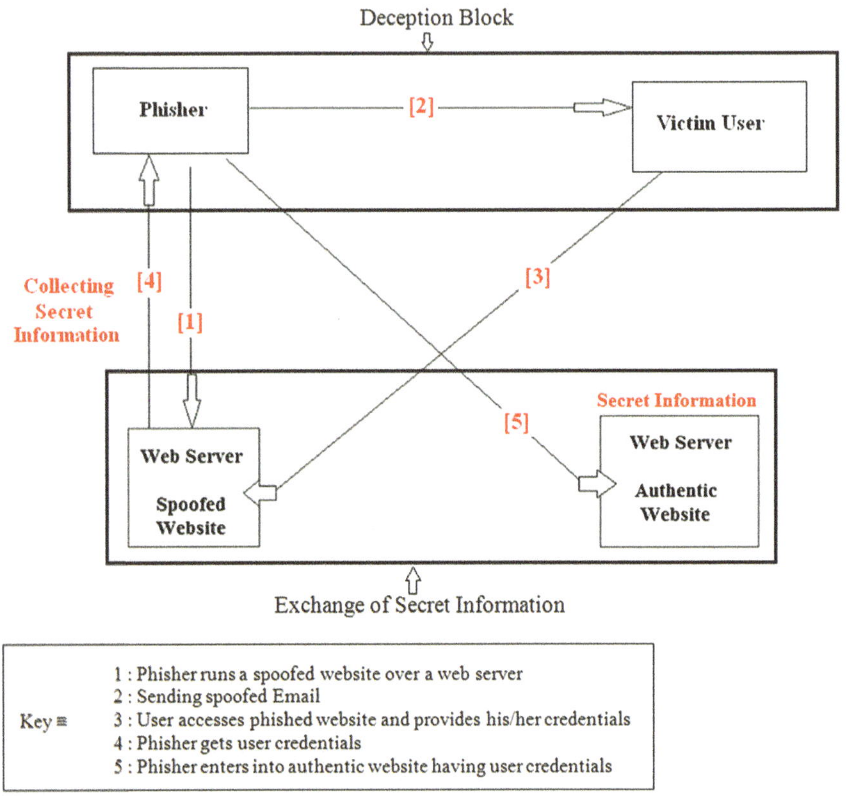

Fig. 3 A practical phishing scenario

- Spear Phishing for an individual or a company.
- Whaling: it targets high-profile users.
- Tabnabbing: in this case, a tab switch is used to load a phishing webpage [8].

Figure 4 shows major web phishing techniques. The targeted users are redirected to the phished websites by the highly developed techniques of Internet fraud, for example, website spoofing, email spoofing, exploitation of web browser, etc. The successful deception is executed by the phisher by using the following techniques:

- By manipulating links (the displayed web link leading to an authentic URL, actually redirects the users to a malicious URL in the background).
- By using images instead of texts which evades the phishing detection filters.
- By using malicious web scripting languages.
- By using pop-up windows.
- By Tabnabbing.

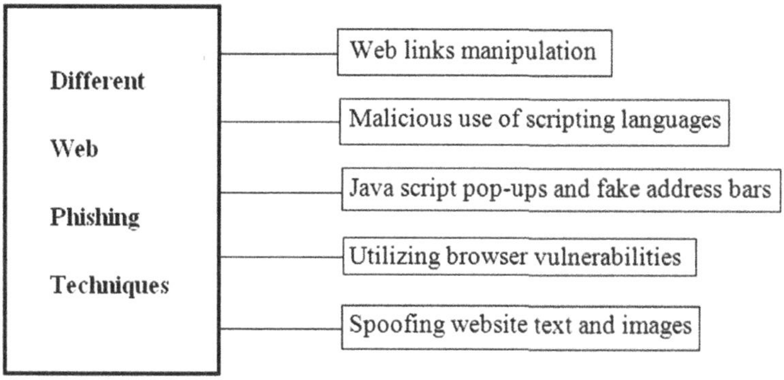

Fig. 4 Web phishing techniques

1.3 Anti-phishing Solutions

Anti-phishing solutions can be broadly classified as.

- Preventing methods for phishing.
- User training approach.
- Attempts to detect phishing.

Some of the proposed solutions for counteracting phishing attacks include [9]:

- Proving the users with proper training on phishing detection, prevention and other activities related to phishing.
- Use of anti-phishing software.
- Adapting new measures of user authentication.
- To monitor and shut down phishing websites in real-time.
- Disabling malicious Java scripts and securing browser developments.

Figure 5 demonstrates the classification of anti-phishing methods.

Fig. 5 Anti-phishing methods

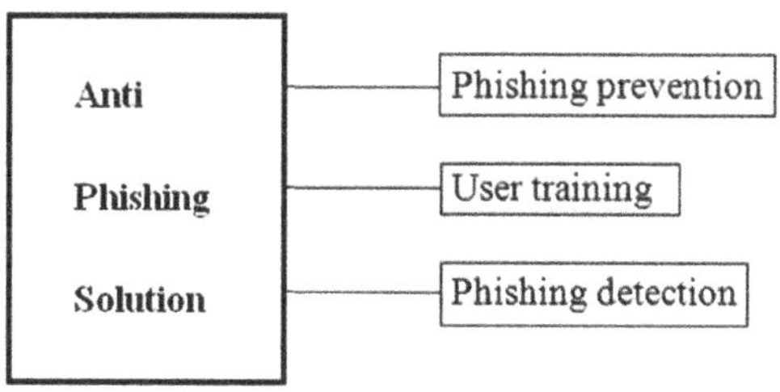

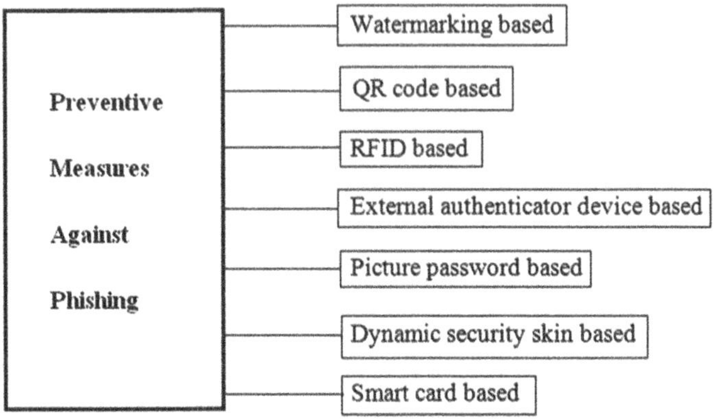

Fig. 6 List of methods for preventing phishing

1.4 Preventing Methods

The aim of phishing prevention schemes is to reduce the chances of a user being deceived by a fraudulent phishing attacker. In order to do that, an effective measure is to provide an extra layer of protection; two-factor authentication or a two-way authentication scheme. Among various approaches of phishing prevention watermarking-based, Radio Frequency Identification (RFID)-based, Quick Response (QR) Code-based techniques are significant. However, the guarantee of the success of these methods depends on support and cooperation from the authentic website and also proper judgment and understanding of the user on this matter. The major drawback of these safety measures are that they may result in a complex user interface, increasing the cost of computation of each authentication. Also, the users may require using additional authentication devices, making the implementation and use of the system more cumbersome. Figure 6 shows phishing prevention methods.

1.5 User Training Schemes

The awareness of the users is necessary so that they can identify phishing attempts and avoid malicious websites. The user training schemes are aimed at this goal. However, this approach may not be proved to be satisfactory since, a large portion of the population lacks the general knowledge to understand the authenticity of Secure Sockets Layer (SSL), certificates, and URLs of the websites.

1.6 Phishing Detection Schemes

A better approach to stop web phishing is to detect and block the access to the probable phishing website by means of various phishing detection schemes; such as, through a web browser on the client-side or via specific software at the host, or solutions that detect phishing at the server-side. These methods guarantee a better success rate because they don't rely upon the user's expertise and user training. The following parameters are used to evaluate the accuracy of the detection schemes:

- True Positives (TP) represents phishing websites as phishing.
- True Negatives (TN) represents legitimate websites as legitimate.
- False Positives (FP) represents legitimate websites as phishing.
- False Negatives (FN) represents phishing websites as legitimate.

Some benchmark datasets are used to evaluate the accuracy of phishing detection schemes. Some of the popular authentic dataset and phishing dataset are given below:

- Authentic Dataset—This dataset contains the list of benign websites as a raw text file.
- Phishing Dataset—It is a community-based system containing a benchmark dataset of phishing websites.

2 Existing State of Art

We conduct a study on phishing sites, which are either fake or having phishing related components.

2.1 Classification-Based Approach

Phishing detection schemes are classified as:

- **Search Engine-based approach**—In this approach features such as text, images, and URLs are extracted from websites. These features are searched using various search engines and the results are collected. The authentic websites have a higher index than phishing websites and hence the valid sites will be among the top of the search lists.
- **Blacklist and Whitelist-based approach**—In this approach two types of lists are used, namely, whitelist for authentic websites and blacklist for phishing websites. The blacklist is prepared either by getting the feedback from authentic users or based on the reports of the third parties who have the dataset of malicious URLs obtained by performing any other phishing detection approach.

- **Machine Learning (Heuristics)-based approach**—This method uses certain rules and generates a set of heuristics-based learning algorithms to detect anomalous websites. These algorithms extract text, image or URL information and then determine thresholds for detection of the phished website.
- **Domain Name System (DNS)-based approach**—Domain Name System (DNS), used to validate the IP address, can also be used as an approach for phishing detection.
- **Visual Similarity-based approach**—If the visual characteristics of a website are found to be similar to that of an authentic website, the system checks whether the URL belongs to the list of authentic URLs.
- **Proactive Method for Phishing Detection**—In this scheme different combinations of URLs are generated from existing authentic URLs; then it is checked whether the generated websites exist or not. If they exist whether they are involved in phishing-related activities.

The merits and demerits of the aforementioned techniques are summarized in Table 1.

Table 1 Merits and demerits of phished website detection methods

Method	Merits	Demerits
SEB	• low complexity • storage requirement is low • may be implemented within browsers • very accurate • perform in real-time and no update required	• accuracy and efficiency are dependent on extracted features • it fails if search engines are toppled to rank phishing webpages at the top
PBWB	• low complexity • may be implemented within the browser	• additional cost to monitor each URL • frequent updates of blacklist or whitelist required
HMLB	• high accuracy • easily modifiable	• frequent training under new features required • time consuming • computational resources required • limitations of resources over browsers
DNSB	• storage requirement is low at client and server end • requires less number of features • implementation of client site is possible	• affected by DNS poisoning • communication cost is high • client-side implementation causes delay
VSB	• easy to detect phishing attacks • can be implemented at client end	• computationally heavy and storage requirement is high • it is not better than existing techniques
PPUDB	• helps in detecting phishing websites before act occurs	• cannot be implemented on client-side • random combination required for URL's to be generated • needs central or distributed server • computationally heavy

Table 2 Performance of phishing attack detection capabilities

Methods	Client-side implementation of client-side	Detection in real-time	Training required	Updates required	Computational/complexity	Storage Cost
SEB	✓	✓	✗	✗	Low/High	Low
HMLB	✗	✓	✓ Training over new features required	✓	High/Medium	High. Store and update training dataset
PBWB	✓	✗	✗	✓ Updates of blacklist and whitelist	Low/High	Medium. Blacklist is stored
VSB	✓	✓	✓		High/Medium	High
DNSB	✓	✓	✗	✗	Low/High	Low
PPUDB	✗	✓	✗	✗	High/High	High. Store probable phishing URL's

Key: Yes ≡ ✓, No ≡ ✗

Table 2 represents the performance of phishing detection.

Among the aforementioned techniques for phishing detection, black and white list and heuristic-based approaches are mostly used. But the problem with this approach is that we manually select some features and weighted equally to all features. But we may have missed some important features and cannot correlate the presence of features in a specific sequence. Since UTLs structure followed some sequence pattern, so we must concern about it.

2.2 *Machine Learning-Based Approach*

Machine learning-based approach detects malicious URLs of all the popular attacks. It also identifies the nature of attacks. Features such as lexical, link popularity, Webpage content, DNS are used. Real life data are collected from DMOZ Open Directory Project, Yahoo's directory, Spam URLs from jwSpamSpy. Statistical method is proposed to discover the lexical and host-based properties of malicious website. In [10] a machine learning-based approach is proposed to detect phishing webpages. They have used Random Forests, Support Vector Machines (SVM) with Radial Basis Function (RBF), Logistic Regression, etc.

2.3 Deep Learning-Based Approach

Deep learning-based approach learns the feature from raw data without using any hand-extracted features. Our objective is to detect malicious URL. Since we use deep networks over lexical features, it is closely related to Natural Language Processing (NLP). Deep learning method is also successfully used in text classification, machine translation, question answering system, etc. Recurrent neural networks (e.g., LSTM) are used to capture sequential information. CNN is an alternative to LSTM. CNN is used for text classification. Little attempt is applied to exploit deep learning for malicious URL detection. Though character levels CNNs are available but it ignores structural information. In this chapter, we discuss both word-level and character-level information. We demonstrate the importance of word-level information to capture large temporal patterns. Character-level embedding is utilized for word-level information to handle too many unseen words and obtain information about unseen words at test time. After obtaining the feature vector from CNN we train our model on Bi-LSTM model to keep this sequence information of URL. Thus, the present model through character and word-level information captures the structural details of URL strings.

2.4 Statement of the Problem

To address the existing problems as stated earlier, we design a system that takes URL strings as input. We apply CNNs for both character and word in the URL. Character-level CNN extracts unique characters. Each character is represented by a vector. Thus the URL is converted to a matrix on which convolution is applied. Character extracted by character level CNN detects the maliciousness of URL. We extract unique words in the training corpus from word-level CNNs and subsequently, convolution is applied.

Word embedding, however, generates certain problems; for instance, it cannot detect new words at the time of testing and in case of many unique words of malicious URL it generates memory constraints during learning. Hence we consider advanced word embedding. Advanced word embedding learns from the character level information of each word. It can also detect sub-word level information. CNNs are optimized for the Bi-LSTM model. We classify the URLS information using the softmax function.

Given a URL we want to classify whether it is malicious or not. This is basically a binary classification. We consider a set of "T" URLs, $\{(u_1, y_1), \ldots\ldots, (u_T, y_T)\}$, where u_t for $t = 1, \ldots., T$ indicates a URL and $y_t \in \{0, 1\}$, is the level of the URL, where $y_t = 1$ stands for a malicious URL and $y_t = 0$ stands for a benign URL. We consider an n-dimensional feature vector $u_t x_t$, where $x_t \in R^n$ of URL u_t. The prediction function is $f : R^n \rightarrow R$. It indicates the classification score. Thus the prediction is $\widehat{y_t} = sign(f(x_t))$, where the function f is represented by a deep neural network. We minimize $\sum_t^T I_{\widehat{y_t}/y_t}$ using a loss function.

2.5 *Contributions of the Present Work*

Our first contribution is to design a real-time, Language independent phishing detection model. Since our model is trained on real-time data, and it understands the anatomy of malicious URL, so it classifies the new unknown test URL. Since we are using the URL of website, we do not use the web-content of the URL. Thus, it is a language-independent model. It also eliminates the time latency, because it does not analyze the web content of URL.

The second contribution understands the "natural" evolution of malicious websites over the time. Since our model is trained on real-time data of URL, so our model understands the strategy of attackers used to modify the URL over time. For instance, initially, phishing URLs is long in length to hide the suspicious information in URL. But researchers built a model that classifies long length URL as phishing. Then attackers become smart and try to trick the user and use some famous domain name in URL and make a length of average size. The user generally ignores it and becomes a victim; but the researcher then uses the host-based feature and detects phishing URL. Since our model learn from real-time data; so it understands the strategy used by attackers in phishing URL over time.

The third contribution is to learn semantic and sequential patterns in the URL. Most of the researchers focus to build a model that extracts lexical features from URL, but forget to keep the sequence information of URL, as URL follows some standard sequence.

3 Data Collection Method

The methodology adapted in this work is described in this section. For implementation, two sources of data are used:

- For malicious URLs the data are collected from (https://www.phishtank.com/);
- For benign URLs the data are collected from (https://www.alexa.com/).

For the detection of malicious websites across different layers of Internet protocol, an automated system is necessary, that can collect contents of application layer and the network layer. Figure 7 is self-explanatory for data collection method. It is basically centered on a Crawler. It uses the URL. The collected data are registration dates of websites and the geographic location of URLs owners.

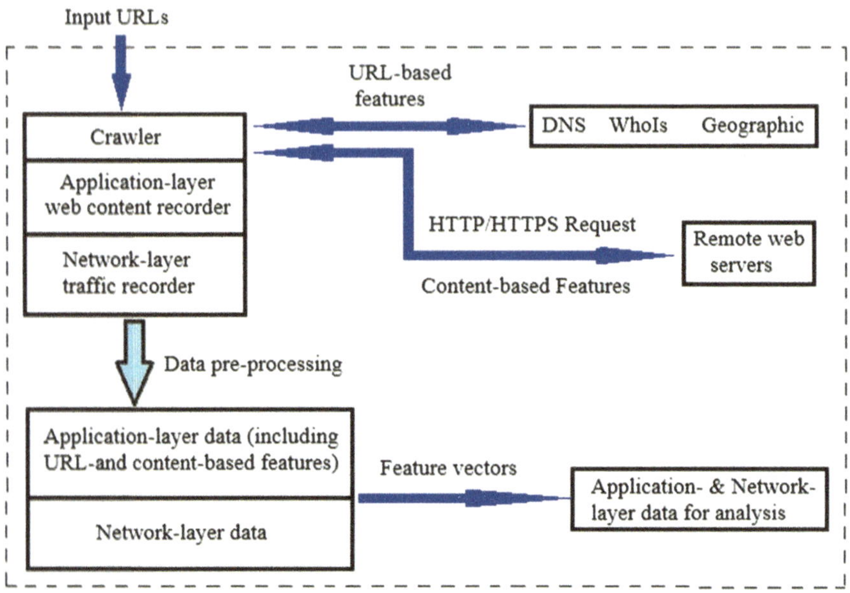

Fig. 7 Data collection system architecture

4 Feature Engineering and Deep Learning-Based Approach for Malicious URL Detection

4.1 Feature Based on the URL Lexical Information

URL dataset is not as simple as text data. So we need to do some feature engineering (use domain knowledge of the data) to extract some feature, we do it manually by using some novel approaches of researchers. Some of the manually selected features are given below:

URL Length: Long URL to Hide the Suspicious Part. So, long URLs (length $\geq$ 54) are treated as phishing.

Presence of @ Symbol in URL: The presence of "@" symbol in the URL leads the browser to ignore everything preceding the "@" symbol. It may be the case that the real address is put after the "@" symbol.

Presence of Redirection Symbol: The redirection symbol "//" within a URL path means that the user will be redirected to another website. An example of such URL is: "https://www.legitimate.com//https://www.phishing.com". The location of the "//" symbol is examined. If the URL starts with "http", that means the "//" should appear in the sixth position. However, if the URL employs "https" then "//" should appear in seventh position.

Prefix or Suffix Separated by (-) to the Domain: The dash symbol is rarely used in legitimate URLs. Phishers tend to add prefixes or suffixes separated by (-) to the domain name. For example https://www.Confirme-paypal.com/.

Sub-Domain and Multi Sub-Domains: The legitimate URL link has two dots in the URL since we can ignore typing "www." However, if there are three dots in the URL, then it is suspicious, as it has a sub-domain. If there are more than three dots, then it has more than one sub-domains and it is malicious.

Presence of IP Address: If the domain name of a URL is replaced by an IP address, for example, "https://125.98.3.123/fake.html", then certainly it is a malicious URL trying to steal user's confidential information. In some cases, to deceive the user, the IP address is encoded in hexadecimal form, for example, "http://0x58.0xCC.0xCA.0x62/2/paypal.ca/index.html".

Using URL Shortening Services "Tiny URL: URL shortening is a method on the \WorldWide Web" in which a URL may be made considerably smaller in length and still lead to the required webpage. This is achieved by "HTTP Redirect" on a domain name. For example, the URL "https://portal.hud.ac.uk/" can be shortened to "bit.ly/19DXSk4".

Existence of protocol in domain part: The web phisher may deceive the user by adding "https" token in the domain part of the URL, for example, https://www.paypal-it-webapps-mpp-home.soft-hair.com/ domains.

Abnormal URL: The WhoIs database can be used to check the authenticity of a URL; since the identity of a benign website is a part of its URL.

Google Index: Generally phishing websites are accessed for a short period of time and therefore, many phishing websites are not indexed by Google. So, to verify the authenticity of a website, we can inspect if the website is listed in Google index or not (Webmaster resources 2017).

Website Traffic: If a website is benign, typically it is listed among the top 100,000 websites, in terms of its popularity and number of visitors. But the phishing websites are active for a short period of time and have a very small number of visitors. In that respect, phishing websites have very low popularity and they will not be identified by the Alexa database. So, in the detection procedure, we may use a rule that if a website's rank is less than 100,000, the website is considered as legitimate; it the rank is >100,000, it is considered to be suspicious and finally, if it has no recorded traffic and Alexa database does not recognize it, then it is marked as Phishing.

Domain Registration Length: The phishing websites are created for cheating the users and to gather some confidential information in a quick span. For this reason, it is expected that phishing websites are paid for a short period, whereas the authentic websites are paid for several years in advance. So the registration period can be used as a feature for detection.

Age of Domain: The WhoIs database can be used to search for the age of a website. If the age is at least six months it can be taken to be a trustworthy website. It is expected that phishing URLs are active for a lesser time period.

DNS Record: The DNS record, stored in WhoIs database, can be used to distinguish between phishing and benign websites, because, there won't be any record of a fraudulent phishing website in the WhoIs database.

Statistical-Reports-Based Feature: The statistical reports, presented by PhishTank and StopBadware and similar agencies, can be incorporated in the feature space. These reports are updates from time to time by these agencies.

4.2 Feature Extracted from CNN

A URL u can be represented by a matrix of the form $u \rightarrow x \in \mathbb{R}^{L \times k}$, where x_i, $i = 1, \ldots, L$ in a sequence is either a character or a word. Character-level CNNs for malicious URL detection captures the properties of characters of URL. This can be achieved by identifying the unique alphanumeric and special characters in the dataset by replacing the less frequent characters like <UNK>.

Another variant is Word-level CNNs. Here convolution operation is done over words. One of the challenges of Word-level CNNs is that the number of unique words depends on the size of the training corpus. This is where Word-level CNNs differ from Character-level CNNs. All unique words are obtained as a sequence of alphanumeric characters. <PAD> token is used makes the lengths of the URLs uniform.

The final URL matrix can be represented as the sum of two matrices denoted by, $URL_w + URL_{cw}$ and shown in Fig. 8. In our model, CNNs are applied for character level as well as word level. Each character is represented as vector. Thus the entire URL is converted to a matrix on which convolution is applied. Character extracted by character level CNN detects the maliciousness of URL. We extract unique words in the training corpus from word-level CNNs. By word embedding, we get a matrix of the URL. Following this, convolution is applied.

Word embedding, however, generates certain problems; for instance, it cannot detect new words at the time of testing, in case of many unique words of malicious URL it generates memory constraints during learning. Hence we consider advanced word embedding. Advanced word embedding learns character, word and sub-word. Bi-LSTM model is trained and classification is done by softmax function.

Given a URL we want to classify whether it is malicious or not. This is basically a binary classification. We consider a set of "T" URLs, $\{(u_1, y_1), \ldots, (u_T, y_T)\}$, where u_t for $t = 1, \ldots, T$ indicates a URL and $y_t \in \{0, 1\}$, is the level of the URL, where $y_t = 1$ stands for a malicious URL and $y_t = 0$ stands for a benign URL. We consider an n-dimensional feature vector $u_t x_t$, where $x_t \in R^n$ of URL u_t. The prediction function is $f : R^n \rightarrow R$. It indicates the classification score. Thus the prediction is $\widehat{y_t} = sign(f(x_t))$, where the function f is represented by a deep neural network. We minimize $\sum_t^T I_{\widehat{y_t}/y_t}$ using a loss function.

After getting features from Character-Level embedding, Word-Level embedding and some manually selected features, we concatenate all together. We give input to the Bi-LSTM model to keep the sequence information of URL.

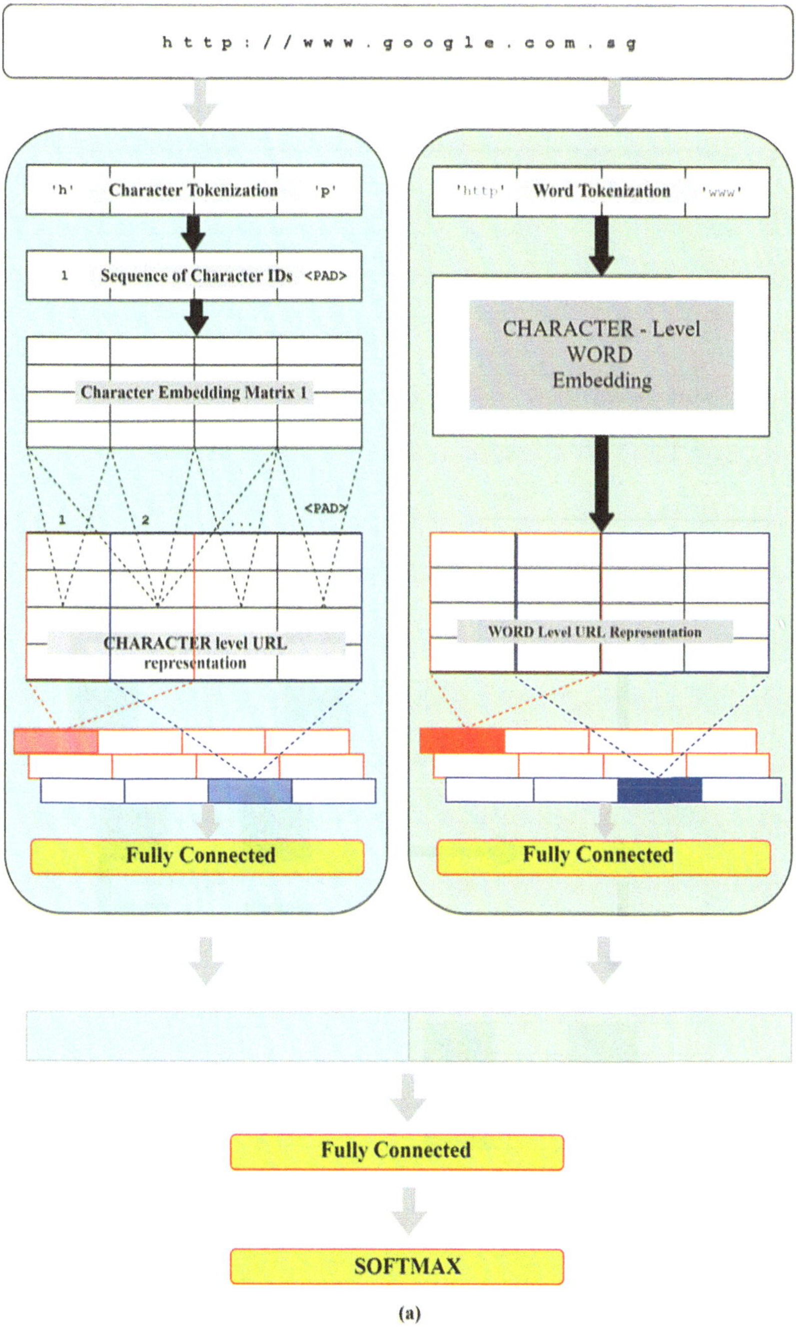
h t t p : / / w w w . g o o g l e . c o m . s g
'h'
Character Tokenization
'p'
1
Sequence of Character IDs
<PAD>
Character Embedding Matrix 1
1
2
<PAD>
CHARACTER level URL representation
Fully Connected
'http'
Word Tokenization
'www'
CHARACTER - Level WORD Embedding
WORD Level URL Representation
Fully Connected
Fully Connected
SOFTMAX
(a)

Fig. 8 URLNet

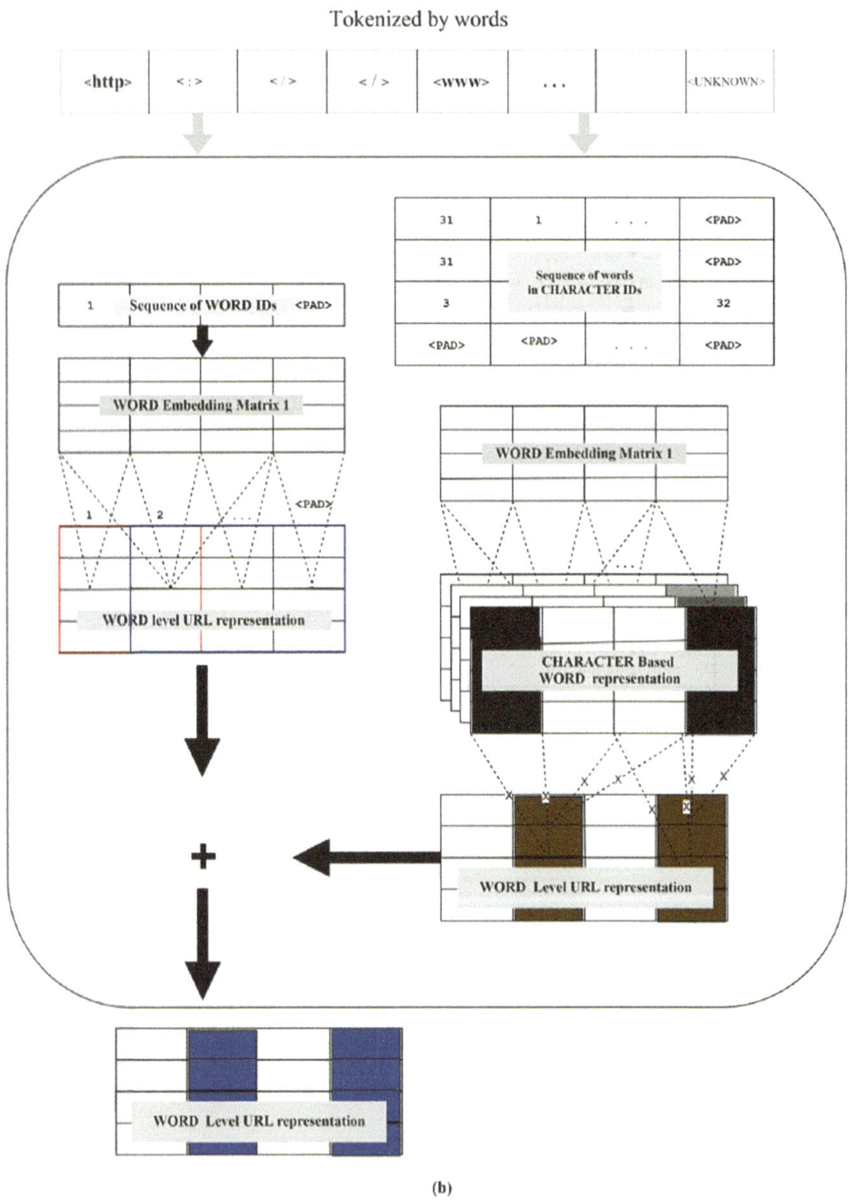

Fig. 8 (continued)

5 Architecture Preliminaries

This section briefly describes some architectures of neural network which we use in our proposed model. The most common application of CNN, the class of Deep Neural Network (DNN), is in the domain of analyzing visual imagery. Presently, the core of the most Computer Vision Systems is built with CNN. This complex feed-forward neural network, used for image recognition and image classification, captures the significant features in the images from raw pixel values automatically. This technology is also used in classifying text in the domain of Natural Language Processing (NLP). In NLP, CNNs capture from the text, the raw values of word or character. In this chapter, we discuss the use Bidirectional LSTMs to track the sequence information of feature vector, and classify the URL.

5.1 Convolutional Neural Network (CNN)

CNNs are analogous to the traditional artificial neural network (ANN). The drawback of traditional ANN is that it always struggle with computational complexity for image data. MNIST database for hand-written characters is suitable for ANN because it can easily handle small image dimension 28×28. Thus for this dataset a single neuron of the first hidden layer considers 784 weights which is manageable for ANN. But for large color image of 64×64, a single neuron of the first layer requires 12,288 weights. For color image network requires larger size than the one used to classify color MNIST dataset. This problem is to some extent overcome by the use of CNNs. CNNs preprocess the input image and extract some necessary features by convolving the input image with proper filters. Then these extracted features are fed to a traditional neural network. The loss function is minimized using backpropagation and modifying the weights of the neural network layer as well as updating the filter weights.

5.2 Long Short Term Memory (LSTM)

Long Short Term Memory networks are a variant of recurrent neural networks (RNN). LSTM is used to learn some context-dependent information from sequential data, for example, some information related to text inputs or video data. It is an enhancement of standard RNNs in terms that LSTM can easily learn long-term dependencies, which is not possible in standard RNN. This makes LSTMs much more human like, since humans also use contextual information to predict future data. LSTMs can remember past information over a long period of time and this property makes it useful for many applications.

5.3 Bidirectional Long Short Term Memory

Bi-LSTMs are bidirectional LSTMs which are a modified and improved version of LSTMs. In this architecture, the signal propagates in both forward and backward direction. For classification problems on sequential data Bi-LSTMs give better performance as compared to ordinary LSTMs; because, LSTMs only use the past data for problem-solving, whereas Bi-LSTM can utilize past as well as future information. In Bi-LSTM architecture there are two parallel LSTMs one of which processes original input information and the second one processes the input in reverse order. This way Bi-LSTM can analyze data from both perspectives which makes learning much faster. It finds its application in text processing, natural language processing and in many more fields.

6 Configuration of the Proposed Model

Figure 9 represents the overview of the developed model. There are three branches in the model to process the input URL string:

- Character-level branch represents the input URL.
- Word-level branch represents the input URL by using two sub-branches, that is, a word-embedding and a character level word embedding.
- Manual feature selection branch uses some host-based features, selected manually from the input URL.

In the procedure initially, convolution operation is applied over the character-level and the word-level branches. Then the obtained output is passed through a fully-connected (FC) layer. Both the branches are regularized with adequate dropout layers to avoid overfitting. Then the output of the fully-connected layer is concatenated with the manually selected feature vector. Finally, Bi-LSTM leads to the output classifier. Training is done by Adam optimizer with backpropagation.

We consider characters and words. Then matrix $u \rightarrow x \in R^{L*k}$ is formed. After finding the feature vector our aim is to find the sequence information of these feature vectors. For this we use Bidirectional LSTM model, which takes feature vector x of size R^n as an input to our LSTM model and classification is based on soft maxfunction. The entire model is trained using backpropagation algorithm with binary cross-entropy loss function [11, 12].

6.1 Combination of CNN and LSTM

In the present architecture, we have combined CNNs (convolutional neural networks) and LSTMs (long short term memory). These two techniques are capable of dealing

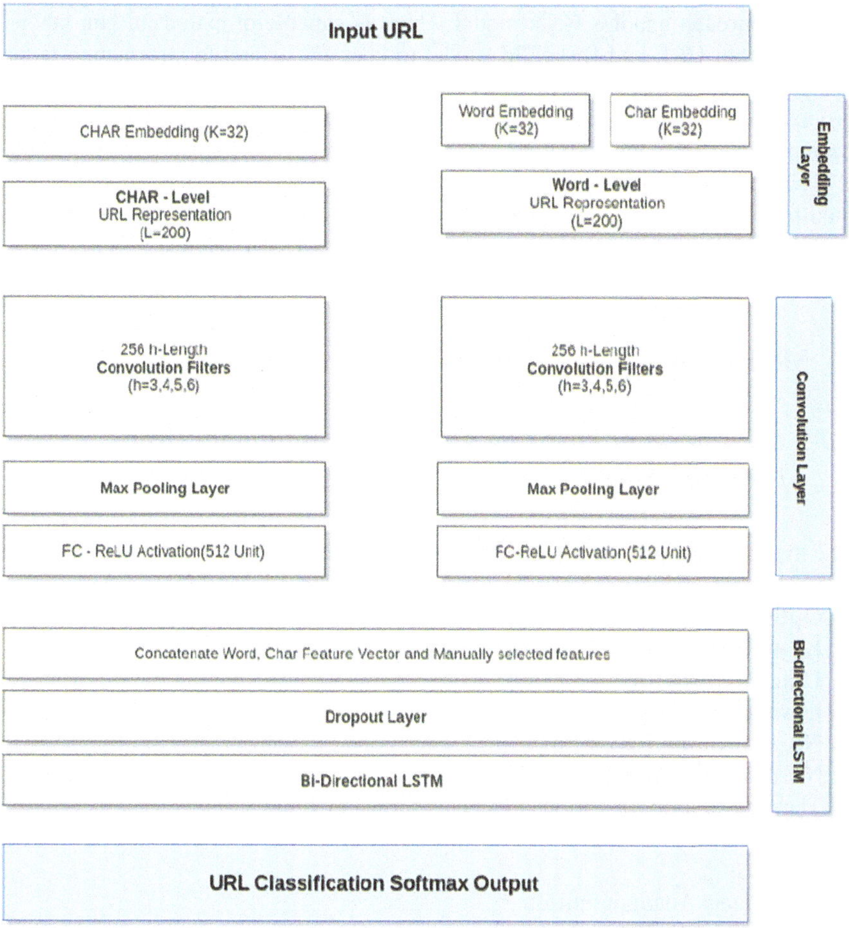

Fig. 9 Model configuration

with completely different problem types which we have merged to detect malicious URLs.

- **CNN** scan effectively handle spatial or hierarchical data and capable of extracting unlabeled features, such as written characters, images, etc. The inputs and produced outputs of CNNs are of fixed size.
- **Bi-LSTM** scan efficiently handle temporal or sequential data; for example, specific letters or words from text, data from stock market or speech recognition. The inputs and produced outputs of Bi-LSTMs are of arbitrary lengths. It can control the amount of prior training data which is supposed to be remembered as well as the amount of forgotten data.

Our approach handles CNN model which is capable of extract in gun labeled features from URL and Bi-LSTM model to keep the sequence information (since every URL followed some standard sequence). Many models have already been proposed in the combination of these tools. In our model, CNNs and Bi-LSTM have been merged so that the output of the CNN can be used as the input of the Bi-LSTM.

The original URLNet architecture is proposed by Le et al. [13] which we have modified. We replace the last Fully Connected convolution layer by Bidirectional LSTM model to maintain the sequence information of URL and classify it.

7 Model Comparisons and Results Achieved

7.1 Accuracy and Loss Plot of Various Traditional Deep Learning Models

The hyperparameters used to train the model by CNN, LSTM and Bi-LSTM are as follows:

Optimizer: Adam optimizer.

Loss Function: Binary Cross-Entropy.

Learning Rate: 1e−4.

Epsilon: 1e−08.

No. of training URLs: 5613.

No. of validating URLs: 1871.

The dataset used for various models is shown in Table 3.

The hyperparameters used to train the emerging model of CNN and Bi-LSTM are as follows:

Optimizer: Adam optimizer.

Loss Function: Binary Cross-Entropy.

Learning Rate: 0.001.

Epsilon: None.

No. of training URLs: 5613.

No. of validating URLs: 1871.

The dataset used for various models is shown in Table 4.

The accuracy and loss plots of traditional deep learning models are shown in Fig. 10.

Table 3 The dataset used for the traditional model

Model	Dataset	Accuracy
CNN	Alexa and PhishTank	0.934
LSTM	Alexa and PhishTank	0.945
Bi-LSTM	Alexa and PhishTank	0.9459

Table 4 The dataset used for emerging model

Model	Dataset	Accuracy
CNN + LSTM	Alexa and PhishTank	0.9738
CNN + Bi-LSTM	Alexa and PhishTank	0.975

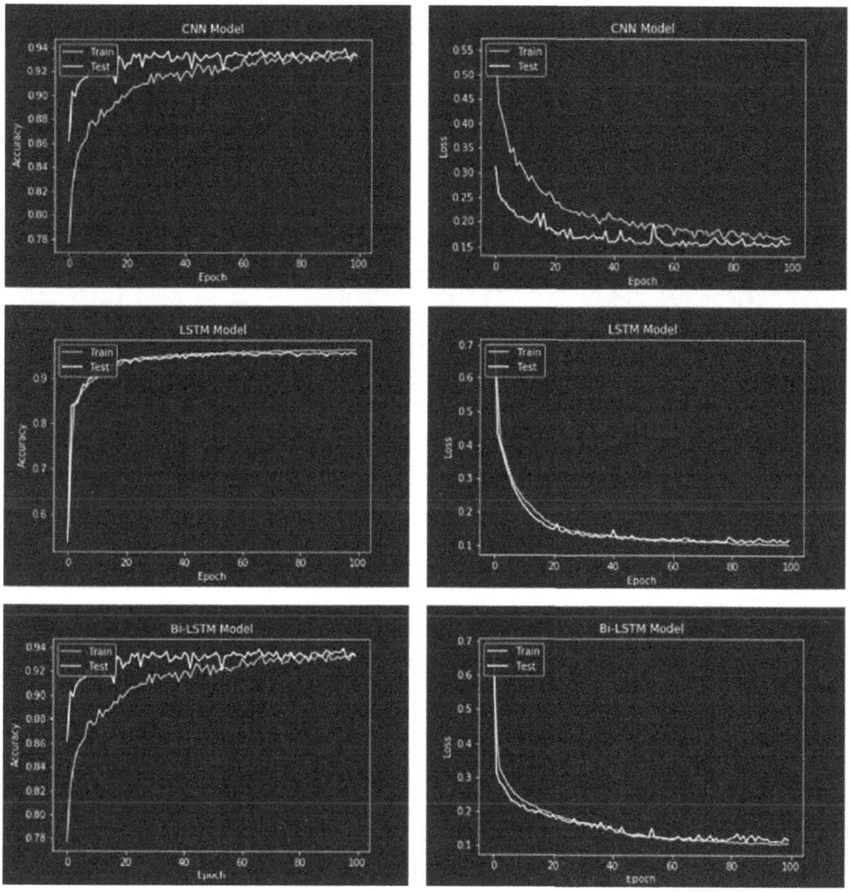

Fig. 10 Accuracy versus epochs of traditional models

7.2 *Accuracy and Loss Plots of Emerging Deep Learning Models*

The accuracy and loss plots of emerging deep learning models are shown in Fig. 11.

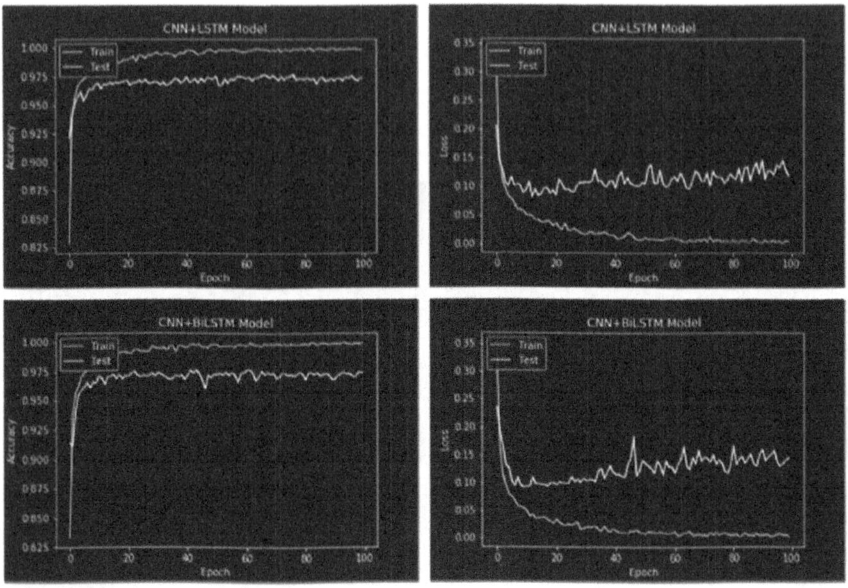

Fig. 11 Accuracy versus epochs of emerging models

8 Constraint of Proposed Model

The constraints of the proposed method are listed below;

- The present approach is incapable to analyze and detect obfuscated Java Scripts in the webpages which is the key reason for attacks such as XSS, drive-by downloads, malware-delivery, etc.
- More research should be done on discriminative spam URL features to differentiate them efficiently from authentic URLs.
- The features of short URLs should be studied further for the successful detection of fake webpages. Recently, the modern age micro-blogging sites such as Facebook, Twitter, etc., are facing these kinds of problems.

9 Concluding Remarks

In this chapter, we propose URLNet for detecting malicious URLs. The functional strategy of URLNet is based on CNN which can be defined a class of deep neural network. Most of the previous methods for the detection of fake URLs use Bag of Words such as features, but this leads to some significant disadvantages. Some drawbacks of these approaches include the incapability to handle unseen features of fake URLs and inefficiency to detect sequential concepts in the strings representing

URLs. To overcome these problems we propose Character CNNs and Word CNNs for joint optimization of the network. The present model incorporates an advanced word-embedding technique that is capable of dealing with rare words used in fake URLs. In addition, the proposed model allows capturing embeddings from unseen words at test time which exploits sub-word information. By comparing the three leading aspects, that is precision, recall, f-score, of CNN, Bi-LSTM and CNN-Bi-LSTM, we can conclude that the model is based on CNN + Bi-LSTM is more advanced than the CNN and Bi-LSTM model.

Our future plan is to design a framework based on the proposed technology which can handle extensive real-world test data. The improved version of the model will be capable to minimize the chances of false positives and negatives. We will compare the efficiency of existing online algorithms, which are capable in detecting malicious URLs, with the present model. Our next aim is to implement an attention network that gives special attention on specific features of URLs. In transaction websites we should consider the protocol of URL and domain name of URL instead of other features. Complexity of attention network is high. Since the Deep learning model learns weights and creates a model, everything inside the model acts, such as a black box. It is proven that in deep learning as we increase the complexity of the model performance is going to reduce.

References

1. Canali D, Cova M, Vigna G, Kruegel C (2011) Prophiler: a fast filter for the large-scale detection of malicious web pages. In: Proceedings of the 20th international conference on world wide web, pp 197–206
2. Carpineto C, Romano G (2017) Learning to detect and measure fake ecommerce websites in search-engine results. In: Proceedings of the international conference on web intelligence, pp 403–410
3. Chiew KL, Chang EH, Tiong WK (2015) Utilisation of website logo for phishing detection. Comput Secur 54:16–26
4. Ding Y, Luktarhan N, Li K, Slamu W (2019) A keyword-based combination approach for detecting phishing webpages. Comput Secur 84:256–275
5. Khonji M, Iraqi Y, Jones A (2013) Phishing detection: a literature survey. IEEE Commun Sur Tutor 15(4):2091–2121
6. https://geekflare.com/find-remove-blacklist-site/
7. Mohammad RM, Thabtah F, McCluskey L (2015) Tutorial and critical analysis of phishing websites methods. Comput Sci Rev 17:1–24
8. Singh A, Tripathy S (2014) TabSol: an efficient framework to defend Tabnabbing. In: 2014 international conference on information technology. IEEE, pp 173–178
9. Sarika S, Paul V (2014) An anti-phishing framework to defend Tabnabbing attack. In: International conference on security and authentication, pp 132–135
10. Sahingoz OK, Buber E, Demir O, Diri B (2019) Machine learning based phishing detection from URLs. Expert Syst Appl 117:345–357
11. Pao H-K, Chou Y-L, Lee Y-J (2012) Malicious URL detection based on kolmogorov complexity estimation. In: 2012 IEEE/WIC/ACM international conferences on web intelligence and intelligent agent technology, vol 1. IEEE, pp 380–387

12. Sahoo D, Liu C, Hoi SCH (2017) Malicious URL detection using machine learning: a survey. arXiv:1701.07179
13. Le H, Pham Q, Sahoo D, Hoi SCH (2018) URLnet: learning a URL representation with deep learning for malicious URL detection. arXiv:1802.03162

Part V
Security Networking

Software-Defined Network Vulnerabilities

Swati Chakraborti, Atrayee Majumdar Ray, Swagata Roy Chatterjee, and Mohuya Chakraborty

Abstract Software-Defined Network (SDN) is gaining popularity day by day in the enterprise cloud and data-centric network. It provides flexibility to manage large scale complex network which needs time to time reconfiguration. SDN proposes a new paradigm of the programmable network having centralized management. In this approach, a software control program named as the controller is responsible for decision making. Hardware such as routers, switches only forward packets to their destination. Thus the control is decoupled from the data plane providing a more dynamic environment. However, this benefit of SDN also brings in new vulnerabilities. SDN architecture mainly consists of data plane, control plane and application plane. Each layer consists of different attack vectors and possible vulnerabilities. The security aspect of SDN is very important because it inherits the security flaws of the classical network and additional security vulnerabilities of the control plane. In this chapter, the different assets of SDN that need to be protected from attack are described. Threat vectors such as Fake Traffic Flows, Switch Specific Vulnerabilities, Control Plane Communication Attacks, Controller Vulnerabilities, Lack of trust between Controller and Management Applications, etc., that are either intrinsic or extrinsic part of SDN are discussed thoroughly. The security mechanisms, both generic as well as SDN specific, that are to be embedded into the network design are also discussed. SDN will bring a breakthrough in the network architecture if properly designed to manage different vulnerabilities to enhance the security of assets.

Keywords Software-defined network · Threat vector · Security · Vulnerability

---

S. Chakraborti · A. M. Ray · S. R. Chatterjee (✉)
Netaji Subhash Engineering College, Kolkata, India
e-mail: rcswagata@gmail.com

M. Chakraborty
Institute of Engineering & Management, Kolkata, India

M. Chakraborty et al. (eds.), *The "Essence" of Network Security: An End-to-End Panorama*, Lecture Notes in Networks and Systems 163,
https://doi.org/10.1007/978-981-15-9317-8_9

1 Introduction

Traditional IP-based networks are complex and difficult to manage. The network operator has to configure each network element separately using commands, which is often vendor-specific to express the required high-level network policies. Automatic reconfiguration and automatic response are also not present in the current network. Moreover, it is vertically integrated. The control plane that takes the decision about handling traffic and the data plane that forwards the traffic are placed together inside the networking elements. This reduces the flexibility of the network and also hinders the evolution of infrastructure. Software-Defined Network (SDN) proposes an emerging paradigm where the control and management part is separated from network traffic forwarding functionalities breaking the vertical integration. So switches and routers become simple traffic forwarding devices. Traffic forwarding decisions are not a destination-based, instead, flow-based forwarding is adapted in SDN. Flow means here a sequence of packets flowing between source and destination. A centralized controller is implemented to provide all control functionalities. The network is programmable. The core configuration and evolution of the network become easier. Policy enforcement is also simple to implement in this emerging technology. The architecture of SDN is described in Sect. 2 [1, 2].

Separation of control and data plane can be implemented through a programming interface between control and data plane. The controller controls network elements through it. OpenFlow is a widely used programming interface for SDN. Switches following this network protocol have one or more flow tables containing traffic forwarding rule. When incoming traffic matches the flow rule, different actions such as forwarding, modifying or dropping can be performed on it. Depending on the flow table installed by the controller a switch can function as a switch, router, firewall, etc., SDN and OpenFlow have gained significant popularity in the industry within a few years. Many big organizations such as Google, Facebook and Microsoft, etc., are funding the Open Networking Foundation (ONF). The logical components of SDN are described in Sect. 3 [2, 3].

SDN can be used to enhance the security aspects of a traditional network. The controller has a global view of the network, which may be considered as the greatest advantage as compared to the traditional network in terms of security. All network elements in SDN collect and report traffic statistics. It makes network-wise intrusion detection (misuse and anomaly detection) and malicious switch detection easier. SDN provides services that are required to improve the security of the network such as security policy enforcement, random host mutation, monitoring of cloud infrastructure and many more. Network forensic analysis also has become better. A flow-based traffic forwarding scheme is adopted by SDN. This offers more control over network traffic compared to the traditional network, which forwards network traffic blindly. This limits the generation as well as the flow of malicious traffic [4].

Before discussing the security challenges of SDN we must be familiar with the terminologies such as threat, attack and vulnerability. Vulnerabilities are the weaknesses that are inherent to a network due to design, configuration or implementation.

A threat is an accidental or intentional incident that hampers confidentiality, integrity or availability of a system. When vulnerabilities are exposed, they must be addressed to reduce the threat. An attack is a harmful action taken to cause some damage to a system. The main difference between threat and attack is that a threat can be unintentional also, but an attack is always intentional. Attributes of a secured communication network are narrated in Sect. 4, which are authentication, non-repudiation, confidentiality, availability, and integrity. Confidentiality ensures the protection of delicate information against unauthorized access. Integrity protects unauthorized alteration of network element and controller information. The availability of information, when needed, is also important. The properties of SDN that attracts malicious uses are programmable network and centralized control. The main threat vectors that exploit the vulnerabilities of SDN are described in Sect. 5. Some of the threats are general and some are specific to SDN. Threat analysis helps to develop a more secure network. Section 6 describes the general design for a secure and dependable SDN architecture. Fault and intrusion tolerance is considered as one of the key factors from the design perspective of a secure and dependable network. The security issues of the SDN data plane and control plane are described in Sects. 7 and 8, respectively. Vulnerabilities of both the planes are identified. Attacks and countermeasures of data plane and control plane are also discussed. SDN applications communicate with the controller through the Northbound Application Programming Interface (API). Therefore applications of SDN can also enact as attack vectors. However, the network controller can check the permissions of a particular application if it is not from a trusted source [5–8]. The security issues of the SDN application plane is described in Sect. 9.

The chapter ends with a conclusion in Sect. 10. The design of a secure and dependable SDN has many challenges as well as opportunities. The challenges will increase gradually with the development of this technology. These challenges need to be addressed proactively to take complete advantages of this newly developed technology.

2 Architecture of Software-Defined Network

The architecture of SDN is based mainly on the client-server model. In this model, the commands from the client are sent to the server. The server sends an acknowledgment and notification to the client. It is a layered network shown in Fig. 1. It focuses on building the network and the computer systems with the amalgamation of open, software-based technologies and commodity networking hardware. This networking hardware separates the SDN control plane and SDN data plane in the networking stack.

Unlike conventional network where the functions are decentralized, SDN [1] separates the control and the data plane centralizing a controller to determine all the data flow in the network.

Typical core SDN architecture broadly consists of three layers or plane, namely application layer, control layer and data plane layer. In the conventional network,

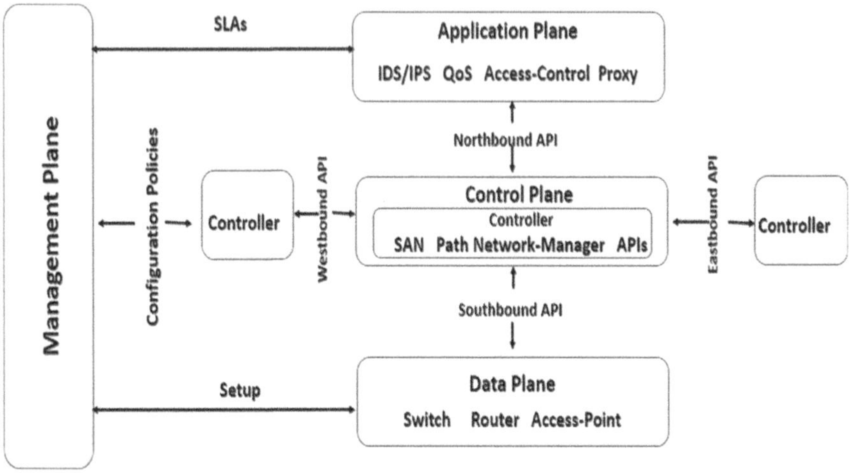

Fig. 1 SDN layered architecture

there was no application layer and the infrastructure layer (data plane) and the control layer were merged together. In SDN, these three layers are separated, which allows controlling the whole behavior of the entire network from single control software.

Application layer—The top of the SDN stack is the application layer. It is an open area for developing innovative applications and also provides an open interface to establish connections with the rest of the layers in the network. The application layer defines a number of rules and offers services such as firewall, access control, routing, proxy service, monitoring balancer, etc. This layer is also responsible for the abstraction of the SDN network control management through the boundaries. By receiving the global view of the network, this layer helps and guides the control layer with the global view information.

Control layer—The control layer is below the application layer. The main component of this layer is one or various controllers that forward the different types of rules and policies to the infrastructure layer through the interface [1].

Data plane/Infrastructure layer—This is the bottom-most layers in the SDN stack and is also called the data plane or the forwarding plane. It consists of all network infrastructure element such as switches, routers, etc., the main function of this layer is data forwarding, fragmentation, local information monitoring and statistics gathering.

Apart from the above mentioned three layers, there is another layer called **Management Layer** that governs these three layers. This layer controls all the network administrative tasks, spreads the policies of the network and allocates the SDN resources.

SDN breaks the traditional physical boundaries that exist between the switches, routers and controllers through well-defined API. There are two main interfaces in SDN architecture:

(a) **Southbound API**—It connects the controller in the control plane and the network infrastructure in the data plane. Southbound API can be explained in two scenarios.

- In-band communication—In this type of scenario, some specified flow rules are followed for the traffic.
- Out-band communication—In this type of scenario, no flow rules are followed for the traffic. VLAN implementation is required to distinguish the traffic flow and the communication that depends on the open flow rule.
- **OpenFlow** is the most common southbound interface, which was standardized by Open Network Foundation.

(b) **Northbound API**—This is the interface between the controller and the network application.

Apart from these two APIs there are eastbound and westbound APIs (ex. HyperFlow) [9], which share control information regarding flow in the data plane.

2.1 Characteristics of SDN Architecture

SDN architecture exhibits the following characteristics

Directly Programmable: Network is directly programmable as it is not coupled with the forwarding functions.

Agile: The administrators can adjust the traffic flow by abstracting the control from forwarding according to the changing needs.

Centrally Managed: The total global view and the control are managed by the software-based centralized controller.

Programmatically Configured: The network managers can easily manage the network resources via dynamic, automated SDN programs which can be written easily as the program is not dependent on proprietary software.

Open Standards-Based and Vendor Neutral: Even when implemented through open standards, the SDN network is comparatively simple as the instruction is provided by the controller and not by the multiple, vendor-specific devices or protocols.

2.2 Few Components of SDN Architecture

This section describes two very important components in the conventional SDN architecture, the controller and the protocols for the communication.

Controller: The controller is said to be the brain of the Software-Defined Network. It is a control point through which the data flow to switch/router via the Southbound API and application and different business logic via the Northbound API. The controller platform runs on a server and with different protocols informs switches where to send the packets. Controllers are classified into two groups: physically centralized controllers and physically distributed controllers.

In a physically centralized controller, the controller is considered as the single entity and it controls all the forwarding devices in the network. NOX is the first centralized controller supporting OpenFlow protocol. There are also other controllers based on different programming languages such as Beacon [10], Maestro Trema, etc.

Initially, in the beginning, SDN was considered to have a single centralized controller, but as the network expanded, the performance was not scaling well. Moreover, due to some drawbacks, such as the single point of failure and scalability problem, physically distributed designs came into consideration.

In a physically distributed design, a number of controllers are used to manage and facilitate the forwarding device. Here the control plane consists of a number of controllers that work together to have a more secured network. The distributed controller approach may again be divided into four types: logically distributed controller approach, physically distributed approach, hierarchically distributed approach and hybrid controller approach.

In a physically distributed controller approach, a large network is divided into a number of small networks and each network is maintained by a local controller. All the local controllers have the same global view of the network. This approach is mainly used to connect multiple domains, e.g., ONIX controller [11].

In a logically distributed controller approach also the entire large network is divided into a number of small domains. Each domain is individually controlled by a centralized controller having different responsibilities. Each controller in spite of having its own responsibilities, must have the total network view. This view should be shared with other controllers to maintain a global view. In this approach, each time as the state changes, the controllers have to update other controllers. DISCO [12] is an example of this type of controller.

In a hierarchical distributed controller approach more than one layer of controllers exists in the control plane. Each controller possesses an individual task and may have a partial view of the total network to take decisions. The upper layer of the control plane acts as the central controller, for example, Kandoo controller [13].

The hybrid controller approach is considered as the combination of all other approaches to omit the limitations and adopt the benefits of all approaches, for example, SOX/DSOX [14].

2.3 OpenFlow Protocol

OpenFlow is the most accepted Southbound open protocol communicating data layer and control layer, which enables the server to dictate the switches about the destination of the packets. This protocol defines packet forwarding rules, processes, keeping track of the process and finally manages the entire process. Before coming in detail in the mechanism, we will discuss a few terms.

Packets are the basic units of forwarding, which is the combination of bytes having a header, payload and occasionally a trailer. All the control information is stored inside the packet as the Packet Header. The series of packets having the same pattern is called Flow. The switch component consists of a number of flow tables and group tables which perform the packet forwarding. The switch communicates with the controller and the controller controls the switches through the protocol. Flow Entry is a rule in the Flow Table to process the packets by matching technique. OpenFlow switches use this Flow Table for message forwarding. The message is forwarded between the SDN switch and the SDN controller using the protocol. According to the network condition, the controllers can add, delete or change any entity in the Flow Table. The network connection between the switch and the controller is called the Open Flow Connection and the main unit that flows within the connection is called the messages. Based on the sender, the messages can be of three types: controller-to-switch, asynchronous and symmetric.

In **controller-to-switch**, the sender is the controller and is responsible for the embellishment of the handshakes, switch and flow table configurations.

In **asynchronous** type, the message starts from the switch by sending a packet in-port status, flow removed message and any error to the controller. An example of this type of message is PACKET-IN.

For the **symmetric** type, there is no restriction for the sender and it is mainly used for lightweight messages such as hello or echo message.

Mechanism of OpenFlow

The Open flow switches forward the packets according to the flow table. The flow table is controlled and manipulated by software in the controller. The Flow Table contains one or more flow entries that determine the matching packets and the process of flow of the matching packets. The flow entries consist of:

(a) The matching rules for matching the incoming packets.
(b) Counters to have the detail of flow such as total packets received, number of bytes and the flow duration.
(c) Set of instructions for handling the matching packets.

During the forwarding process, the packets go through the flow table following pipeline architecture. The packets are numbered, starting from 0 according to their movement. Pipeline process takes place in two stages—ingress processing and egress processing. Within every pipeline, a sequence of different table lookups is performed on different flow tables. The detailed working flow is described in Fig. 2 [15].

It is shown in Fig. 3 that when a packet reaches the OpenFlow switch, the header field is taken out from the packet and is matched against the matching field of the flow table entries. If there is a matching entry, the switch applies appropriate instructions that match with that flow entry. If there are a number of matched entry packets, they are matched according to priority. The counter of that flow table entry is then updated and the switch forwards the packet to the respective port. If there is no match found, the packets are sent to the SDN controller, which is later on buffered according to the switch capability. To that end, the unbuffered packets or the first byte of the buffered

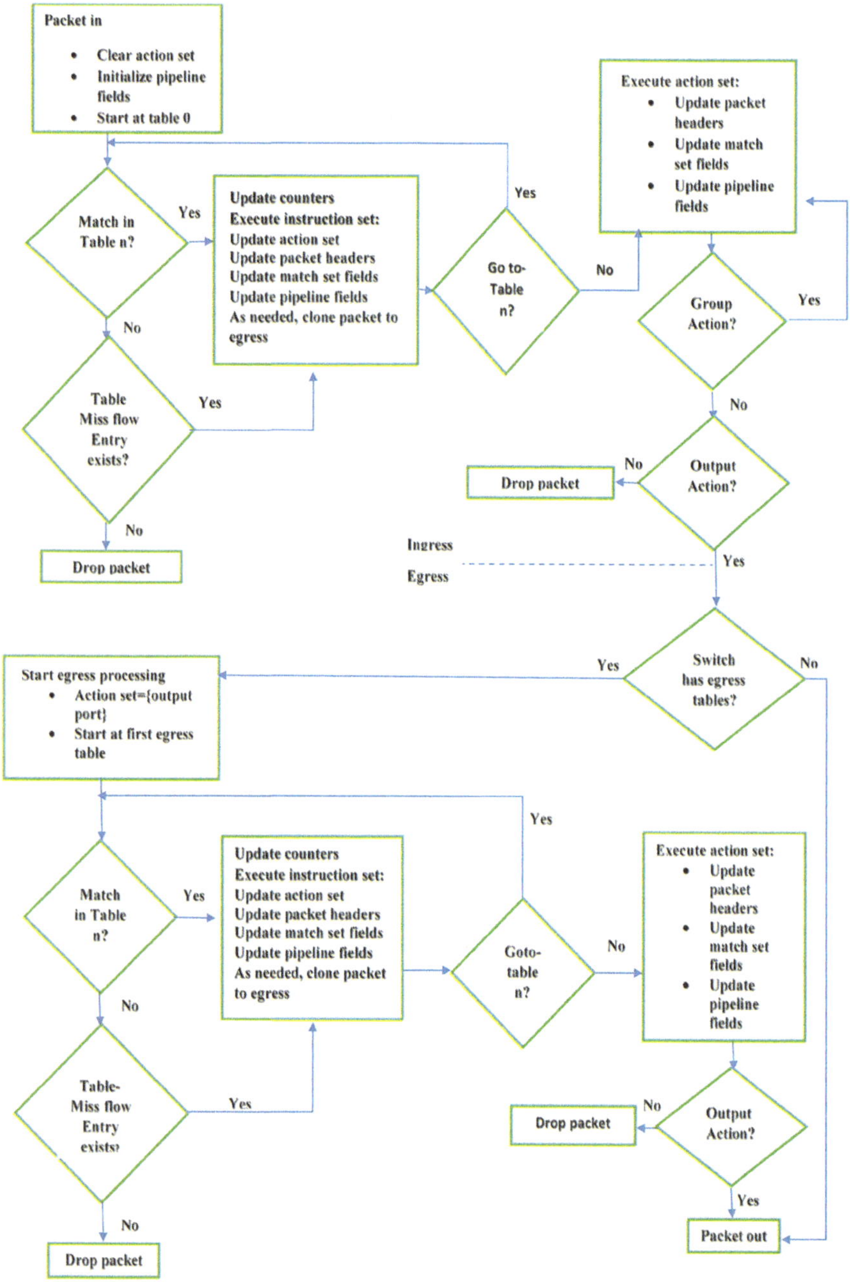

Fig. 2 Detailed working flow of the OpenFlow switch

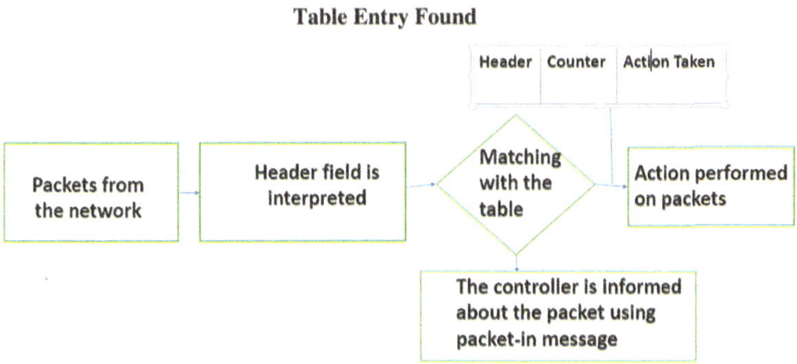

Fig. 3 Open flow process

packets are enclosed using a PACKET-IN message and are sent to the controller. After receiving the PACKET-IN message, the controller takes necessary actions for the packets and finally, the buffered packets are forwarded using the PACKET-OUT message. In the OpenFlow protocol, the remote controllers have also access to modify the flow entries from the switch's flow table.

3 Classification of SDN Logical Components

The SDN consists mainly of three layers. The application layer is placed at the top of the architecture. It includes many services such as firewalls, load balancing, VoIP, etc. The control layer controls the hardware infrastructure and also provides network data and the state of the application layer. Interaction between the control and application layer is through Northbound APIs (A-CPIs). The main function of the data plane layer is forwarding traffic. The interaction between the data plane layer and the control layer is through southbound APIs (D-CPIs) [2, 3]. The logical components of SDN are listed in Table 1 and shown in Fig. 4.

Data Plane Layer Entities:

The data plane layer forwards and processes data traffic. This layer consists of network switches. Data Controller Programming Interface (D-CPI) allows the sharing of data plane resources by the controller program. D-CPI Agents execute controller instructions. Information about data plane resources is stored by the network element Relational Database (RDB). Data plane resources are data source, data sink and data processing engine. Data traffic is handled and delivered by the data sources. Data processing engine processes the data. Data sink stores data coming from data source and data processing engine.

Table 1 Asset components of SDN network

Sl No	Asset component	Location
1	D-CPI agent	Data plane
2	Network element RDB	Data plane
3	Data source	Data plane
4	Data sink	Data plane
5	Data processing engine	Data plane
6	D-CPI	Control plane
7	C-agent	Control plane
8	C-API	Control plane
9	C-RDB	Control plane
10	Controller function	Control plane
11	Controller content	Control plane
12	A-API agent	Application layer
13	Application content	Application layer
14	Application function	Application layer
15	Application controller	Application layer

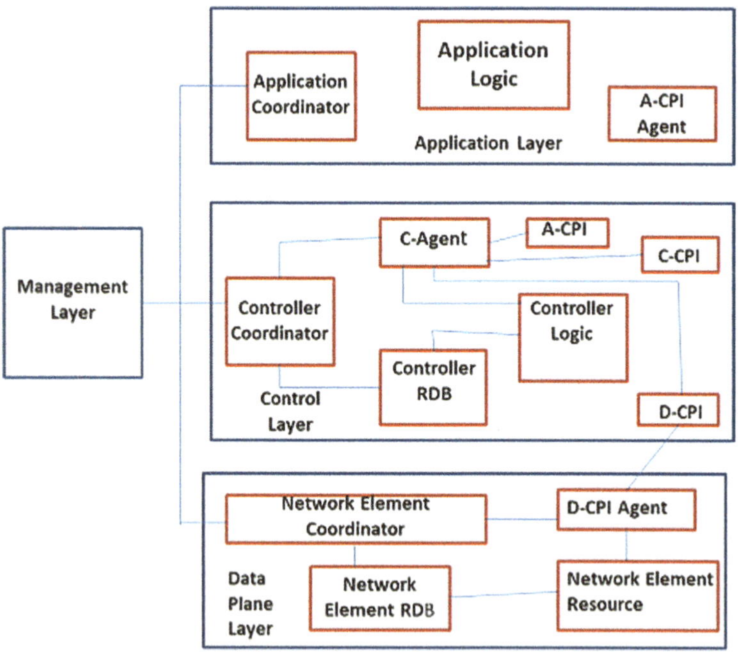

Fig. 4 Logical components of SDN

Control Plane Layer Entities:

The control layer controls the data plane layer elements. It also translates the application layer's instruction, which is in a high-level language, into the low-level language to program the data plane layer. C-API provides information about controller resources to other controllers. C-Agent describes the controller's functions to other entities. C-RDB stores information about the resources in a controller. The controller function refers to all control operation of the control plane. Controller content is the data produced or used for control operation.

Application Plane Entities:

The application plane consists of A-CPI agent to communicate the information from the controller. The application function is responsible for the operations of the application layer and the content that is processed is the application content. The application coordinator coordinates with the management layer.

4 Attributes of Secured Communication Network

The National Security Agency (NSA) has identified the security requirements of the communication system to protect transmitted data, as well as to guarantee the service requirements. They are specified considering the stakeholder's requirements, assets, threats and attacks. Stakeholders are users of the network, the network provider or any government or public authority. Assets are the component of the network. Any potential violation of security or disruption of service is the threat. When a threat actually occurs, it is an attack. The identified attributes of a secured communication network are confidentiality, availability, integrity and non-repudiation [16].

Confidentiality: Confidentiality means protection against unauthorized access of delicate data at the network element as well as controller level. Techniques such as spoofing, scanning or hijacking can be used by an attacker to get access to important information at the switch level. Spoofing is the creation of a fake IP address to hide the source of the attack. At the controller level confidentiality is very essential. Data, policies and rules need to be protected from unauthorized access. Also, safe communication between switches and controllers and among switches is essential for maintaining confidentiality.

Integrity: Integrity means protection against unauthorized alteration of network elements and controller information by a third party. The modification may be intentional or unintentional. At the switch level, information components can be flow tables and switches. At the controller level also controller data and policy need to be protected from modification by unauthorized third parties. Integrity must be ensured at the communication channel level as well.

Availability: At switch level, availability means the authorized users must be able to access network elements as well as data when required. At the controller level, it is very necessary to ensure that approved parties such as switches or other controllers should be able to access a controller and related data. Information is only valuable if they can be accessed at the right time.

Non-repudiation: Non-repudiation is a legal concept where a party cannot deny after doing something. Repudiation can take place at the controller level from an app or any other controller.

5 Security Threats of SDN

The SDN is programmable, so the entire network can be controlled by software programming. Control has been separated from network elements and centralized. These two features of SDN attract malicious users. This section discusses potential risks (threat) of the SDN network and its effects. Threat source can be an SDN or a non-SDN component, a malicious controller, a malicious application or a management resource. Analyzing threat will help in developing a more secure network which will preserve only the numerous benefits provided by SDN network [6, 7].

Threat Vector 1 Fake or Forged Traffic Flow: This type of DoS Attack can be generated from either non-malicious faulty machines or from malicious users. The targets of this attack are switches and controllers. A DoS attack is launched to exhaust the capacity of Ternary Content Addressable Memory (TCAM) and controller resources. The problem can be avoided by using an authentication mechanism. For the cases where an attacker is able to get the control of an application server that contains user information, this problem cannot be mitigated so easily. At that time, an attacker can use the same authenticated ports, as well as the source MAC address, to produce forged authorized flow into the network. The intrusion detection system, along with the mechanism to control switch behavior dynamically, can support the detection of abnormal flow in the network.

Threat Vector 2 Vulnerabilities in Switches: The switches of SDN are vulnerable to certain kinds of attacks. For example, a switch can be used to slow down traffic in the network. Clone switches can be used to deviate the path of packets for data theft purpose. Forged request can be injected into the system to overload the controller or other switches. Monitoring and detection of abnormal behavior of network elements can help to solve this problem. Software attestation can also be a mechanism that mitigates this problem for software components.

Threat Vector 3 Control plane Communication Attacks: Many researchers have shown that TLS/SSL does not guarantee secure communication of the control data plane even in the presence of public key infrastructure (PKI). For example, vulnerable application and libraries, a compromised certificate authority or a self-signed certificate can expose the control plane channel of SDN to the attackers. If an attacker gains access to the controller, it can also use switches to initiate DDoS attacks. However, secure communication may be possible by using threshold cryptography. The oligarchic trust model can also be used with multiple trust anchor certification authorities.

Threat Vector 4 Vulnerability of Controller: This type of threat may be considered as the most severe one because the most important component of SDN is the controller. If it is faulty or malicious, the entire network would be vulnerable. Also, the intrusion detection system will not be able to find which combinations of events have stimulated a particular behavior or to identify it as malicious. For example, an old version of the SDN controller can be a cause for remote DoS attack. This type of threat can be avoided by using replication, employing a diversity and recovery technique. Also, the security of all elements inside the controller, which are sensitive, is very important for this purpose.

Threat Vector 5 Lack of Mechanism for Ensuring Trust between Controller and Management Application: Similar to Threat Vector 3, controller and management applications may not establish a trusted relationship. Applications that are integrated to the SDN network may not be completely secure, which can be a threat to the entire network. However, autonomic trust management mechanism may ensure the trust of an application during its lifetime.

Threat Vector 6 Attack on Administrative Station Vulnerability: This threat vector is also present in the traditional network. But the severity is more in the SDN network since the entire network may be affected by a single compromised machine that can be used to reprogram the entire network. One of the possible solutions to this type of threat can be using protocols that will require a double credential verification to access the network controller.

6 Mechanism for Designing Secure and Dependable SDN Environment

This section presents the general design for a secure and dependable SDN architecture. Fault and intrusion tolerance is considered as one of the key factors from the design perspective of a secure and dependable network. Crash and Byzantine are the two main fault models. Crash fault tolerance mainly models benign fault due to the crash of machine or OS. Byzantine fault tolerance model can be used for any abnormal behavior. Security mechanism such as replication, diversity, self-healing, etc., illustrated in Fig. 5, can be used to deal with the different kinds of threat vectors described in the previous section. Replication helps in masking fault. Self-healing helps in removing errors. Diversity helps in reducing common vulnerability [16]. Different threat vectors and security solutions related to them are listed in Table 2.

Replication: Replication of controller and application helps in improving the dependability of the system. The key element to guarantee a highly robust system is fault and intrusion tolerance, which can be masked by using replication. It helps in dealing with situations such as controller or application failure due to software vulnerability or high traffic volume. Figure 5 shows multiple controller versions to provide replication. Few Apps are also replicated. This helps in dealing with both

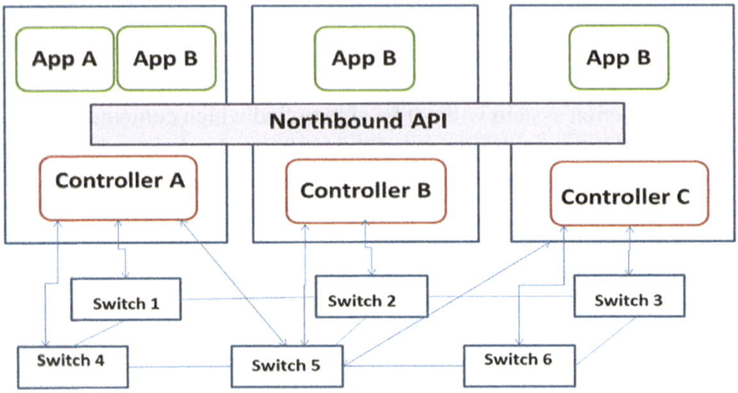

Fig. 5 Secure and dependable SDN

Table 2 Design solution to threat vectors

Sl. No	Threat vector	General/SDN specific	Security solution/Mechanism
1	Threat vector 1	General	Replication, trust between controller and switch
2	Threat vector 2	General	Self-healing, trust between controller and switch
3	Threat vector 3	SDN specific	Diversity, trust between controller and switch, dynamic device association
4	Threat vector 4	SDN specific	Replication, self-healing, trust between controller and switch, dynamic device association
5	Threat vector 5	SDN specific	Replication, trust between controller and app, security domains, secure components
6	Threat vector 6	General	Self-healing, reliable and quick software update

hardware as well as software failure. Malicious applications can also be masked without disrupting the service.

Diversity: When only a particular kind of software or a fixed operating system is used throughout the network, it becomes easier for the attackers. If a diverse set of OS is used, intrusion tolerance will increase. If replication is used along with diverse controller dependability and security will be more. Since there exists only a very few intersecting vulnerabilities among different OS or software, diversity will help in avoiding common fault and vulnerabilities. Using a diverse controller will also restrict the lateral movement of an attacker preventing cascaded system failure.

Dynamic Device Association: The association between a controller and a switch must be dynamic in nature. If a particular controller fails to operate for some reason, the switch must be allocated to another backup controller in a secured manner.

Before allocating a switch dynamically, the backup controller must be checked using threshold cryptography or any other proper authentication mechanism. This kind of dynamic association increases the fault tolerance of the control plane. Also, using several controllers increases control plane throughput (load balancing) and reduces control delay. Dynamic device association requires a general purpose CPU inside Switches instead of a custom ASIC. It can be implemented in another way by placing a proxy element attached to the switch.

Self-healing: A security attack may cause service disruption. A proactive and reactive recovery system helps in maintaining virtual normal functioning of the SDN network. When a particular compromised component is replaced, the replacement should be a new and diverse version of the component. For example, when replacing an SDN controller where the software component is OpenDaylight, an alternate version such as ONOS or Floodlight may be used. Exploring diversity in the process of recovery will strengthen the security against a few specific vulnerabilities of the system.

Trust between Controller and Switch: Fake flow of traffic is often inserted into the SDN network by faulty or malicious switches. Therefore building trust must be between the switches and the controller is a very important requirement. A controller can maintain an authenticated white list of trusted switches that can send messages specific to the control plane. This mechanism lacks the desired flexibility of SDN. Alternatively, a controller can trust all switches. Any anomalous or malicious behavior of a particular could be reported based on an anomaly detection algorithm by other switches or controller. When the trustworthiness of a switch is reported, it will be put in quarantine mode by all switches and controller.

Trust between Controller and App: The behavior of software component changes due to environmental change. The aging of software may also add security vulnerability. Therefore autonomic trust management mechanism is required to ensure trust between controller and application plane component. It is based on mutual trust and delegated trust. The trustworthiness of a component is measured based on qualitative attributes such as confidentiality, reliability, integrity, maintainability, availability, etc.

Security Domains: Isolated security domain technique helps in the trust-based segmentation of the network. It helps in restricting threat to the only affected section of the SDN network. For example, an operating level subsystem cannot access kernel level subsystems. Similarly, a web server application can only interact with database backend application and not allowed to interact with any other application of the same network. The security domain can be implemented in SDN using virtualization and sandboxing techniques. It restricts minimum communication between different domains.

Secure Components: Secure components are the important building blocks of a dependable and secure system such as TCB (trusted computing bases). TCBs are a set of very secure hardware and/or software components that cannot be tampered. Sensitive security information can be stored on TCB to assure confidentiality even when the system is compromised.

Reliable and quick software update: All software have some bugs which require a regular and secure update to reduce vulnerability. A control platform should have the proper mechanism to ensure a fast, smooth and reliable way of updating software.

7 Security Issues of SDN Data Plane

The concept of SDN is to decouple control logic from devices of data plane, such as SDN switches and routers. The controller function is implemented centrally and the function of SDN Switches is thus limited mostly to forwarding traffic. This new paradigm has brought programmability and flexibility in the network. Network operators will get better control of the network, making automation and optimization easier. Also, the risk of security policy collision reduces because of the centralized nature of SDN. Updation of security policies becomes easier. In SDN, separation of data plane and control plane have made the switches, router and virtual switches simple because forwarding policies are provided by the controller instead of the switch hardware or firmware. The minimum requirement of SDN switches (i) flow table which is used to maintain flow entries for forwarding traffic, (ii) secure channel for communication with the controller and (iii) OpenFlow protocol for communicating with the controller. The first security challenge is, therefore, differentiating between genuine and malicious flow rule. SDN switches have limited memory. The second challenge is therefore avoiding a saturation attack by generating a huge number of flow rules [17–20].

7.1 SDN Data Plane Vulnerabilities

Security Vulnerabilities are the weak areas of a network, which may be the possible doors of attacks in the system. Since most of the attacks exploit the vulnerabilities, it is very important to identify them. Attacks can be prevented by the elimination and correction of vulnerabilities. In SDN, the identified source of vulnerabilities is shown in Fig. 6.

i. Flow information leakage via Side Channels of data plane: An attacker can use this vulnerability to get information about the network configuration. The time overhead for inserting a new flow rule by the controller can be used by malicious users to observe the switch flow table, network policies, OpenFlow protocol version and controller software version.
ii. TCAM Memory, which is exhaustible: Each switch has limited memory. When a packet is received, it is forwarded to a specific switch port according to the flow rule. When a packet is received, which is not matching to any of the existing flow rule the packet is buffered. A new flow rule is requested from the controller.

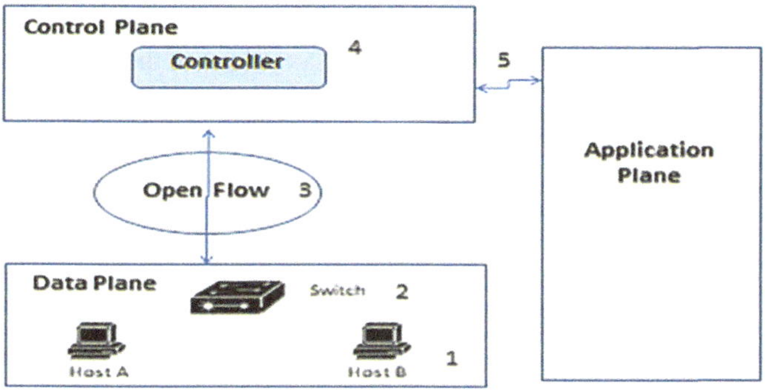

Fig. 6 Security vulnerabilities of SDN

When a large number of flow entries are inserted, TCAM memory of the switch can be exhausted.

iii. Topology Poisoning: SDN switches have controller ID through which they can make a connection to the controller. A controller after getting connected to a particular switch sends a Feature_Request message. The switch in response sends a Feature_Reply message containing its Data Path Identifier (DPID), which uniquely identifies an OpenFlow instance on a switch and a list of active ports. The controller creates Link Layer Discovery Protocol (LLDP) packets containing DPID and port number. The switch sends it to the mentioned port. When an attacker gets access to LLDP packet, the invariants of the packet can be modified giving fake link information to the controller. Thus the topology gets modified. It can be used by an attacker.

iv. Switch Controller Bandwidth: If a packet is received, which is not matching to any of the existing flow rule, a new flow rule is requested from the controller. When a large number of table miss packet are sent to the controller, it will be saturated. The switch controller bandwidth can be exhausted.

7.2 *SDN Data Plane Attacks and Countermeasures*

DoS Attack and Countermeasures: In SDN, the control and the data plane are separated. The control plane controls the behavior of the network. The communication between control and data plane is through the southbound API. Control plane install flow rule on the data plane. The data plane obeys those flow rules to manage network traffic. A table miss flow or data is the one that does not match with the existing flow rule. On arrival of a new table miss flow, the data plane sends a packet_in message to the control plane. The bandwidth between the control and data plane is limited, less than 10 Mbps. An attacker can exploit this vulnerability and launch a data-to-control plane saturation attack by generating forged table miss packet by modifying some fields of data or flow rule. This will trigger the switch to generate a huge number of packet_in messages consuming the communication bandwidth as well as the memory of Switches. Therefore the challenges against DoS attack at the data plane level are identified as

- Efficient handling of table miss packets
- Distinguishing between benign traffic and attack traffic.

For the first challenge, a low packet loss rate and short packet delay have to be maintained. If the number of table miss packets is dropped, then packet_in messages from benign sources will also drop. It is important to identify attack traffic otherwise to filter them from normal traffic. A connection migration was introduced on the data plane by AvantGuard to identify attack traffic. It is done after verification of the TCP handshake for every new flow. Statistical analysis of flow and flow classification also helps to distinguish between attack traffic and benign traffic. Since switch controller bandwidth is limited, a rate limit if imposed on the packets sent to the controller, will help to prevent the attack on it. In this process, a meter is attached with a switch element for measuring and controlling packet flow. If the packet rate exceeds, a threshold limit "RateLimiter" is triggered.

Side Channel Attack and Countermeasures: SDN switches are controlled by the SDN controller. It delivers flow rules to the Switch to handle the flow of traffic. When a switch receives a packet for which there exists no matching rule for packet forwarding, the controller installs a new flow rule for that packet. The delay suffered by a packet for which no matching flow rule exists is much more than the delay encountered by a packet for which there is an already existing flow rule. Flow rules that are delivered to the switch are stored in switch memory until and unless these two situations arise-

- Flow Rule cache filled up
- Rule expired for not matching with any of the flow within the timeout duration of the rule.

A side-channel attack is the one where an attacker can submit a probe request. The delay in response can be measured. The attacker can determine if a rule for the probe flow already exists or not. This helps in detecting the use of SDN in a network, size of the flow table, versions of software, etc. This issue can be solved by implementing a timeout proxy at the data plane. When the controller is unable to respond within a particular time interval, the default forwarding instruction is sent by this proxy.

Topology Poisoning Attack and Countermeasures: As we have already discussed, when an attacker gets access to the LLDP of switches, it can modify the network topology. There are two methods by which topology poisoning can be done.

I. Fake LLDP injection
II. LLDP Relay.

For the first case, an attacker can generate fake LLDP packets by varying invariants of the original LLDP and launch the Link Fabrication Attack. The attacker also could

Table 3 Summary of data plane attacks and countermeasures

Sl No	Attacks	Attack plane	Countermeasures
1	DoS	Data plane	AvantGuard to identify attack traffic, timeout proxy, topology update checker, self-healing, dynamic switch association
2	LLDP spoofing		
3	Side channel attack		
4	Topology poisoning		
5	Data modification		

adjust the injection rate of LLDP in a way to avoid anomalous traffic detection. In the second case, an attacker does not inject a forged LLDP packet. Instead, the attacker, upon receiving the LLDP packet from a switch, repeats it to another target switch without any alteration. Thus a fake topology view is created to the controller as if there exists an internal connection between the two switches. When the controller notices this connection, the Shortest Route is computed again on the basis of this modified topology information. An attacker can exploit this fake route for initiating an attack. It is known as a man-in-the-middle attack. Fake LLDP injection can be avoided by adding extra authentication such as Type Length and Value (TLV). If a controller signed TLV is added in the LLDP packet, it will be very difficult for an attacker to modify it. But it cannot defend the Link Fabrication attack. Another approach for providing security against topology poisoning is TopoGuard. In this security mechanism, a Topology Update Checker is designed. It validates any updation in the network topology. The data plane attacks and their countermeasures are described in Table 3.

8 Security Issues of SDN Control Plane

We have already discussed that in SDN the control plane is decoupled from each individual network element and implemented in a centralized external entity known as the controller. The controller has complete information about the entire network topology. It programs the traffic forwarding table of the network elements. A conventional network has control functions embedded in each network element. So compared with the conventional network, SDN has some distinguishing properties [20–25].

I. Control and data plane separation
II. Network programmability.

These capabilities of SDN also have brought some new security problems. There are three elements of the control plane whose security is important.

- The controller
- The communication channel that exists among controllers (East West bound API)
- The channel between controller and switch (Southbound API).

8.1 Vulnerability of Control Plane

1. The vulnerability of the Controller: Controller can be considered as the heart of SDN. Since the control is decoupled from the data plane and the centralized failure of a controller will lead to a single point failure of the network. A compromised controller can spoil the entire network. It is also vulnerable to DoS attack and also can generate fake flow rule. Also, based on forged information, a controller may create a false routing path.
2. Southbound API: Southbound API communicates between control and data plane. It follows OpenFlow protocol. Transport Layer Security (TSL) or Secure Socket Layer (SSL) is often downgraded to a lower version for communication compatibility of two clients. In this process controller, TLS version can be downgraded to launch a man-in-the-middle attack.

8.2 SDN Control Plane Attack and Countermeasures

DDoS Attack: DDoS attack is generated by multiple malicious hosts that are distributed in the network. They may try to exhaust the memory and other resources of a controller, making it unavailable for genuine users. This happens when a controller cannot differentiate between normal traffic and attack traffic. We have already discussed that if a packet is received for which the OpenFlow switch does not have an existing flow rule, a Packet_In the message is sent to the controller. During a DDoS attack forged table, miss packets can generate huge amounts of Packet_In messages causing exhaustion of memory of the controller and other resources. Replication of controller may be one possible solution for such problems where instead of a single controller, a number of controllers manage the network. Though these controllers are physically distributed, they are logically centralized.

Attack due to compromised Controller and Countermeasures: When the access to the controller is gained by an attacker, it poses a serious threat for the entire network. Such a compromised controller can use a switch to launch an attack on any target. Replication of the controller in such a case cannot be treated as an efficient solution because the controllers are equally vulnerable when such an attack occurs. Diversity of controller may be useful in such cases if switches, instead of communicating with any master controller, get a response from all the controllers. A majority voting technique can be used for the selection of flow rules.

8.3 Security Vulnerabilities in OpenFlow

It is known that in SD, the whole control plane function is carried out by centralized program logic at the controller. Though this centralized nature has a number of benefits but the major vulnerability in the OpenFlow protocol is due to the lack of authentication and authorization required. The security lies in the protection of the switches, controller and channel as they are very prone to several attacks such as DoS, spoofing, hijacking and so on. Security vulnerabilities of OpenFlow and their countermeasures are listed in Table 4 [26]. Some attacks and their mitigation is given below.

- Denial of Service Attack—The controller is the main point for the DoS or flooding attack. For an example, in some cases, while receiving unknown network packets, the control plane receives the request from the data plane or the forwarding plane for the flow rules, but the attackers can make the control plane unavailable for accepting the request from the data plane by injecting high amount of traffic in the network and making the controller down. Thus DoS Attack is launched on the OpenFlow switches. To mitigate this threat, Intrusion Detection Systems (IDS) and dynamic control of traffic flow may be used.
- Host Hijacking Attack—In this case, the Host Tracking Service (HTS) in the network sometimes gets disrupted through host attack, man-in-the-middle attack or DoS attack [27]. The controller is aware of the whole network information and the attacker, through the scanning attack, track the total network information and modify the valuable information slowing down the SDN information. Moreover, if the controller is hijacked, then the password details can be changed and communication mode can be altered.
- Tampering Attack—In this attack, the API messages are spoofed by inserting harmful flow rules towards the network devices, making network misbehavior. Dynamic Flow Tempering is an example of this type of attack.
- Spoofing Attack—In this type of attack, the attacker gets control of the controller. The entries of the flow table may be changed by creating or updating the entries and the network specialist may not have the view of the change. As a result, the attacker has the control of the network completely.

Table 4 Countermeasures of SDN [26]

Author	Method	year	Attack	Description
Tootoochina et al.	Hyperflow	2010	DoS attack	Minimizes the response time of the control plane to oppose the DoS attack
Suh et al.	CONA	2010	DoS attack	It is the countermeasures against resource-exhaustive attacks
Yao et al.	VAVE	2011	Spoofing attack	Verifies the source address of the external packets preventing the spoofing attack
Matias et al.	ARM	2012	ARP spoofing attack	Helps in tracking the MAC address to prevent spoofing attack
Wen et al.	PermOF	2013	Potential attacks	Proposed to give minimum privilege to the OpenFlow application to avoid attacks
Fichera et al.	OPERETA	2015	DDoS attack	Implemented in SDN controller to reject useless connection request
Wang et al.	FloodGuard	2015	DoS attack	Used to reject requests for data to control planes saturation attacks
Hong et al.	TopoGuard	2015	Poisoning attack	Useful for fronting automatic and real-time disclosure of poisoning attack
Kuerban et al.	FlowSec	2016	DoS attack	Restricts the number of incoming packets to the controller
Buragohain et al.	FlowTrApp	2016	DDoS attack	Helps in detecting DDoS attack and mitigate for the data center
Dridi et al.	SDN-Guard	2016	DoS	Scheme proposed to protect the SDN controller from DoS attack

(continued)

Table 4 (continued)

Author	Method	year	Attack	Description
Nguyen et al.	SDN Extension	2016	Host impersonate attack, man-in-the middle attack DoS attack	Protect the controller from host tracking Service

9 Security Issues of SDN Application Plane

SDN provides high-level abstraction and programmability of the network by decoupling control and data plane. This exposes high-level abstraction to the application layer. Deployment of new applications on any OpenFlow controller to implement a new security mechanism becomes easier. But malicious applications can be the cause of security threat for network elements by spreading through the network. Also, the variety of applications developed from different vendors can create security policy collision [28]. Some of the security challenges are

Authentication and Authorization: Authenticating applications is a major issue in SDN. In OpenFlow most of the applications are developed by other parties, but they are running on the controller. They access network resources without any proper security for the protection of those resources. The mechanism is therefore required to establish trust between controller and applications. A centralized mechanism is required to certify these applications.

Access Control and Accountability: Proper access control mechanism is needed for ensuring network security. A malicious application can bypass the procedure in SDN. There are also nested applications that are unaware of SDN. Therefore access control of such application is difficult. PermOF is a permission system that enforces permission control. Permission sets can be categorized into read, notification, write and system permission. For example, in case of sensitive information, read permission is provided to an application. If an application has notified permission, it can notify in real-time. An application with write permission can modify the switch or controller state. System permission provides access to local resources. Also, a consistent view of the network must be available to applications. There are several debugging and verification programs for this purpose. Languages such as VeriFlow is used to know the flow rules in run time.

10 Conclusion

SDN improves network security by implementing a global view of the network. The centralized control plane and programmability of the network provide strong security enforcement compared to the traditional network. These advantages also have brought in new security challenges with it. In this chapter, the security weaknesses of SDN are highlighted. The vulnerabilities of control, data and application plane are discussed. Security solutions are also presented. Due to centralized control, the most vulnerable asset of SDN is the controller. The security aspects of program interfaces for communication between controller and switches are also discussed. The threat space will grow with the gradual deployment of technology. Therefore security mechanisms need to be developed for fast detection and quick response of security threats.

References

1. Kreutz D, Ramos FMV, Veríssimo PE, Rothenberg CE, Azodolmolky S, Uhlig S (2015) Software-defined networking: a comprehensive survey. In: Proceedings of the IEEE, vol 103, pp 14–16
2. Open Networking Foundation (2014) SDN architecture overview, ONF, Palo Alto
3. Zerkane S, Espes D, Le Parc P, Cuppens F (2017) Vulnerability analysis of software defined networking. In: Cuppens F, Wang L, Cuppens-Boulahia N, Tawbi N, Garcia-Alfaro J (eds) Foundations and practice of security. FPS 2016. Lecture notes in computer science, vol 10128. Springer, Cham
4. Dabbagh M, Hamdaoui B, Guizani M, Rayes A (2015) Software-defined networking security: pros and cons. IEEE Commun Mag 53(6):73–79
5. Scott-Hayward S, O'Callaghan G, Sezer S (2013) SDN security: a survey. In: Proceedings of the software defined networks for future networks and services (SDN4FNS 2013), pp 1–7
6. Kreutz D, Ramos FMV, Verissimo P (2013) Towards secure and dependable software-defined networks. In: SIGCOMM HotSDN, pp 55–60
7. Huang D, Chowdhary A, Pisharody S (2018) Software-defined networking and security: from theory to practice
8. Ahmad I, Namal S, Ylianttila M, Gurtov A (2015) Security in software defined networks: a survey. IEEE Commun Surv Tutor 17(4):2317–2346
9. Tootoonchian A, Ganjali Y (2010). HyperFlow: a distributed control lane for OpenFlow. In: Proceedings of the 2010 internet network management conference on research on enterprise networking USENIX Association, p 3
10. Erickson D (2013) The Beacon OpenFlow controller. In: Proceedings of the second ACM SIGCOMM workshop on hot topics in software defined networking. ACM, pp 13–18
11. Poutievski L, Zhu M, Ramathan R, Iwata Y, Inoue H, Hama T, Shenker S (2010) Onix: a distributed control platform for large-scale production networks. In: Proceedings of OSDI, vol 10, pp 1–6
12. Phemius K, Bouet M, Leguay J (2014) Disco: distributed multi-domain SDN controllers In: IEEE network operations and management symposium (NOMS)
13. Fu T, Hu L, Yu X, Hu J, Zhao K (2016) Rolebased intelligent application state computing for OpenFlow distributed controllers in software defined networking. In: Soft Computing
14. Luo M, Tian Y, Li Q, Wang J, Chou W (2012) SOX—a generalized and extensible smart network Openflow controller (X) In: The first SDN world summit, Germany
15. OpenFlow specification. Version 1.5.1 (Wire Protocol 0x06) (2015) Open Networking Foundation
16. Dharma NIG, Muthohar MF, Prayuda JDA, Priagung K, Choi D (2015) Time-based DDoS detection and mitigation for SDN controller In: 17th Asia-Pacific network operations and management symposium (APNOMS), Busan, 2015, pp 550–553
17. Yoon C, Lee S, Kang H, Park T, Shin S, Yegneswaran V, Porras P, Gu G (2017) Flow wars: systemizing the attack surface and defenses in software-defined networks. IEEE/ACM Trans Netw 25(6):3514–3530

18. Shang G, Zhe P, Bin X, Aiqun H, Kui R (2017) Flood defender: protecting data and control plane resources under SDN-aimed DoS attacks. In: IEEE INFOCOM 2017—IEEE conference on computer communications, Atlanta, GA, pp 1–9
19. Kuerban M, Tian Y, Yang Q, Jia Y, Huebert B, Poss D (2016) FlowSec: DOS attack mitigation strategy on SDN controller. In: IEEE international conference on networking, architecture and storage (NAS), Long Beach, CA, pp 1–2
20. Dayal N, Maity P, Srivastava S, Khondoker R (2016) Research trends in security and DDoS in SDN. Wiley Online Lib 9:6386–6411
21. Celesova B, Val'ko J, Grezo R, Helebrandt P (2019) Enhancing security of SDN focusing on control plane and data plane. In: 7th international symposium on digital forensics and security (ISDFS), Barcelos, Portugal, pp 1–6
22. Scott-Hayward S (2015) Design and deployment of secure, robust, and resilient SDN controllers. In: Proceedings of the 2015 1st IEEE conference on network softwarization (NetSoft), London, pp 1–5
23. Natarajan S, Ramaiah A, Mathen M (2013) A software defined cloud-gateway automation system using OpenFlow. In: IEEE 2nd international conference on cloud networking (CloudNet), San Francisco, CA, pp 219–226
24. Hu H, Han W, Ahn GJ, Zhao Z (2014) FLOWGUARD: building robust firewalls for software-defined network. In: 3rd ACM SIGCOMM workshop on hot topics in software defined networking. In: Proceedings of HotSDN 2014
25. Abdou A, Oorschot, Wan T (2018) Comparative analysis of control plane security of SDN and conventional networks. IEEE Commun Surv Tutor 20(4):3542–3559. Fourthquarter
26. Mutaher H, Kumar P, Wahid A (2018) OpenFlow controller based SDN: security issues and countermeasures. Int J Adv Res Comput Sci 765–769
27. Nguyen T-H, Myungsik Y (2016) Attacks on host tracker in SDN controller: investigation and prevention. In: International conference on information and communication technology convergence (ICTC)
28. Lee S, Yoon C, Shin S (2016) The smaller, the shrewder: a simple malicious application can kill an entire SDN environment. In: SDN-NFV security 16 proceedings of 2016 ACM international workshop on security in software defined networks and network function virtualization, New Orleans, LA, USA, pp 23–28

Demystifying Security on NDN: A Survey of Existing Attacks and Open Research Challenges

Madhurima Buragohain and Sukumar Nandi

Abstract Named Data Networking (NDN) is a clean slate Future Internet Architecture that incorporates the visionary idea of focusing on contents instead of its location/host. It is mainly designed to address long-standing problems in current IP architecture such as security, mobility, and content distribution inefficiency. NDN introduces one specific layer called "security layer" in the NDN protocol stack, which undertakes the responsibility of securing contents instead of securing the communication channels and hosts in IP. The design features in NDN such as stateful forwarding plane, in-network caching helps to counter many existing problems prevalent in IP; however, at the same time, it brings new security challenges. In this chapter, we survey the current literature in security in NDN and present a classification of the existing attacks based on the layers in the NDN protocol stack explicitly link layer, strategy layer, network layer, and application layer. We review the attack scenarios and discuss the pitfalls of the existing solutions to these attacks. The chapter concludes with insightful findings and open research challenges.

Keywords Named data networking · NDN security

1 Introduction

The Internet that we are using today was designed and developed during the late 1960s to solve resource sharing problem as, during that time, the cost of hardware resources was too high. On the other hand, if we observe today, we are using the same Internet, mostly for content distribution and retrieval. According to CISCO visual networking Index report, by 2021, 80% of the total traffic will be video traffic [1], and it indicates that the usage of the Internet has been shifted from resource sharing to distribution of

---

M. Buragohain (✉) · S. Nandi
Computer Science and Engineering Department, Indian Institute of Technology Guwahati, Guwahati, India
e-mail: madhurima.2015@iitg.ac.in

S. Nandi
e-mail: sukumar@iitg.ac.in

M. Chakraborty et al. (eds.), *The "Essence" of Network Security: An End-to-End Panorama*, Lecture Notes in Networks and Systems 163,
https://doi.org/10.1007/978-981-15-9317-8_10

contents; still, the fundamental architecture remains the same. The mismatch between the architecture and the usage has given rise to many problems in IP, including security. During the design of the current TCP-IP architecture, nobody has considered the aspect of security. Security was taken into consideration once the problems started to appear. Due to the Host-to-Host communication model, the security model in TCP-IP is channel-based. The existing solution such as Transport Layer Security (TLS) secures the communication channel between two hosts rather than contents. Once the contents get out of the channel; there is no guarantee of the packets. Though a lot of security solutions are developed to counter security attacks, the number of ongoing attacks does not decrease. For example, WannaCry Ransomware attack captured global attention in 2017. The researchers have realized that ad-hoc solutions increase the complexity of network architecture. To make it simple and secure, researchers have paved the way for finding new Internet architectures.

Van Jacobson, a leading TCP-IP Architect, has proposed the visionary idea of focusing on content, not its location. This idea has led to the design of named data networking. When we download any movie or access any information from the Internet, we do not care about from which server/location it is coming. Our major concern is content as long as it is the right content that we want. The tremendous growth of secured applications such as online banking, healthcare applications has forced NDN researchers to consider "security as primary design goal" from the very beginning. Now, the idea of channel-based security will not work as communication is not host-to-host. Therefore, NDN focus on securing content directly instead of the communication channel through which data transfer takes place. Each content carries a digital signature signed by the original producer. A consumer, when receives the data, verifies it before consumption. Once the verification is successful, there is no need to concern about where and how it is retrieved. If the application requires confidentiality, the original producer can encrypt the content so that only the intended recipient can decrypt the data.

Though the architectural design and presence of digital signature help us secure NDN from many existing attacks prevalent in TPC-IP such as ICMP flooding attack, reflection attack, TCP SYN flood attack, etc., however, the inherent features in NDN such as stateful forwarding plane, in-network caching helps to emerge new kinds of attacks. This chapter classifies the existing attacks in NDN based on the layers in the NDN protocol stack. We discuss the current mitigation approaches and investigate the drawbacks of the proposed schemes.

This paper provides the answer to many questions that may strike in the reader's mind, which may be as follows:

(i) How inherent features in NDN helps to mitigate the existing attacks in TCP-IP architecture and make it secure?
(ii) How attackers take advantage of the architectural features of NDN and launch new kinds of attacks?
(iii) How NDN security is different from TCP-IP security?

The rest of this chapter is organized as follows: Sect. 2 covers a brief review of NDN and Sect. 3 discusses the inherent features which provide security. In Sect. 4, we present the classification of some major attacks in NDN and discuss the existing countermeasures. Section 5 provides a brief discussion on TCP versus NDN security and some research challenges. We conclude the chapter with the lessons that we have learnt during the survey in Sect. 6.

2 NDN Overview

2.1 NDN Protocol Stack

NDN inherits the hourglass shape from TCP-IP architecture. Figure 1 shows the hourglass shape for both the TCP-IP protocol stack and the NDN protocol stack. The upper and lower portion signifies the independent innovation of upper and lower layer technologies without worrying about the thin waist. The significant difference between both the protocol stacks is that the narrow waist in IP represents host-centric delivery (delivering packets to a particular destination). In contrast, in NDN, it is content-centric (retrieve contents from the network based on names). In addition to that, the NDN stack introduces two new layers: security and Strategy layer. The security layer holds the responsibility of securing content. The strategy layer decides whether, when, and where to forward NDN packets. As we notice in Fig. 1 that there is no transport layer in the NDN stack. The reason behind it is that in NDN, content names carry all the required information for transport. No port and the sequence number are required.

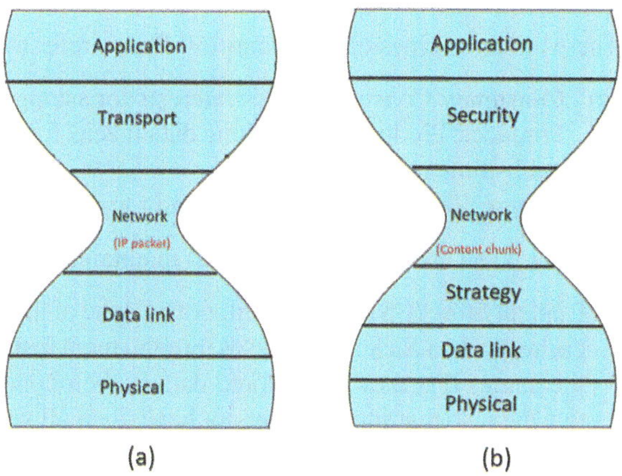

Fig. 1 **a** TCP-IP protocol stack, **b** NDN protocol stack

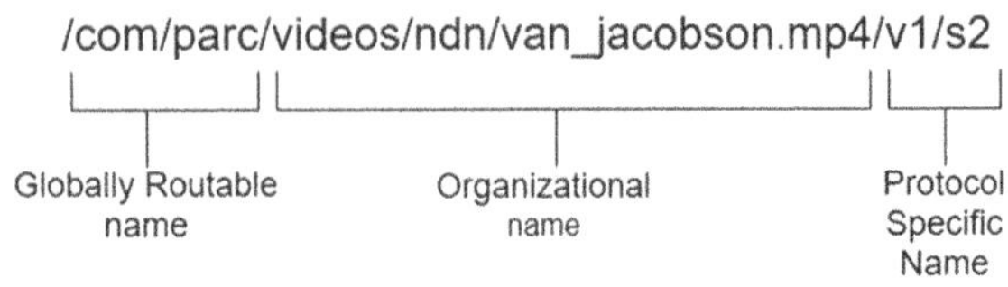

Fig. 2 NDN naming

2.2 *Naming*

Each content in NDN has a human-friendly name. It is used both for content identification as well as for routing the packets. Each name can have many components. "/" is used to separate two components. There is no limit to the number of components and the component length. Similar to HTTP URLs, the names are hierarchical, and therefore the names can be aggregated when required. Application designs the naming rule according to their requirement. For example, in IoT applications where devices are resource-constrained, the packet size should be limited, and therefore the number of components must be less. A search application helps to generate the name depending on the typed keywords by the user. We can see in Fig. 2, there are three parts in the name. The globally routable prefix part is necessary when we want to make the data globally available. In other cases, there is no need for a global prefix; just it needs customization inside the local networks such as my computer or my home, etc.

2.3 *Packet Forwarding in NDN*

There are three types of NDN entities. Please refer to Fig. 3 for packet forwarding.

(a) Consumer: Consumer sends interests for asking for data.
(b) Producer: Producers publishing data listens for interests with the name that it is publishing and send back data.
(c) Router: Routes interest packets and forwards corresponding content packets.

Each NDN entity (not just routers) maintains three major data structures:

- Content Store (CS): CS is a temporary cache of the data packet. Forwarding nodes cache the data packet that passes through them in their CS. Any interest that match the names that they have stored data in their content store, they will reply with that data, so that interest may not have to go all way back to the producer to find the data.
- Pending Interest Table (PIT): It holds all not-yet-satisfied Interests that have been sent upstream towards potential data sources. Each PIT entry contains one or multiple incoming and outgoing physical interfaces: multiple incoming interfaces indicate the same Data is requested from multiple downstream users; multiple outgoing interfaces indicate the same Interest is forwarded along multiple paths.

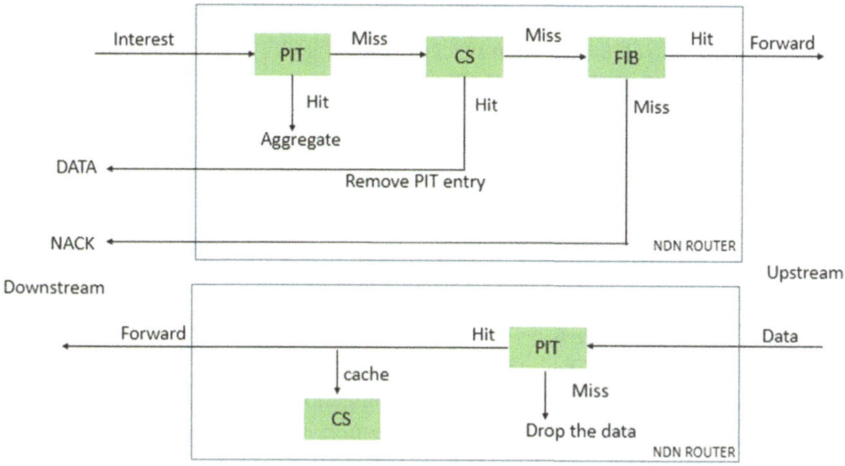

Fig. 3 Packet forwarding in NDN

- The Forwarding Interest Base (FIB): It is similar to the IP Forwarding table. The main difference is that it stores name prefixes instead of IP prefixes. Each FIB entry may provide multiple interfaces instead of a single best interface for each name prefix. FIB is built by the name-based routing protocol, for example, NLSR.

Interest packet forwarding: When an Interest arrives at an NDN node, it extracts the name of the interest packet and uses it to look up the CS table if there is a matching Data packet. If it has the matching data, the data packet is forwarded to the same face from which it arrives. Otherwise, PIT is looked up to determine whether there is already an entry in the PIT for similar interest. There are three possible outcomes;

- If the same name is already in the PIT and the arrival interface of the present interest is already in the set of arrival interfaces of the corresponding PIT entry, then the interest is discarded.
- If the PIT entry for the same name exists and the arrival interface is new, the router updates the PIT entry by adding the new interface to the set. The interest is not forwarded further.
- Otherwise, the router creates a new entry and records incoming interface in the PIT entry. It further looks up the interest name in the FIB using the longest prefix match. If a matching FIB entry is found, the Interest is forwarded by a forwarding strategy module. Otherwise, the router cannot satisfy the Interest and may send a NACK back to the incoming interface of the Interest. Outgoing interfaces are also recorded in the PIT entry.

Data packet forwarding: When the data packet arrives at an NDN node, the first PIT is checked if there is any matching name. It has two results;

- If there is no entry, then the packet is dropped since it is either unrequested or no longer wanted.

- Otherwise, the data packet is stored in CS (depends on the caching policy) and then forwarded to all the incoming faces recorded in PIT entry. The corresponding PIT entry is then removed.

3 Inherent Features Providing Security Support in NDN

In this section, we will discuss the in-built features in NDN, which helps to provide security and make it resistant to many IP-based attacks.

(i) *Signature in Data packets*: In NDN, each producer needs to put its signature on the NDN Data packet. A consumer verifies the packet before consumption. The signature provides data origin authentication, integrity, and correctness.
(ii) *No information about source and destination on NDN packets*: It makes the packet anonymous. If an attacker captures any packet during transmission, attackers will not know who has requested the data.
(iii) *Receiver driven communication model*: Communication in NDN is receiver-driven. A consumer receives any data only in case it has requested it before. It makes NDN resistant from TCP-IP-based attacks, such as TCP SYN flooding, ICMP flooding, and UDP flooding. These attacks are possible in IP as any node can flood packets to any IP address and port number.
(iv) *Stateful forwarding plane*: Forwarding in IP is stateless and to inspect ongoing traffic, IP has to do extra state. On the other hand, a stateful forwarding plane in NDN gives accurate insight on network traffic and helps in prevent and mitigate some existing attacks. Due to the stateful forwarding plane, non-requested data packets are dropped.
(v) *Symmetric Path*: A data packet in NDN traverses back along the same path traversed by its corresponding Interest packet. This symmetric nature makes NDN resistant to a reflection attack where attackers forge an IP packet with a victim's IP address and send it to other secondary victims (reflectors). Responses are routed back to these reflectors.
(vi) *In-network caches*: In a scenario where attackers send requests for existing contents, in-network caches reduce the effect of the attack. Static contents in data packets (e.g., image, HTML files, CSS files) are cached in CS of the router to satisfy future Interest packets. However, we cannot ignore the effect of an attack where attackers send a large number of Interests, and the cache has limited capacity.

Though the design of NDN helps to prevent some of the existing attacks in TCP-IP, attackers exploit some of these features and make these features an advantage for them. They can launch new kinds of attacks. In the section below, we review the existing attacks in NDN.

4 Classification of Attacks in NDN

4.1 *Attacks in the Application Layer and Its Countermeasures*

Major attacks in the Application layer are Interest flooding attack, Cache pollution attack and Content Poisoning Attack.

(i) *Interest Flooding attack (IFA)*:

In this attack, attackers issue a large number of interest packets to exhaust the PIT or dissipate computation resources of the content producer. Figure 4 shows a scenario of IFA where the attacker sends a large number of Interests and it causes the PIT filling up. The requests from the legitimate consumer (Alice) are dropped. IFA can be classified into three types based on the type of Interest packet: (i) existing (ii) non-existent, and (iii) dynamically generated content. The impact of type (i) is limited due to the presence of in-network caching as some contents can come from in-network caches. In type (ii) attack, attackers target a specific namespace (/com/overleaf) by sending fake interests so that the requests can never be satisfied. Fake interests can be easily generated by appending some random keywords to the targeted name prefix. Therefore, PIT entries corresponding to those requests remain in routers until the expiration, which eventually helps to make the PIT full and helps in dropping legitimate interests. Type (iii) attack mainly overloads the producer because each time producer has to sign the generated content, which involves a significant amount of computational resource [2]. Layer-wise attacks in NDN are provided in Table 1.

Mitigation approaches of IFA

Afanasyev et al. [3] propose three IFA mitigation techniques for IFA: (i) Token bucket with per-interface fairness (ii) Satisfaction-based Interest acceptance, and (iii) Satisfaction-based Pushback (SBP).

The first method is adapted from a well-known TCP congestion control mechanism called Token Bucket. For each interface, each router sets a limit on the number of pending Interests. The value of the limit depends on the capacity of the interface

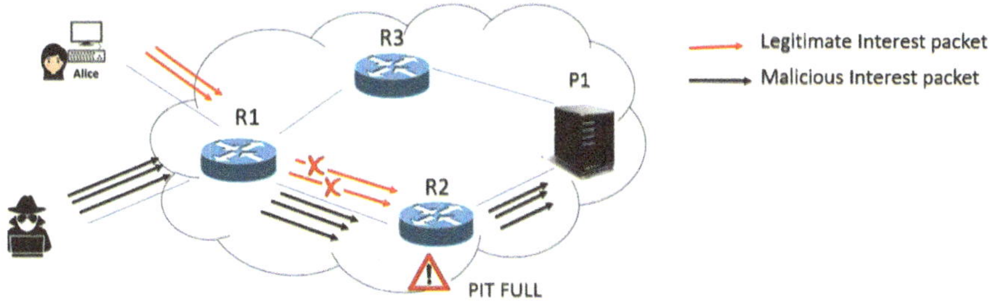

Fig. 4 Interest flooding attack

Table 1 Layer-wise attacks in NDN

Layers	Possible attacks
Application layer	Interest flooding attack Cache pollution attack Content poisoning attack
Network layer	Prefix hijack attack
Strategy layer	Timing analysis attack Cache monitoring attack Object discovery attack Flow cloning attack
Data-link layer	Selfish attack

(bandwidth-delay product). The authors extended this token bucket idea with per-interface fairness. To fairly distribute the outgoing link capacity among the incoming interfaces, they have added a separate queue per incoming interface. It prevents malicious interest from consuming the entire outgoing link capacity. The drawback of this approach is: It cannot distinguish between malicious and legitimate interest, and therefore satisfaction rate of legitimate consumers is reduced.

In Satisfaction-based Interest acceptance, an incoming interest is accepted or rejected based on a probabilistic model. The model considers the value of Interest satisfaction ratio (ISR) of the incoming interface. The major drawback of this scheme is: it affects the forwarding of legitimate interests due to the independent decision of acceptance or rejection at each router. In Satisfaction-based Pushback, an interest limit is maintained for each incoming interface. The value of the limit depends on the ISR of the interface. Routers announce these limits to their downstream neighbors so that they can adjust the forwarding rate. Though approach (iii) is effective among all three, still legitimate Interests suffer.

Wang et al. [4] propose an approach named'Disabling PIT Exhaustion' (DPE) to counter a specific IFA attack where attackers request for non-existing contents. DPE reduces the stress of PIT by decoupling malicious Interests from PIT. For that, each router maintains a list of malicious prefixes (*m-list*). A name prefix is inserted in *m-list* in case the number of expired Interests for that name prefix exceeds a predefined threshold. Moreover, to prevent a legitimate name prefix staying in *m-list*, an expiry timer is added to each entry. For each new interest whose name prefix in m-list, the expiry timer is reset to the initial value. When a router receives an Interest, it checks whether its name prefix is in the *m-list* or not. If yes, it adds a name component called *Interface-list* to the Interest name and adds the incoming interface to *Interface-list*. This kind of interests is named as *m-interest*. When this *m-interest* moves towards the data producer, the router adds the incoming interface to the interface-list one by one. The data packets corresponding to m-interests are forwarded back to the requested consumer based on the interface list. Table 2 lists the countermeasures of IFA.

Table 2 Countermeasures of interest flooding attack

References	Type of content	Mitigation approach	Utilized information	Drawbacks
[3]	Existing	Token bucket with per-interface fairness, satisfaction-based interest acceptance, and satisfaction-based pushback	Interest satisfaction rate and link capacity	Legitimate interests suffer
[4]	Non-existent	Reduces load on PIT by decoupling malicious interests	No. of expired interest per name prefix	Forwarded malicious Interest may cause congestion
[5]	Non-existent	Traces back the origin of attacking interest by sending spoofed datapacket	PIT size	Does not work well when the edge router is compromised
[6]	Non-existent	Limit interest rate after observing per-face statistics	Interest satisfaction rate and PIT usage per interface	Affects legitimate Interests
[7]	Existing, dynamic and non-existent	Send NACK packets towards downstream, and rate-limiting is done at edge routers	Fake interest name list	Performance degrades in case of compromised edge router

The drawbacks of this approach are:

- Malicious interests are still forwarded, which can contribute to network congestion.
- Additional processing on routers.

Dai et al. [5] propose an approach called "Interest Traceback" where originators of attacking Interests are traced back. Here each router monitors the PIT size, and once it exceeds a predefined threshold, the router generates a spoofed data packet for the longest unsatisfied interest packet. This data packet helps the edge router to identify the malicious consumer. Then the edge router limits the incoming Interest rate by dropping Interest packet.

The drawbacks of this approach are:

- It can cause a negative impact on the legitimate clients as they may mistakenly request a non-existent content.
- An edge router may itself compromise.

Compagno et al. [6] propose Poseidon, a framework named Poseidon for IFA detection and mitigation. Each router monitors two values for a time interval: (1) Ratio of the number of incoming Interests to the number of the content packet for each Interface (R) (2) Number of bytes used to store Interests for each Interface (B). Poseidon detects an attack for a particular interface when both R and B exceed their respective thresholds. Once detected, to mitigate the attack, the router sends an alert message to its downstream routers. After receiving such alter messages, a router decreases the threshold values to half and reduces the rate of Interests.

The drawback of this approach is: The interests of legitimate consumers may coexist with the interest of attackers. In that case, due to the limiting of interest rate, legitimate consumers are highly affected.

PAP (Producer-assisted pushback) [7] propose a DDoS mitigation approach where the producer pushes back the traffic to the source of the malicious traffic. The idea is based on the assumption that the producer can judge whether it is under attack or not. It can also identify the prefix under attack. Moreover, it can distinguish between a valid and fake Interest. To mitigate the attack, it sends a NACK packet towards downstream. A NACK packet has the following information: reason code (Fake or valid Interest attack), prefix under attack, Fake/valid Interest receiving rate that producer can handle under prefix P, and a fake Interest name list (FL). When a router receives a PAP NACK, first it checks the reason code. If it is a fake Interest attack, it checks FL and removes the corresponding PIT entries. In both cases, the router sends a new NACK to all its interfaces along with the newly calculated T/C and pruned FL. Finally, the gateway router, after receiving the PAP NACK, will perform the rate-limiting. Rate-limiting is allowed by edge router so that legitimate clients do not get affected.

The drawback of this scheme is: It will not work well for compromised edge router.

(ii) *Cache Pollution Attack*:

In this attack, the attacker sends a large number of unpopular content. It results in the degradation of cache's performance as popular contents will be found less frequently in caches. Therefore, more requests will be forwarded towards the original producer, which decreases network good-put. Figure 5 shows the scenario of a cache pollution

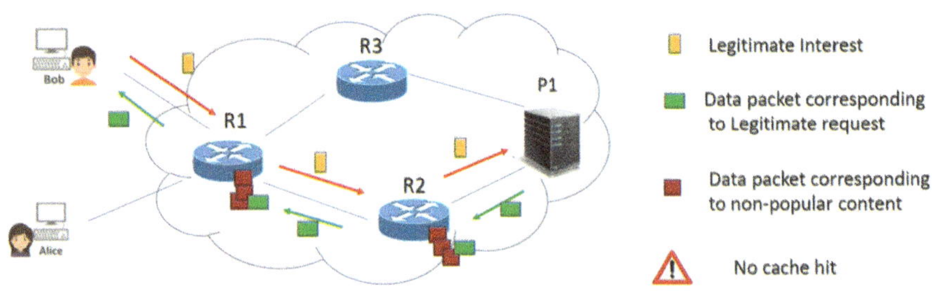

Fig. 5 Cache pollution attack

attack. Due to the presence of unpopular contents in router's caches, Bob's request for contents is forwarded to the producer P1.

Cache Pollution attack can be classified into two subcategories as below:

- Locality disruption attack: Here, attackers continuously send interests for new unpopular contents to disrupt the cache locality. As a consequence, the effectiveness of cache degrades as original popular content may get replaced.
- False locality attack: Here, attackers request the same set of unpopular contents repeatedly with the help of compromised nodes. These selected set of contents will occupy the cache replacing the original popular contents and thereby induces a false locality of contents.

Mitigation approaches:

Xie et al. [8] propose an approach called "CacheShield," which aims to handle unpopular requests. It has two significant components: a cache shielding function and a list of content names and their corresponding request frequencies. After receiving a request for a content chunk, the router first checks whether it is in its cache or not. If it is yes, then it sends the content from the cache. Otherwise, it will forward that request. When the router receives the content corresponding to that request, the cache shield function is evaluated. $\psi(t) = 1/(1 + e^{(p-t)/q})$ where p and q are the system parameters, and t is the tth request for that particular content object. With probability ψ, the router stores the chunk in its CS.

If the chunk is not stored in CS, then the router checks whether its name is in the content name list or not. If yes, then request frequency is updated. Otherwise, the name of the chunk is included in the list.

The major drawbacks of this scheme are:

- Storage overhead as CS needs to store information for each content name and its corresponding request frequency.
- This approach suffers from the fact that the shield function's parameters p and q are constants and can be easily deduced (if not known), and hence an attacker can easily calculate the value of t. Then the attacker has to just ensure that it requests the unpopular contents more than t times.

Karami et al. [9] propose a cache replacement policy to mitigate cache pollution attack. The policy is based on the Adaptive Neuro-Fuzzy Inference System (ANFIS), which consists of three parts: extraction of input–output data and construction of ANFIS structure, accuracy verification of the constructed ANFIS structure, integration of the structure as a cache replacement policy. In the first step, the ANFIS structure is created based on the properties of cached content such as content's time duration in the cache, request frequency, and standard deviation of the request frequency. All these feed into a non-linear system that returns a value (range 0–1). 0 indicates false-locality, 0.5 indicates locality-disruption, and 1 indicates a valid content.

The system iteratively evaluates the goodness of the cached contents that have been cached beyond a predefined time period. The system selects the contents with goodness values less than a goodness threshold, ranks them, and applies cache replacement over the content with low goodness values. The authors showed the advantages of their proposed mechanism over CacheShield [8] in terms of hit damage-ratio (proportion of hits that cannot occur due to the attack), percentage of honest consumers receiving valid contents, and communication overhead. The drawbacks of this scheme are:

- High storage overhead. Needs to store historical and statistical information for each cached content.
- The iterative computation of statistics undermines scalability.

(iii) *Content Poisoning Attack (CPA)*:

In this attack, the attacker fills the routers' caches with fake content with a legitimate name. Fake content may be of the following types.

- Content with an invalid signature. The malicious producer does not have valid signing information to sign the content, and therefore, verification fails.
- Content with a signature signed with the wrong key, that is, not the key of the original producer.
- Content having fake payload. The malicious producer has valid information to correctly sign the packet, which leads to successful verification. Therefore detection of such content is difficult.
- Malformed Signature field.

To insert fake content into routers' cache, the attacker has to take control of one or more intermediate routers on the path between the consumer and the original producer. These controlled routers inject fake content in their routers caches. Moreover, fake content can also be inserted in routers with the help of collaboration between malicious producer and malicious consumer. The attacker leverages caches to spread fake content in the network. To escalate the outcome of the attack, the attacker is likely to forge poisonous data with popular content names. Figure 6

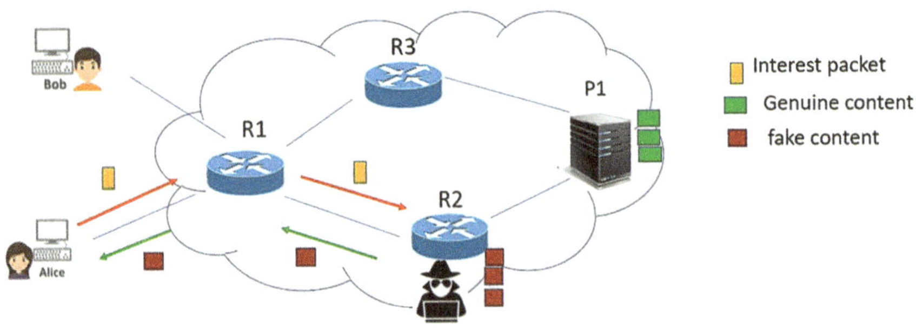

Fig. 6 Content poisoning attack scenario

illustrates one specific scenario of CPA where Alice and Bob are consumers and P1 is the producer. Alice sends Interests packets requesting for some content in P1. Its requests get hit in R2 and contents are returning from the router R2. However, the important point is that router R2 is under the control of the adversary and its cache is filled up with fake contents. Therefore, instead of genuine content from the original producer P1, the invalid content is returned from the router R2.

Mitigation approaches of CPA:

Theoretically, routers in NDN can avoid this attack by simply verifying the signature of the contents. However, it is impractical as verifying at wire speed requires a massive amount of computational power, which is unrealistic.

Gasti et al. [2] is the first paper to discuss CPA. They have proposed a series of solution to counter this attack. However, they did not present the effectiveness of those solutions mathematically or through simulation.

"Self-certifying Interest and content" is the first concept that is used to efficiently validate the received contents by routers. For the static content, the Interest will carry the hash along with the content name, which is computed over data, name and signature. Now, a router can verify by comparing the hash of the received content to the hash included in the Interest. Though it seems effective as compared to existing signature verification schemes, this approach assumes that consumer knows the hash of the desired content beforehand. They have provided a solution where each chunk can carry the hash of the next chunk within it. However, the authors accept that it boils down to a problem where getting the hash of the initial content chunk is a challenge.

This idea does not work for dynamic content. Therefore authors proposed the concept of inclusion of hash of the producer's public key. Though it seems reasonable, authors did not ignore the possibility of an attack in case the attacker knows the public key of the producer. In addition to these, the authors proposed three collaborative approaches.

(1) Probabilistic Disjoint Verification, (2) Neighbor Verification Feedback, (3) Consumer Feedback. In (1) Routers verify the cached contents in a distributed way, so that load of verification reduces. To make an efficient distribution, each router R_i checks the following condition: $h^k{}_{CO} mod n = i$. Where h_{CO} is the least significant 32 bit of the content hash and n is the total number of routers. If the result is yes, then only it verifies.

In (2), routers probabilistically decide to verify the cached content. If the verification fails, it issues a warning Interest, and it is sent to all faces. This type of Interest is restricted to one hop only. Later, when a router receives such type of Interest, first, it finds whether such type of content is in its cache. If not, it simply discards the request; otherwise, it verifies that content with probability p, which depends on its current load. If the verification fails, it sends its warning message. To prevent the network from fraudulent warnings, the adjacent routers can share symmetric keys.

In (3), verification is done with the help of consumer feedback as a consumer always verifies before consumption. Though it looks simple, there are few challenges, for example, lack of trust between router and consumer, compromised consumers,

Table 3 Countermeasures of content poisoning attack

References	Mitigation approach	Utilized information	Drawbacks
[2]	Self-certifying interest and content, collaborative signature verification	Consumer feedback, neighbor verification	The consumer can be itself compromised
[10]	Routers verify content only if there is a cache hit	Cache hit	Malicious consumers can send interests to increase cache hit to overload the routers
[11]	A cached content ranking algorithm to differentiate between valid and fake content	Exclusion feature in the Interest	The algorithm works on the pattern/number of exclusions issued by the consumers who themselves may be compromised

detection of consumers, which issued false earning. The authors introduce the concept of trust value associated with each content. The trust value of new content is 0.5. For verified content, it is 1, and this value is decreased each time the router receives the negative feedback. The value relies on consumer feedback. In the case where an Interest can be satisfied by multiple contents, in that case, a high ranked content should be sent. Table 3 lists the countermeasures of CPA.

Kim et al. [10] propose a scheme named "Check on cache-hit" to reduce the verification load in routers. Here every content is stored in the CS without signature verification. However, if there is a cache hit, then the content is sent into the network only after verification. To avoid multiple verification of the same content, verified content is marked. It saves a considerable amount of computational resource. The drawback of this approach is: The malicious consumers can send a large number of Interests to make cache-hit to increase the load on the router.

Ghali et al. [11] propose a cached content ranking algorithm to distinguish between valid and fake content objects. The initial idea considers the fact that after the detection of fake content, a consumer issue a new Interest that excludes previously received fake content by specifying hash in the exclusion field of the new Interest. Each cached content has a rank value between 0 and 1. Initially, 1 (highest value) is assigned to each content, and this value gradually decreases over time. Contents with higher rank are more likely to be served to users. The rank of cached content depends on three features:

- Number of Exclusions: A content with more number of exclusions has a lower rank value.
- Time Distribution of Exclusions: A lower rank is assigned to content with much recent exclusion over fewer old ones. The reason behind it is that sometimes it is normal that valid content objects might not satisfy a consumer interest. In that case, little exclusion can occur. Besides, a consumer might exclude certain content since it represents a wrong (e.g., old/stale) version.

- Excluding Interfaces Ratio: The content which has been excluded by interests arriving on multiple interfaces will be penalized more.

  The significant drawbacks of this scheme are:

- The content ranking depends on the number of exclusion issued by consumers, and it is more likely to be compromised.
- Exclusion is a part of content exploration in NDN. Therefore, good content may be possibly marked as bad content.

4.2 Attacks in Network Layer and Its Countermeasures

Prefix hijack Attack:

Here, the attacker advertises the victim's prefixes. So, Interests intended for the victim will move towards the attacker. After receiving the Interest packets, it silently drops and thus creates a black-hole.

In NDN, due to the symmetric nature of communication, each node can know about the information of unsatisfied Interests packets. This information may help to infer about hijacked prefixes. In addition to that, nodes can keep track of the performance of each interface w.r.t name prefix. Moreover, NDN's multi-path forwarding helps to decrease the effect of this attack by forwarding the packets through different paths. For the complete elimination of this attack, routing advertisement/updates need to be signed by the advertised router. Hence, each router requires a public key, which must be provided by the network operator. Furthermore, each interface of a router should also have a public interface key, which is signed by the public key of the router. All updated information needs to be approved by the interface key for providing authentication.

4.3 Attacks in Strategy Layer and Its Countermeasures

(i) *Timing Analysis attack*:

In this attack, attackers probe caches and use data retrieval times to infer whether the content was received from the original data producer or the cache. The difference in response time between cached and non-cached content helps to gather information about past communication. Thus one user can leak information about another user, and thus use privacy can be breached.

Let's take an example scenario (refer to Fig. 7) where Alice and Adv are connected via routers R0, R1, R2, and R3 to an original producer holding a content C. Suppose, Adv is an adversary while Alice is an honest consumer. If Alice issues an interest for content C that resides in the original producer, the interest traverses the path: *Alice* $\rightarrow R0 \rightarrow R1 \rightarrow R2$. Due to symmetric communication, the response is sent along

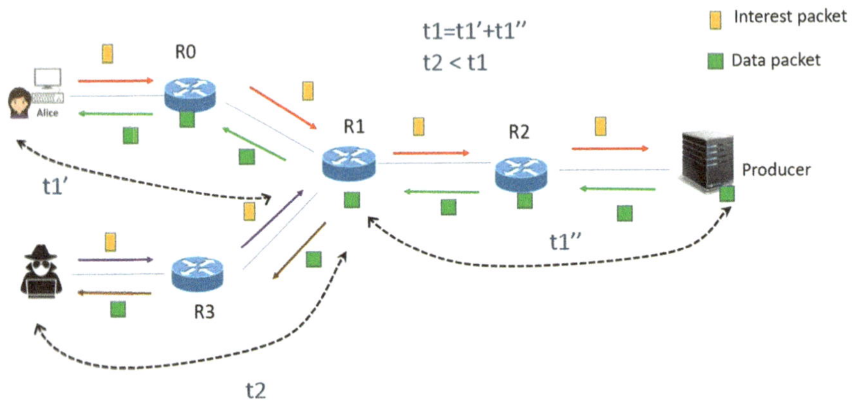

Fig. 7 Timing analysis attack scenario. The attacker can infer whether the user Alice has accessed a content based on the difference in response time for cached and uncached content

the path: *producer* → *R*2 → *R*1 → *R*0 → *Alice*. The total response time required for sending the request until receiving data packets on the returning path is t1. On the other hand, when Adv request C, the path that the interest would traverse is *Adv* → *R*3 → *R*1, and the contents would return on the reversed path and would require a time t2. The time t1 is greater than t2, which an adversary Adv can use to infer that Alice has accessed the content C before.

Mitigation approaches:

Acs et al. [12] categorize the traffic into two types: interactive and content distribution traffic. For interactive content, the authors proposed the addition of a random number to the content name; the number is mutually agreed upon by the requester and the content provider. This prevents the attacker from successfully probing the cache for this content if the precise content name matching approach is employed. However, this approach does undermine caching–cached content can no longer be reduced. The authors suggested that the requester and producer mark privacy-sensitive interests and content as private. The intermediate routers do not cache this marked con-tent, thus preventing privacy leaks. The authors also suggested the emulation of a cache miss at a router, with the router applying a random delay before satisfying a content chunk request. But, a delay undermines the user's Quality of Experience (QoE).

The authors reduced the impact on QoE by using a popularity threshold. The premise of the model is that the privacy-sensitive contents are usually unpopular and that increased popularity generally results in a reduction of the privacy need. With this addition, the router randomly delays satisfying a content for the first-timers it is requested, and deliver the content as soon as possible for the subsequent requests. This model reduces the latency for popular contents, but clients experience the extra delay for the first-interests, and this mechanism also requires an additional state for maintaining the number of requests.

Chaabaneet al. [13] discussed collaborative caching and random caching to preserve cache privacy. Collaborative caching increases the anonymous clients set by increasing the number of clients that share a set of routers; thus, it implicitly helps to preserve privacy. The authors provided no analysis of the caching approaches. We believe collaborative caching is a good direction for further exploration.

(ii) *Cache monitoring attack*:

In this attack, the attacker monitors the accessed data of other consumers connected to the same router [14, 15]. For that, it continuously probes the cache with random names. In the case of encrypted data packets, the attacker can still get substantial information from data-name and data size. The attacker mainly targets the routers at the edges of the network. The reason is: the cache of an edge router serves only a limited amount of consumers. In contrast, the cache of a core router serves a large number of consumers. In that case, there will not be many benefits. In most cases, the attacker mainly targets a particular victim, and they try to find out the requested contents by the victim.

Mitigation approaches

To counter this attack, Lauinger et al. [14] have proposed two approaches: selective caching and selective tunneling. The former method works on the assumption that privacy-sensitive contents are non-popular, and the later one assumes that Internet Service provider (ISP) is trustworthy. In selective caching, a node decides to cache a content only when it crosses a particular popularity threshold. In selective tunneling, contents are classified into privacy sensitive and privacy non-sensitive content. To make a clear distinction, a flag bit is added in case of privacy-sensitive interest/data. Whenever a consumer requests for a privacy-sensitive content, it sets the flags in the interest packet. The corresponding content having a flag bit set will not be cached in any intermediate while traversing from producer to consumer.

The drawback of the selective caching are:

(i) The assumptions may not always be valid.
(ii) Attackers can create fake popularity of content by frequently requesting non-popular content.

One future direction for proposing an efficient solution: Designing an efficient tunneling mechanism for partial content caching.

(iii) *Object discovery attack*: In an object discovery attack, an adversary exploits the exclude feature in the Interest packet for getting the information of other contents in the victim's cache. The attacker takes advantage of the longest name prefix matching as the snooper does not need to know the full name. For handling object discovery attack, an exclusion filter can be disabled, and name prefix matching should be limited.
(iv) *Flow cloning attack*: In this attack, attackers replicate an entire flow of packets by predicting future requests. To implement this kind of attack, at first, attackers need to perform the object discovery attack to know the name of the requested

content. A straightforward solution for this attack is the encryption of the content name. For those scenarios where we cannot perform encryption, in that case, we should design the name such that it is difficult to predict or understand.

4.4 Attacks in Data Link Layer and Its Countermeasures

(i) *Selfish Attack*:

Due to the limitation of computation and storage capability in an NDN IoT environment, the nodes may act selfishly. They may not forward Interest or Data packets of other nodes, which result in DoS attacks. In addition to that, if an attacker wants to stop providing service for specific names, they can sniff the name of the Interest packet and can filter out those name components and can prevent further forwarding.

One naive mitigation approach for Selfish attacks is to provide an incentive to motivate each node for packet forwarding.

5 Discussion and Open Research Challenges

5.1 TCP-IP Security Versus NDN Security

The significant difference between TCP-IP and NDN security centers on the fact that TCP-IP addresses the content locations, whereas NDN addresses the content.

The widely adopted security protocols in TCP-IP, such as TLS and IPsec, secure the communication channel between two hosts. Therefore, in an application where multiple parties are involved, communication overhead increases significantly. The reason is we need to secure each channel between two hosts. Even though a channel is secured, a receiver cannot guarantee whether the received data packet is authentic or unaltered. An application only cares about the security of the content, not its underlying channel. On the other hand, data are protected in NDN as it carries the signature in it. Therefore a consumer as long as the signature verification is successful, there is no need to worry about how the data packet is retrieved or from where it has come (original producer/in-network caches).

In current TCP-IP-based security protocol such as HTTPS, a signature is accepted only when a trusted Certifying Authority directly signs it. For establishing trust, signature verification is not sufficient. NDN utilizes name semantics to build trust policies that further prohibit in blind signature verification of contents and allow its applications effectively reason about trust. In addition to this, a naming convention can also be helpful in the distribution of keys, and it will enhance NDN's usability.

5.2 Research Challenges

(i) *NACK and its security aspect*:

A router or producer sends NACK to inform its downstream routers to inform about forwarding failure or in-existent content request. Sending NACK is helpful for many reasons. Without NACK, a consumer can not differentiate a packet loss or other reasons. Consumer or routers need to wait until the expiry time. NACK helps to flush PIT states, which release valuable resources. There is no good reason for sending packets to alternative paths after receiving a NACK with reason code "non-existent content." However, these advantages come with a price, and it can open the door for another DoS attack with insecure/fake NACK. When these fake NACKs are cached in routers, it will poison the caches, and subsequent interest will be satisfied with these fake NACKs. The straightforward solution is the use of the signature. It can be simply argued that since NACK is a special type of content and exemption of signing violates the basic tenet of its architecture. However, attackers can take advantage of the signing NACKs. As we know, signing requires huge computational power; producers may exhaust its resources due to signing. It results in yet another DoS. Therefore, from the observation of the pros and cons of NACK, we believe that an in-depth study is required and a strong guideline is necessary on the use of NACK.

(ii) *Negative impact on legitimate consumers*:

Though various countermeasures are already proposed to mitigate existing attacks, we observe a lack of concern on the impact on legitimate consumers. Some mitigation approach looks simple and feasible, but legitimate consumers also need to pay a considerable price. For example, the rate-limiting approach no doubt can reduce the impact of DoS attacks such as Interest flooding attack, but along with that legitimate consumer also suffers from it. Therefore there is a need for an in-depth analysis of the proposed countermeasures on the impact on legitimate consumers.

(iii) *Detection of a collude attack*:

In collude attack, attackers aim to flood routers in collaboration with malicious producers, and due to saturation of PIT, legitimate interests are dropped. If it is a low rate collude attack, the detection becomes very difficult. The reasons are:

- From the network perspective, there are no fake interest packets as they have a valid signature.
- Detection parameters, such as Interest satisfaction rate/Interest expiration rate, may not work.
- Network-layer NACKS itself can cause a NACK-flooding attack.

Although it seems that congestion control schemes can help to reduce the impact of the attack, however, it cannot differentiate between malicious and legitimate interest packets. Therefore an efficient collude attack detection scheme is necessary.

(iv) *A requirement of a unified solution framework*:

We have observed that most of the works in NDN security deal with a particular kind of attack and its possible countermeasure. We suggest NDN research community to develop a unified solution framework which covers major attacks in NDN. The reason behind this suggestion is that to make NDN secure, it has to be resilient to all possible attacks. There is a possibility that a small group of solutions can help to resist most of the existing attacks. We also keep in mind that the fundamental principles of NDN should not be violated, and there should be very less overhead on NDN entities.

6 Conclusion

This chapter sheds some light on the inherent features on NDN, which make it secure from many TCP-IP-based attacks or makes it less effective. We classify the existing attacks on the basis of layers on the NDN protocol stack. We discuss the proposed countermeasures and also highlight their shortcomings. During the review, we have learnt a few lessons, which we have pointed out below:

- NDN's architectural features help in achieving security, along with mobility and efficient content distribution. However, adversaries take advantage of some of those features such as in-network caching and launch various kinds of attacks.
- As compared to TCP-IP-based attacks, some of the attacks in NDN are entirely new; some are similar but have different impacts.
- Most of the attacks discussed so far targets two security requirements: availability and privacy. The reason is that the digital signature included in the data packet already helps to satisfy the security requirements like integrity, origin authentication, and correctness.

References

1. Cisco (2017) Cisco visual networking index: forecast and methodology, 20162021. Technical report
2. Gasti P et al (2013) DoS and DDoS in named data networking. In: 2013 22nd international conference on computer communication and networks (ICCCN). IEEE
3. Afanasyev A et al (2013) Interest flooding attack and countermeasures in named data networking. In: 2013 IFIP networking conference. IEEE
4. Wang K et al (2013) Decoupling malicious interests from pending interest table to mitigate interest flooding attacks. In: 2013 IEEE GLOBECOM workshops (GC Wkshps). IEEE
5. Dai H et al (2013) Mitigate DDoS attacks in NDN by interest traceback. In: 2013 IEEE conference on computer communications workshops (INFOCOM WKSHPS). IEEE
6. Compagno A et al (2013) Poseidon: mitigating interest flooding DDoS attacks in named data networking. In: 38th annual IEEE conference on local computer networks. IEEE
7. Zhang Z et al (2018) Producer-assisted pushback. Technical report NDN-0065, NDN

8. Xie M, IndraW, Haining W (2012) Enhancing cache robustness for content-centric networking. In: 2012 Proceedings IEEE INFOCOM. IEEE
9. Karami A, Guerrero-Zapata M (2015) An ANFIS-based cache replacement method for mitigating cache pollution attacks in named data networking. Comput Netw 80:51–65
10. Kim D et al (2015) Efficient content verification in named data networking. In: Proceedings of the 2nd ACM conference on information-centric networking. ACM
11. Ghali C, Gene T, Ersin U (2014) Needle in a haystack: mitigating content poisoning in named-data networking. In: Proceedings of NDSS workshop on security of emerging networking technologies (SENT)
12. Acs G et al (2013) Cache privacy in named-data networking. In: 2013 IEEE 33rd international conference on distributed computing systems. IEEE
13. Chaabane A, De Cristofaro E, Kaafar MA, Uzun E (2013) Privacy in contentoriented networking: threats and countermeasures. ACM SIGCOMM Comput Commun Rev 43(3):25–33
14. Lauinger T et al (2012) Privacy risks in named data networking: what is the cost of performance? ACM SIGCOMM Comput Commun Rev 42(5):54–57
15. Chaabane A et al (2013) Privacy in content-oriented networking: threats and countermeasures. ACM SIGCOMM Comput Commun Rev 43(3)

Anonymous Traffic Networks

Anand Raje and Sushanta Sinha

Abstract Anonymous traffic networks help anonymizing the identity and privacy of the user and their uses. There has been a great amount of interest in the research community for these implementations. Tor network, based on onion routing, has been a big volunteer-driven network and is being used worldwide. Other than Tor, implementations such as Garlic, I2P, Bitcoin, etc., use the principal concept of anonymous network. There has been plenty of implementations which tried to address the issue of anonymity and privacy has been in practice and evolving. The evolution of the decentralized web, zero-knowledge authentications, in a way, advocate the work toward anonymity and privacy. The usage of the Internet is evolving and so the traffic patterns of a user. This chapter will try to address the Idea of Anonymous Traffic Networks (ATN), Types of implementations of ATN, Challenges of ATN, Tools and Technologies around ATN, Relevance of anonymity and privacy of users for ATN, ATN—Projects and Opportunities

Keywords Anonymous traffic network · Tor · Onion network

1 Introduction

The chapter gives an overview of the Anonymous Traffic Network (ATN) and tries to explain some of the implementations in use. It briefly describes the history and the driver for being anonymous and how the user community is showing their growing interest in the subject. It also covers various personas of users and their interest areas. The learner can also get a fair idea about the fundamental concept of an anonymous network and how the ecosystem works.

The chapter highlights the batching strategies of mix networks, their benefits and vulnerabilities. It also touches upon high-latency and low-latency systems. The chapter describes few scenarios that ensure or otherwise pose as threats to the security

A. Raje · S. Sinha (✉)
India Internet Foundation (IIFON), Kolkata, India
e-mail: sinhasu@yahoo.com

M. Chakraborty et al. (eds.), *The "Essence" of Network Security: An End-to-End Panorama*, Lecture Notes in Networks and Systems 163,
https://doi.org/10.1007/978-981-15-9317-8_11

of information from the perspective of anonymity. The learners get to know about a few browsers that work on these principles. It also mentions tools and platforms for the implementation.

1.1 Idea of Anonymous Traffic Network (ATN)

One of the prime concerns while exchanging information (in any form, be it in digital or physical) is keeping the information secured. It can only be achieved if the right information is sent by the right sender to the right receiver. Now, in order to keep the exchange of information secured, more and more details are captured for the sender and receiver. That raises another concern on Privacy. People want to maintain privacy and often like to remain anonymous, especially when sensitive information is exchanged. With the advent of Internet and its penetration to the daily life of an individual, the issue of anonymity is becoming more and more relevant. Balancing these two, security and privacy, happen to be the most challenging issue for the system designers.

Now, we need to understand why people like to remain anonymous. It is mostly driven by their hesitation to share personal information, knowingly or unknowingly. Let us take an example from the field of statistics and see how a statistician tries to address the challenge of personal information and privacy. Let us take the example of calculating the average age of a group of people, say from a large residential community. One way to collect the information is taking inputs such as Date of Birth (DoB) or simply the actual "age." However, people may have a reservation to share the actual age. So, you may try this way—asking people to choose a number between 0 and 100 (all inclusive) and then add it to the actual "age" (of years) and share. So, you are not revealing the actual age. For a sufficiently large number of people, after calculating the average age, you subtract "50" (since you know the average of randomly chosen supplementary addition is 50). While statisticians argue on their confidence level of accuracy—you protect the Privacy of the Personal Data. Nowadays, programmers use Data Obfuscation (DO) techniques [1], where data are purposely scrambled to prevent unauthorized access to sensitive information.

Apart from personal information, which may be important for authenticating their identity in a secured environment, people may also have their own feelings, sentiments and opinions toward social issues that they do not want to personify. So, it is a normal human tendency to look for anonymity systems, tools and infrastructures available over Internet. Such growing need evolved into the development and popularizing ATN.

The term "Anonymity" has originated from the Greek word meaning "without name" or "name-lees." In such cases the identity of the object or the person cannot be traced back. The whole purpose of being anonymous is that the message or information is more important than the sender, or the receiver, or both [2].

However, there is another term called "Pseudonymity" which has also originated from a Greek word means having a "false-name" or "disguised-identity" where the real identity is veiled under one posing as a different object or person. There are Pseudonym Servers (commonly known as nym servers) that can communicate through e-mails being untraceable to reveal the actual identity of the sender or receiver [2].

There are some situations where people in general want to avoid any kind of censorship from governments or administration by going too far with freedom of speech over the Internet. Some popular Anonymizing networks like "Tor" provide such facilities to remain anonymous. These types of networks hide the actual source and destination address used over the Internet. Tor uses a chosen set of proxy servers that start with an entry node and ends up to the final destination through an exit node. In between, a series of relays passes the traffic through Tor network and makes it difficult, if not impossible, to trace back the originator of the information. At the same time, it works in conjunction with end-to-end encryption of data through HTTPS. More than the privacy of information, the basic need of freedom of sharing views and opinion, being untraceable, is making these networks more and more popular.

1.2 History and the Driving Force Behind

In 1995, David Goldschlag, Michael Reed and Paul Syverson started the research on Onion Routing. They were military scientists at Naval Research Laboratory Anacostia-Bolling military base in Washington, D.C., USA [3]. The primary objective of their work was to separate the trace of identification from routing. It was meant to divert the traffic and bounce it randomly into a peer-to-peer network before sending it to the actual destination. Their work was more driven toward an anonymous routing rather than complete anonymity of any user while browsing the Internet. You may refer to the history of Onion Routing for interesting details [4].

The first formal publication on onion routing was released by mid-1996 [5]. It described the architecture that provides the layered approach of anonymous socket connection by means of proxy servers. Toward the end of 2003, Tor network was deployed and Tor code was released for public consumption under open and free license from MIT. By the middle of 2005, there were over 160 Tor nodes on spread across 5 continents. As we speak in April 2020, there are over 6000 Tor relays inside the Tor network, serving over 2.0 million users (Refer Fig. 1a and b). You may get much such metric information from tor project site [6–8]. Figure 1b shows only directly connecting users and not including those who are connected via bridges.

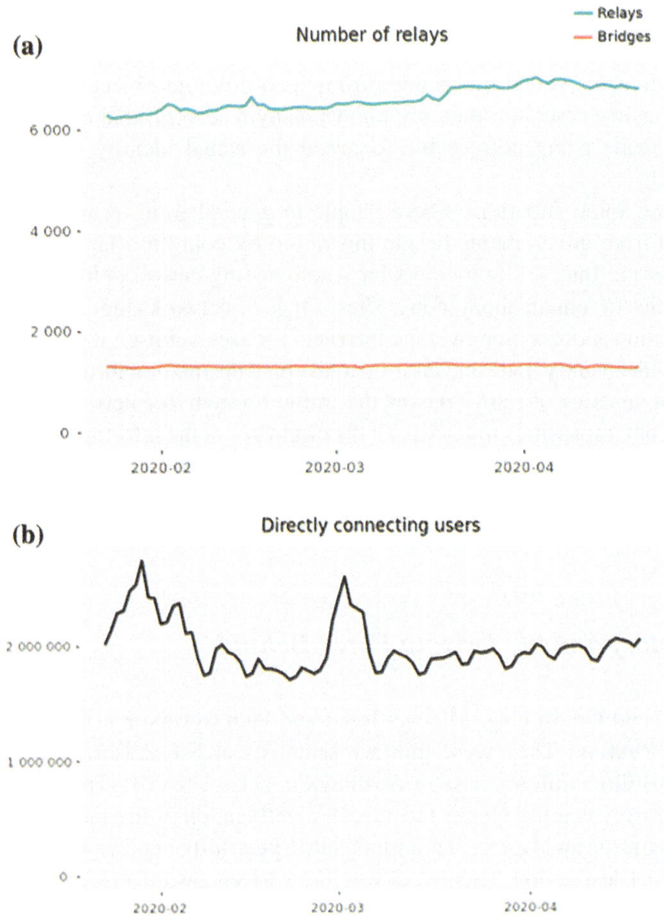

Fig. 1 **a** Number of relays and bridges [6]. **b** Directly connecting users [7]

1.3 Which User Community is Showing Interests Toward ATN?

Today, there are various Tor users having their own point of interest and objective to achieve through anonymous network. One common objective is to surf Internet anonymously, however, for different intent. Keeping the information safe, secured and private is a must-be requirement for every user, and many services assure that. Then why people are drifting toward ATN like Tor? The simple reason is, there is a scare in the minds of the users that they are subject to surveillance by the government and administration or people with vested interest, and.their freedom is curbed due to this. However, not everything is very straight with a clean and good purpose.

Common Citizen with Genuine Concerns

Common citizens are concerned about their identity theft. It is a common practice that user's browsing history and logs are used by marketers for their business promotions. Often ISPs fail to safeguard or protect user's personal information that may lead to harassment or even financial loss. Location tracking is a key component for revealing the identity and prime concern for the common citizen.

On one hand, parents may be concerned that their young kids were falling prey to people with malicious intent while they are studying online or using entertainment sites. On the other, young adults are concerned that their parents are tracking their private conversations. In an extreme situation, on the other hand, in some countries, some popular social networking sites such as YouTube or Facebook are blocked, which is also not desirable since it is violating personal freedom. So, everyone is for an ecosystem where their identities are not revealed to unintended person and Tor-like networks can address the common concern [9].

Not so Common Citizen, but with Genuine Concerns

Those people who hold some positions of public interest or due to their positions or popularity always get media attention, their public comments may influence positively or adversely. So, they, at times, like to share the information being anonymous or restrict their conversation only to the intended audience and do not want someone to track the identity to influence their private life.

Researchers and innovators, while exchanging their thoughts over Internet are worried that the messages and their identities may not be protected from the groups who can steal the idea and spoil the purpose of innovation.

There are journalists, activists and whistle-blowers who may like to bring out untold stories, unlawful activities, environmental disparity and reveal the actual facts. They often try to give voices to voice-less. However, this can directly challenge the ruler or the authority and may pose as a danger for their existence. So, they may like to be anonymous and use such networks for their purpose [9].

Law enforcement authorities, for some genuine reason to protect the interest of the citizens, need to hide their identity. They may like to penetrate into some unlawful sites, not being tracked and get safe and private access to information and its source.

Not so Common Citizen, with not so Genuine Intentions

When something is prohibited or banned, it creates more curiosity and a large scale of such users like to bypass the surveillance by the government and want to visit these banned sites. What else but the Anonymous Traffic Networks can help them doing this?

Some enthusiastic and curious minds, on the other hand, like to explore the wonders of technology and crack the defensive barriers and protective walls of Internet to expose the privacy of un-mindful users. They pose to remain anonymous in order to bypass surveillance. However, sometimes this motivation is not limited to adventurers only but may be driven by business gains, quick money, and ransom or for an even bigger crime.

Figure 2 shows the anonymity and privacy works for various types of users of Internet.

The above network shows a number of users/nodes with different roles: [10]

- Internet User: Common citizen using the network for the genuine purpose
- Attacker/Hacker: Not so common citizen with malicious intent
- Police: Not so common citizen (Law Enforcement) with genuine intent
- ISP: Internet Service Provider, the common platform
- Site: Internet resources, such as the website, application.

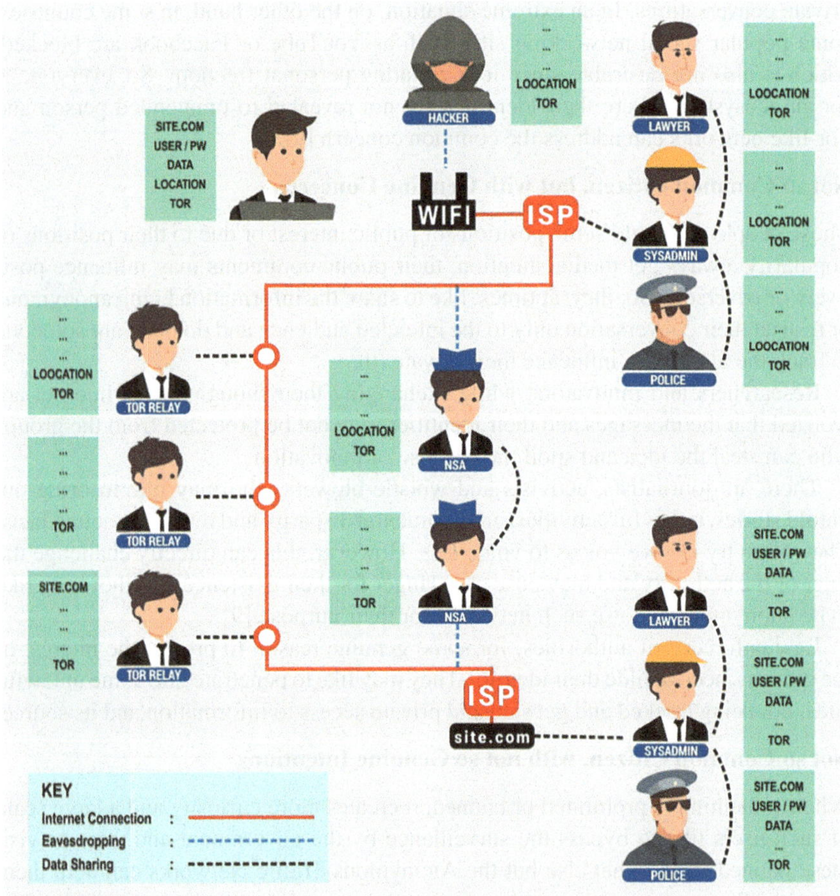

Fig. 2 Combination of anonymity and privacy illustrated using tor and https

1.4 How Does ATN (Like Tor Network) Work?

In the earlier diagram, we see a glimpse of user-spread who are impacted because of anonymity or not having it. It shows how using Tor, the common citizen who uses Internet to access common Internet resources such as websites, and those entities who share data with the website, can be able to see all the information. However, the attacker with malicious intent, the intermediary nodes who share data with the ISP, and the law enforcement officers can only identify the location of the Internet user. It is not possible to identify the user credentials, the Internet resources the user is accessing or the information being exchanged. As a normal precaution against information-stealing Tor encrypts the information before it is transmitted over the network. The entry node is not aware of the final destination and the final exit node of the Tor network is not aware of user location information, hence the identity is protected. Normally when the traffic exits the Tor network, it is not encrypted. However, Tor, along with HTTPS, ensures that the information is encrypted once it leaves the network.

We already know that people use private networks to protect the leakage of information outside the intended audience. Virtual Private Network (VPN) serves a similar purpose [11]. So, why people feel more interested in ATN over VPN? If we observe closely, ISP or any attacker who monitors the user's activity can only see traffic pointing toward the VPN. No destination information is revealed. However, the VPN can see all the traffic movement of the user. So, the privacy is maintained within the purview of integrity and security of the VPN ownerships. It does not truly give you that level of freedom, which ATN can assure. The diagram in Fig. 3a describes the VPN.

If we take a step further, and add another level of VPN [12], we can think of better architecture for anonymity because the source information and the destination information can be separated out into two different compartments, the way it is shown below in Fig. 3b.

In a true sense, the ATNs (such as Tor networks) are built on similar concepts by introducing multiple layers in the traffic path.

Tor network bounces the information across a dynamically formed sequence of Tor nodes. So, from the source node to the destination node, they remain independent and make it difficult, if not impossible, to trace the entire traversed path of the information and trace back the link of the source or the destination. It randomly selects Tor nodes from within to confirm the source and destination data. This is so dynamically done that it is very difficult to trace back. Every now and then, it also changes the pathway for the same user by choosing different Tor nodes. In one word, it simply misdirects the one who likes to run surveillance on the activities.

However, looking at some pattern, one may conclude to a binary question of "Is the client 'A' communicates to destination 'B'?"—with some level of confidence. (Refer Fig. 4) Over a period of time, if someone monitors the traffic pattern of sending and receiving packets from a client to a distinct destination, and if it shows some pattern of synchronization, there is a possibility that the identity can be revealed. The

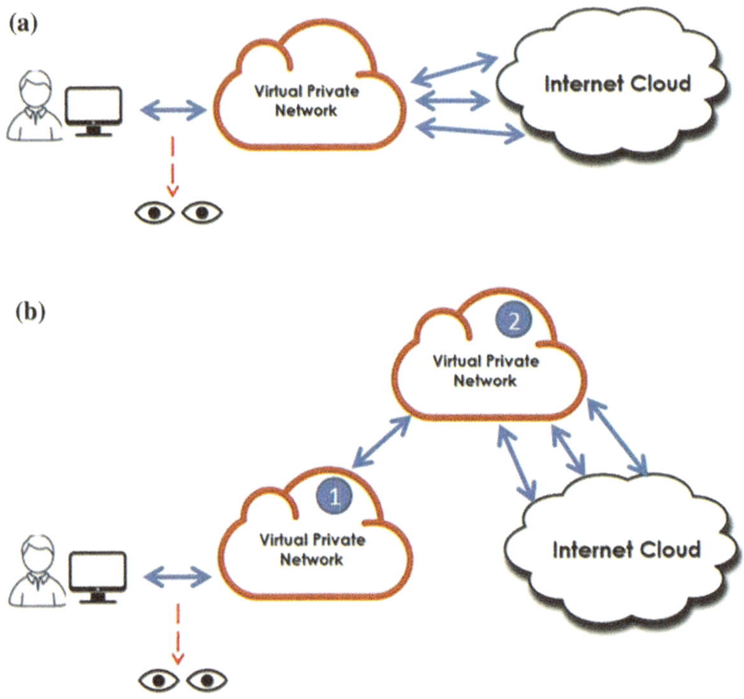

Fig. 3 **a** How VPN can provide a level of privacy to users. **b** Multiple level of VPN can assure better anonymity

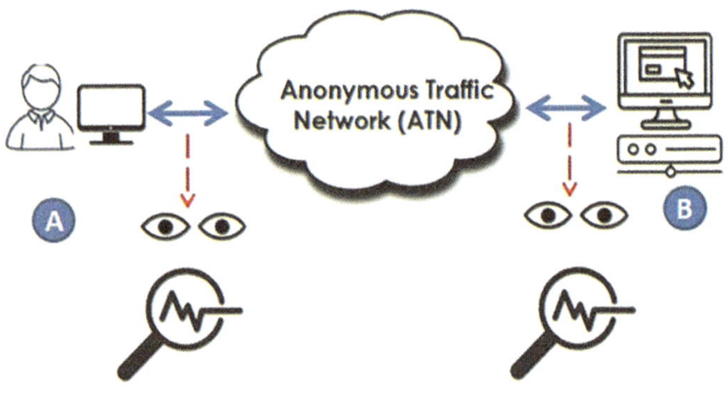

Fig. 4 Analyzing input and output data beyond ATN

following diagram demonstrates the same. However, with a large number of client base and numerous destination resources, it is very difficult to trace.

2 Types of Implementation of ATN

2.1 Anonymity Ecosystem

Anonymity ecosystem normally establishes a platform or infrastructure that assures anonymity over the Internet. It allows messages to navigate from one server to other without revealing the identity of the sender or receiver. Key components of the Anonymous set include nodes capable of being the sender or receiver and the communication channel that transfers the most valuable message [13].

Anonymous Sender: Let us think of a situation in which some citizen wants to send a public complaint to the civil authority or the law enforcement department, but he or she wants to remain anonymous to avoid any further harassment. The fear of being recognized by anyone compels them to remain within Anonymity Sender Set. So, Anonymous Sender refers to that ecosystem that ensures the person originating messages to some intended recipients or any observers remain unrecognized. No passive observer can expose the identity of the citizen.

Anonymous Receiver: If we can think of a situation where a citizen raises his/her voice to multiple group of authorities, or a range of law enforcement agencies in the Anonymity Receiver Set, then any passive observer cannot make out who the real intended recipient of the message. A job-seeker similarly may send job requests to such a series of the potential employer. In all such cases, an ecosystem that provides such a set of Anonymous Receivers ensures that any passive observer cannot identify the real destination from the entire set.

Anonymous Sender-Receiver: Let us think of a situation where both the sender and receiver communicate as anonymous, making the entire activity untraceable end-to-end. When citizens participate in general elections to exercise their franchises, we get to know the outcome in terms of the number of votes polled in favor of a candidate but never know the one-to-one relation from the sender to the receiver. An ecosystem can work in the similar way over Internet while exchanging information.

High-latency and Low-latency Anonymity: Currently large amount of anonymity ecosystems are categorized as either High-latency anonymity or low-latency anonymity [14]. High-latency systems such as Mixmaster and Mixminion works on the principle of introducing significant delay—3 to 4 h, on average. On the other hand, Low-latency anonymity ecosystems such as Crowds [15], Tor [16], AN.ON [17], Invisible Internet Project (I2P) [18], etc., look forward to limit the bandwidth overhead as well as the delay in transmission. They are considered as connection-based systems and more Proxy-based and capable of working on a real-time basis.

Before we get into the detail offerings related to Internet anonymity, it would be helpful to study the basic elements of anonymous communications.

Fig. 5 Mix network concept

2.2 Mix Networks

The objective of every anonymous communication scheme is to allow users to communicate while concealing information about who communicates with whom. The notion of anonymous communication schemes was first introduced by an American computer scientist and cryptographer David Lee Chaum [19], who proposed sending messages through a "Mix server" [19]. In Mix networks, messages are taken from multiple senders and then shuffle them through a series of proxy servers (called mixes) and push them back to their next destination randomly. (Refer Fig. 5).

The destination can be another mix node, as well. In this way, the natural link between the source and destination is broken, making it difficult to trace back. In the process message is encrypted through public key cryptography of series of nodes along the path, only knowing the information of adjacent nodes. Here the point of interest (also the cause of concern) is the way the order of mixes is done. However, the topology is asymmetric and can be either user-defined "free-route" or system defined "mix-cascade."

Each message is encrypted to the respective proxy using the conventional security standard of public key cryptography; complete encryption is built such as a Russian doll of the same size with a clear message at the innermost core. Each server peels off its own layer of encryption to reveal where to send the message next.

Encryption and decryption done in a simple mix network are shown in the Fig. 6. Messages are encrypted through a sequence of public keys. Each mix node peels off a layer of encryption using its own private key. A set of messages are shuffled and then transmitted to the next node [20].

2.3 Batching Strategies

By now, we can understand that Mix Network must be using some batching strategies that, in turn, re-orders, decrypts, and in the process, introduce delays before relaying the messages toward their destinations. Batching strategies determine what and how many messages can be transmitted to their desired destination, what time delay they can maintain and factors like these.

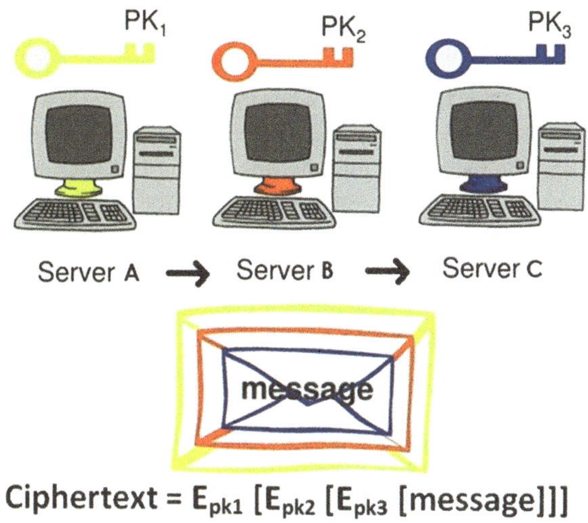

Fig. 6 Simple decryption of Mix network

We can validate the strategies against the effectiveness of the possible attack or act of adversary. A passive adversary can monitor the network passively till the time a target message gets into its scanner, whereas an active adversary is capable to tweak the network to take control and isolate the message via blending attacks.

While in asynchronous batching, messages can enter and leave the network at any point in time, a network that uses synchronous batching uses a fixed batch period, t_{batch}, which is related to the maximum desired latency. Normally it is considered to be 3–4 h. With synchronous batching messages reaching outside the defined time window are considered dropped, so the attacker cannot delay messages without destroying them. Few approaches described below are among the popular basis for determining the strategies.

Simple Mixes: In simple mixes all of the messages in the mix are sent each time it fires. With this simple approach involving low cost, we consider few assumptions like, (a) the mixes have limited physical memory (b) mixes take fixed time to send the message (c) messages may or may not enter at a uniform rate (d) only consider sender anonymity [21].

Threshold Mix: This is the original and fundamental mix, as described by Chaum. The parameter for batching here is "n," the threshold. In this approach, the messages are forwarded in a random order to their next destination when the mix collects all these "n" messages as set by the mix. The minimum delay occurs when the target message arrives at a time when there are already (n − 1) messages in the mix. The maximum delay can be infinite for the target message that arrives at the mix at a time when no other message has arrived. If we assume a constant rate of arrival of messages as "r," we can arrive at a reasonable mean delay of n/2r. Here the minimum anonymity set is "n." However, this approach is vulnerable to flooding attack, when the attacker may occupy n-1 message slots and waiting for the target message to arrive that can be compromised.

Timed Mix: In this approach, the message is forwarded in a random order to their next destination after a stipulated time of "t" second where "t" is defined by the mix. The parameter for batching here is "t." The mix flushes the messages every "t" seconds. The minimum delay is the time when the message arrives just before the entire set is ready to be flushed out. The maximum delay occurs when the message arrives just after the set of message is delivered, and the mean delay can be taken as "t/2" if we consider a constant rate of arrival of messages. This approach can expose identity for low traffic situation where there is a possibility of isolating one message in a stipulated time threshold of "t" seconds. This approach is vulnerable to trickle attack.

If we assume a constant rate of the incoming message, the properties of these two approaches are the same.

Threshold Mix or Timed Mix: In this mix network approach, the message is flushed out to their next destination, if either mix has received "n" packets or "t" seconds have already passed. The parameter for batching here is "n," the threshold and time "t." After sending the message, mix resets the timer. This approach gives the worst combination of threshold and timed mixes. The attacker may choose to perform a trickle attack, or a flooding attack, or a mixture of the two depending on whether it decides to wait or send messages.

Threshold and Timed Mix: Correspondingly, in this combined approach, the mix flushes out all the message if "t" seconds have already passed and mix has received at least "n" packets. Essentially it ensures to satisfy both the parameters threshold and time. The parameter for batching here is "n," the threshold and time "t." Similar to the previous approach, this one is also vulnerable to blending attack. However, it is evident that to carry out an attack on this mix. The attacker must have both the capabilities to insert messages as well as to introduce delay.

Pool Mixes: In all the above cases, the attacks are low cost, but they are certain and exact. In case of Pool Mix, there are uncertain attacks, and it involves high cost [21].

Threshold Pool Mix: Here the parameter for batching is "n" the threshold, and "f" the pool. In this approach, mix flushes out "n" messages to the next destination, after it accumulates "n + f" messages in mix, retaining a pool of "f" in the mix. These "f" messages are chosen randomly but uniformly every time. If we observe minutely, there is a probability, however low, that a message remains in the mix for an infinite time, even though there is a constant flow of messages. The attacker can use this opportunity to single out the target message with some blending attack behavior.

Timed Pool Mix: In this approach, mix flushes out all other messages to the next destination after every "t" seconds, provided there are "f" packets remaining in the pool of mix. They are chosen uniformly but randomly. If there is only "f" or less number of messages left in the mix, it does not fire. Essentially this is a combination of Threshold and Timed Pool mix, where the threshold "n" is zero (0). Here the parameters for batching are "t" time period, "f" the pool and "n" threshold, which is "0."

Timed Dynamic Pool Mix: In this approach, mix instead of sending "n" messages, randomly chooses a fraction of the whole messages (m**fraction*) in the pool, and flushes it out to the next destination, after "t" seconds. But it follows the rule, that mix fires only when there are $(n + f_{min})$ messages remain in the mix.

Pool mixes are vulnerable to blending attack.

2.4 High-Latency Anonymity Systems

One way to offer stronger anonymity is by delivering messages after a significant delay, as in High-latency systems such as Mixmaster and Mixminion. It used to be applied for exchanging e-mails where longer time delays were accepted.

A remailer server redirects mail messages. Anonymous remailer gets a message from a source with some underlining coded instructions to pass on the message to the target or destination server. It sends the message without revealing the identity of the source. So, anonymity is maintained. In contrary, a pseudonymous remailer keeps a pseudo name for the source and connects back to that user with the return message.

There are 3 types of remailer systems commonly used [22].

Type—I: Cypherpunk Remailers

Cypherpunk remailers, often called Type I remailers, receive encrypted messages, and use its private key to decrypt the message, peel the header away, and transmit it to its next remailer. User can request for a remailer public-key, and then import the public key to encrypt through Pretty-Good-Privacy (PGP) and GNU Privacy Guard (GPG). Then the user can compose the message, encrypt it and send that to remailer for the transmission to the next destination address. These remailers are still in use but not having very strong encryption.

Type—II: Mixmaster Remailer

Mixmaster remailers are often called Type II remailers as primarily meant for one-way communication unless the reply address is embedded into the message body. It introduces random padding and its batching strategy follows a dynamic pool mix approach. The message is relayed through remailers using SMTP protocol till it reaches the destination. It is better placed against "replay attack" [23].

Type—III: Mixminion Remailer

Mixminion remailer tried to address the limitations of Mixmaster remailer in the areas of dummy traffic flow, autonomous replies, prevention of replay, rotating the key, exit node policies, forward anonymity, to name a few. Though it makes better use of tiny and small scale synchronized redundant directory servers, it still finds a long way toward integrated Directory Servers. It uses Single Use Reply Block (SURB) to secure reply messages. For forwarding messages, it uses TLS.

2.5 Low-Latency Anonymity Systems

While high-latency anonymity works on batching and delays, low-latency anonymity systems work on a real-time interaction through a connection-based system [24].

The onion routing (Tor) and its way of working are discussed in detail. There are other tools and platforms also available on low-latency anonymity systems such as anonymizer.com, PipeNet, Tarzan, Crowds, Java Anon Proxy (JAP), Morphmix, and so on.

3 Challenges of Anonymous Traffic Networks (ATN)

Though ATN ensures protecting privacy by concealing the identity of the users, it is still been seen facing various challenges. Some of the challenges are, of course, related to technology, its limitation or advancement, but there are huge social stigma also attached to it.

- Drawing the line between privacy and security is always a tug of war. Many countries, due to their concern for internal security, like to ban ATN or want to get a control to trace-back the identity of users. One side may advocate for freedom speech and sharing information, the other side also ready due showcase the legitimate reason to capture the rough gangs spread over Internet with evil intentions.
- Hacking appears to be one of the biggest concerns. Attackers can misuse security holes in any IT product and reveal the identity of the users using it. In many cases, these ATNs are used for illegal activities such as arms selling, drug trafficking, child pornography, which are not allowed in most places. So, normal users suffer when they like to use it for the genuine purpose.
- Sometimes for some specific network, or for some specific activities, the pattern followed by the user is highly predictable and be traceable and the identity of the individual can be revealed. Correlation can be used to preciously identify someone based on the analysis of data points such as timing of Internet usage, location, and other such factors.
- Entry and Exit points of the ATN can be compromised. Attackers can take a chance to get their nodes selected as those vital points. ATN is a constant threat of being exposed to its vulnerable spots. There are numerous types of cyberattacks these networks faced and continuously facing. However, in the process, new enhancement and fixes are emerging continuously.

- Like many other threats, a DNS leak is one among them. Even for an anonymous network for a wrong request to the default DNS server, a trail can be established for real identity through requester IP. A log of activities can capture the information and break the anonymity.
- Sometimes, because of some intentional leaks administered by attackers, can reveal important personal information, which was otherwise kept anonymous.
- Maintaining huge metadata is always a challenge.
- Since the concepts and implementation of Anonymous Traffic Networks are still considered as the brain behind the dark web, it is still not encouraged to bring into the conventional educational curriculum, and stigma is still attached to it.

4 Tools and Technologies

4.1 The Onion Router (Tor)

Tor is a three-letter acronym that got its name from the project—The Onion Router. It is a free and open-source platform that directs Internet traffic through a distributed, anonymous network.

Its initial release was way back on 20th September 2002 and today, it is maintained by The Free Haven Project with all its resources donated by the Electronic Frontier Foundation [25]. Like any open source, anyone can contribute to find out flaws and vulnerabilities as well as its fixes. The primary objective is to provide transparent low latency connections to end-user, with high resistance against network traffic analysis and other attacks to protect individual identity.

Tor network is dominated by some typical kind of Tor nodes. Directory Server is a dedicated server that keeps the updated list of server nodes. Information of such directory authorities are kept in trusted servers, located worldwide. Once the Client node fetches the required information about server nodes, it builds three-hop circuits involving three other server nodes. However, this circuit is not built in one shot, instead in three steps, one hop at a time. It uses Transport Layer Security and the conventional Private Key concept. Every server node only has the information of its earlier node. So, in short, when a user wants to visit some website through Tor networks, the client node does not interact with the website directly. Tor directs the traffic and bounces it back using three Tor nodes randomly, and then transmits the information through this circuit. The diagram below shows how it finds the first entry or guard node in the circuit, then comes to the middle node and finally exit the node. This exit node sends the required information to the destination website's server. An attacker, snooping on the destination website, can only get the information that the request came from an exit node. Similarly, the attacker snooping the Client node can only retrieve the information that it is pointing to the entry/guard relay node. No further way to determine where the information is going after that. (Refer Fig. 7).

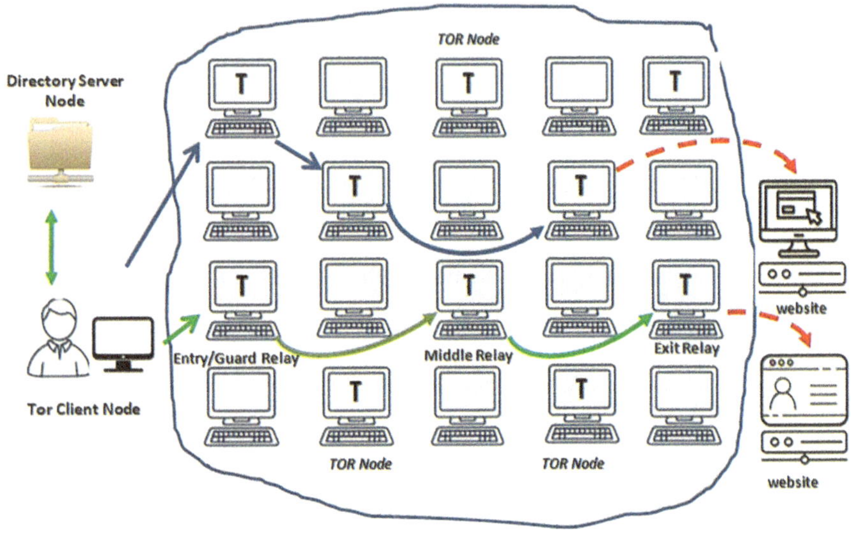

Fig. 7 Tor nodes and the flow of information through its network

Setting up the Tor Circuit

Tor network secures the identity of its users and ensures that no one except the parties exchanging information over the network is capable of extracting the IP addresses of client or the destination node together as a matching pair.

Tor network is tracked and publicized by a handful (around 10) trusted Directory nodes. Each of these Directory Nodes is administered by a unique person or organization. It ensures redundancy as well as less chance of getting compromised. The correctness of the information from Directory nodes ensures the integrity of the Tor network and hence it plays a very critical role in the entire ecosystem. Directory nodes declare the current state with the Relay nodes willing to come online or updating their settings. Over the years, Tor came up with a better approach for resolving the disputes among the Directory nodes and coming to a consensus by shifting the focus to the Directory nodes from the clients to reduce the number of iterations, resource usage as well as risk of being trapped by rouge client node [26].

How Secure is Tor Network?

Tor network ascertains that it secures the identity of its user and no one except the sender and receiver through its network will be able to extract the IP addresses of the client node and the destination node.

Now, there is a question that creates a doubt in the minds of the users. If an attacker poses as the Tor node, and gets a chance to pass on the message as an insider of the network, can it retrieve the information? The answer is "no," since none of the nodes have complete information, as stated earlier.

For a Tor client node, the first entry or guard node has the information about the user's IP address, but it does not know user's destination details. The guard node does not communicate directly with the destination node server. It only bounces the request to the middle node, with zero knowledge on what the middle node does with the request. Only knowledge that the middle node has is IP address of the guard node and the exit node with no knowledge on source or final destination. Exit node knows the final destination website but not aware of who is the original requester.

However, there is a remote possibility that if both the entry node and the exit node happen to be the attacker's malicious node, then the unlucky user is bound to expose all vital information to the external threat and reveals the identity.

What if the entry guard node emerges as a malicious evil node in disguise? Can it steal the vital routing information and IP addresses from the guard node before it is bounced to the middle relay node? Again, the answer is "no" [26].

Let us now try to understand how the message is constructed and retrieved over a Tor network. The user at the client node uses Transport Layer Security (TLS) certificates while handshaking with respective chosen nodes to finally get the symmetric encryption key, which is a contract between the user and the respective node. TLS Security ensures that no other node knows any information about the symmetric keys of others.

The picture (Refer Fig. 8) shows how three layers of encryption works between the nodes. The client node constructs the message, and encrypts it with the help of the symmetric key during TLS negotiation and as the mutual agreement with the exit node. In the next step, it repeats the same—by taking the earlier encrypted message, and re-encrypts it using the symmetric key as the mutual agreement with the middle node. At the last step in the process, it uses the already twice-encrypted message, and encrypts it once more with the help of the symmetric key it mutually agreed with the entry or guard node.

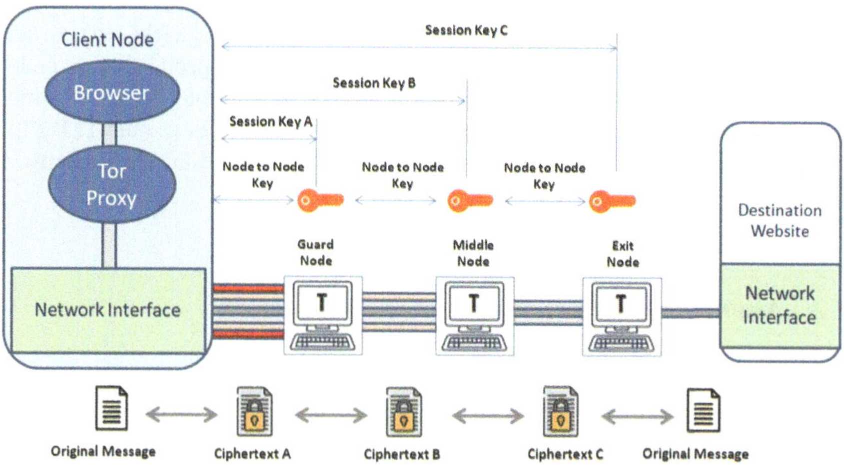

Fig. 8 Encryption-decryption of Tor Network is illustrated

The picture also depicts to and fro movement of the message triply-encrypted and subsequently decrypted securely. Tor client node sends the message to the entry node, commonly known as the guard node. Then the guard node decrypts only the outermost layer of the onion by using its mutually agreed symmetric key. What comes out is a doubly-encrypted message, instructing it to pass on the message to the middle node, whose identity is revealed to it only at that time. So, coming back to the question of an evil guard node, it cannot decrypt the entire message layers, because it does not have the necessary TLS-negotiated keys or the agreed symmetric key of the other nodes.

The process continues further as the message reaches to the middle node. The message is doubly-encrypted at that point. The middle node applies its agreed symmetric key to extract the message from the second layer. What comes out now is a singly-encrypted message with the address of the exit node. The exit node then decrypts the last layer of the encrypted message. Further communication with the destination server is outside the scope of Tor Network. They use conventional security norms to protect the message being read or extracted.

A Bridge Node

Sometime, a repressive state sees Tor as a threat and tries to block it. We know, the complete list of Tor relay nodes is available in public and anyone can download it. Using this list, one can block and censor all traffic sent to these Tor relays. Under this situation, Tor offers unpublished list of proxies that can be used to bypass the censorship [27]. The list is dynamic and can be shared over mail also. It simply re-routes the traffic in such a way that the state also will not know that it is using Tor network. The diagram below shows the traffic diversion through the Bridge Node. (Fig. 9).

Hidden Services

Even for authorized users, web services are exposed to spoofing, DDoS, and other various physical attacks. To avoid and to protect from such attacks, logical and physical location of the service can be kept hidden. Tor has a special concept called "rendezvous points." The user chooses a router as "rendezvous point" to which a tunnel is set up to avail such hidden services. Hidden services use the TLD (Top Level Domain) ".onion," and the hostname has to be looked up using the Tor network [28, 29].

4.2 Anonymous Browsers

Tor Browser

The Tor browser is built on the belief that everyone should be able to explore the Internet with privacy. The browser isolates websites the users visit from any third-party trackers and defends from surveillance. No advertisement can disturb the user.

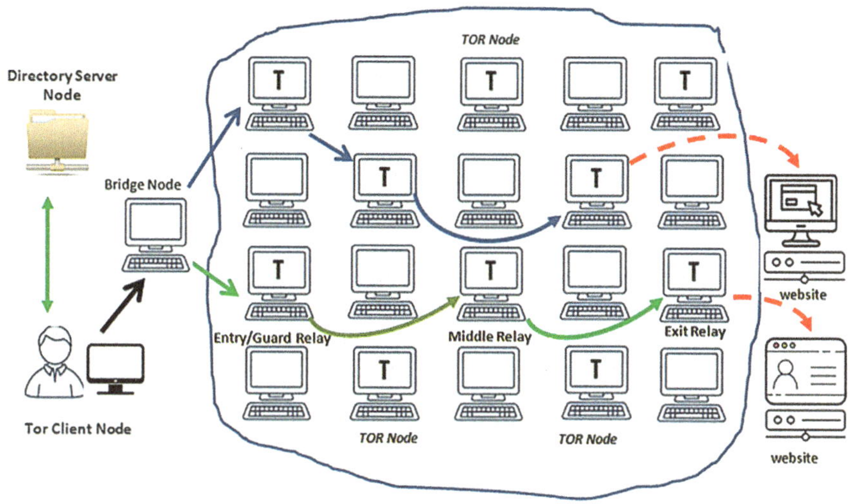

Fig. 9 Bridge node

It aims to make every user looks the same by resisting fingerprinting and secure them through multi-layered encryption.

It can be used on Apple MacOS, Microsoft Windows, or GNU/Linux and it is easy to install and configure. Browsing the web over Tor browsing is comparatively slower than others and some major web services may block Tor users. Tor Browser is also illegal in some states and authoritarian regimes. You may refer to the Tor installation manual to install it in your machine [30].

Epic Browser

"Epic" browser [31] is built on the belief that browsing online should always be private. "Epic" web browser based on Chrome, unlike Tor that is based on FireFox; its appearance is very similar to that of Chrome's incognito window option.

It is one of the world's few private and secure web browsers that block ads, trackers, fingerprinting, crypto-mining. It also offers free VPN (servers in 8 countries).

Its main characteristics are the following:

- It does not store any error reports.
- It does not have any record of installation timestamp.
- It stops tracking the URLs visited by users, hence no suggestion in the address bar.
- There are no automatic updates.

Brave Browser

Brave browser [32] is built on the belief that they have re-imagined what a browser should be. "Brave" offers to give users a safer, faster and better browsing experience.

It is more than a browser. Brave is a new way of thinking about how the web should work. It is open-source, and built by privacy focused, performance oriented pioneers of the web, founded by the inventor of JavaScript and co-founder of Mozilla.

It is available as mobile apps (Play-Store) to be used on the Android platform, for free, fast and easy. It blocks pop-up windows and tabs targeted for ads. It claims to optimize the use of mobile data and saves battery life.

SRWare Iron Browser

SRWare Iron, called as "The Browser of the Future" [33] is an implementation of Chromium by SRWare of Germany. It earned a lot of respect from the world community for its work on identity protection based on Google Chrome. It is a free web browser that offers to eliminate usage tracking and other privacy-compromising functionality that the Google Chrome browser includes. It works with Windows, Mac OS X and Linux. Some of the salient features are as follows:

- It does not track any download or installation (No installation ID).
- It does not offer any updates in the background.
- It includes a customizable user agent as an option.

5 Relevance of Anonymity and Privacy of Users for ATN

The importance of anonymity has long been of interest to social psychologists and other social scientists. It has been suggested that there are really two broad categories of anonymity: technical anonymity and social anonymity [34].

Technical anonymity refers to the removal of all meaningful identifying information about others in the exchange of material.

Social anonymity refers to the perception of others and/or one's self as unidentifiable because of a lack of cues to use to attribute an identity to that individual. Often it appears that one is not truly anonymous in a social context, but the individual perceives him or herself to be anonymous to others [35, 36].

Being anonymous has become a kind of necessity today, wherein hundreds of online services are trying to capture personal information that potentially reveals one's identity. It is paramount to protect privacy in order to better serve its users. There has been a rising concern of overexposure of identity, raising privacy and security concerns.

Anonymous identities, zero-knowledge authentications, virtual IDs are the way forward. In a society where the ATN concept started with anonymous routing, its implementation to integrate service endpoints with user services is having a big role in the mainstream service delivery models over the Internet.

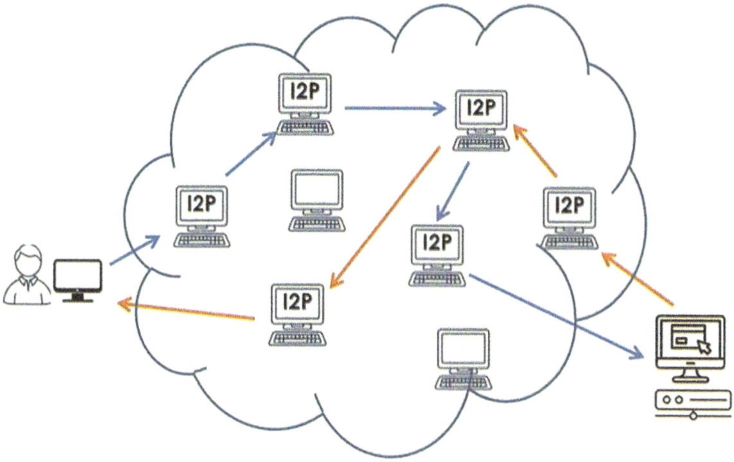

Fig. 10 I2P tunnels for inbound and outbound traffic

6 Anonymous Traffic Networks—Projects

6.1 Invisible Internet Project [I2P]

Like Onion Router Network, Invisible Internet Project (I2P) is also an anonymous network. It has multiple nodes used to transmit the information as data packets. I2P is primarily focused on internal services when it allows its users to access web more for the purpose of applications or features available in it. This includes items such as Torrents, E-mail services, IRCs.

I2P uses layered encryption in unidirectional connections. Each I2P has different incoming and outgoing points and they build inbound and outbound tunnels (Fig. 10).

I2P content is passed using layered encryption. The layers are garlic encryption. I2P can be installed easily by downloading the software from the site [37]. Once the installation over, uses "Start I2P" and after the proxy is set, the user can surf the web using this browser anonymously. I2P Router Console gives a comprehensive interface for applications and configuration [38].

6.2 PipeNet

PipeNet is another anonymous network created by Wei Dai. It routes encrypted messages such as other networks. The principle difference is the synchronization and coordinated flow of the packets. At each cycle of the clock, all the computers connected in the network receive a packet, perform the encryption, and then pass the message.

This approach restricts the access of an omniscient attacker and prevents it from watching the flow of all of the packets in the intention to find out the source and the destination of the packets. Ideally, a large user base would help in providing wide coverage [39, 40].

The PipeNet ensures that each link between nodes carries an equal amount of information at each step. This process has a huge limitation. If one packet is somehow destroyed or one node in the network is locked up, then the whole chain shuts down.

6.3 Freenet

It is one of the earliest networks and is known for P2P file-sharing features. The applications can be installed after downloading from the site [41].

After the installation is done, Freenet homepage page will link to indexes of freenet websites called "freesites." They are similar to the Tor wikis. Freenet allows us to connect with known people using URL https://localhost:8888/friends/—just by exchanging file called "noderefs" Freenet has its own wiki https://wiki.freenetproje ct.org which lists information related to different features and operational details for freesites setup.

Apart from these mentioned networks, there are also some other networks that provide similar functionalities, but Tor, I2P, and freenet are the most popular ones.

7 Conclusion

The chapter made an effort to explain the Anonymity Network and its implementation. Across the world, there is a growing need for using Internet as anonymous to protect privacy. Respecting the demand coming from users, the research community is busy ensuring that the platform is full-proof and good to use in a large scale. Community is also coming forward to collaborate through participation and valued contribution in terms of knowledge sharing.

There is always a stigma attached to the ATN. While the ethical use of ATN is still in question, the social acceptance of the activities carried out using this network is still considered as suspicious. Civil administration or law enforcement department are probably ready for giving legitimate space for freedom of speech, till it is not turning into offences and creating chaos. Balancing both the interests with more trust on ATN can open up a huge opportunity in Internet domain. Usage of Blockchain technology with a pseudo-anonymous public ledger concept can be considered a step toward building an inclusive system with an extension of ATN. Efforts are being made for more awareness to people on the benefits of ATN and bringing it within formal education can be a big step.

If we can analyze the ATN usage for a different segment of communities, it is a wonderful tool to safeguard traffic routing and identifying that can becomes a necessity in future. This is a good area for research as well. Research projects can be considered for innovating better next-generation framework of ATNs.

References

1. What is Data Obfuscation. https://www.talend.com/resources/data-obfuscation/
2. Anonymity. https://en.wikipedia.org/wiki/Anonymity
3. Privacy spooks: almost everyone involved in developing tor was (or is) funded by the US government. https://yashalevine.com/articles/tor-spooks
4. Onion Routing-Brief Selected History. https://www.onion-router.net/History.html
5. Hiding Routing Information. https://www.onion-router.net/Publications/IH-1996.pdf
6. Tor Metrics: Relays and bridges. https://metrics.torproject.org/networksize.html
7. Tor Metrics: Directly connecting users. https://metrics.torproject.org/userstats-relay-country.html
8. Tor Metrics. https://metrics.torproject.org/
9. Anonymity Network Tor and Performance Analysis of 'ARANEA'—an IOT Based Privacy-Preserving Router. https://arxiv.org/ftp/arxiv/papers/1906/1906.01276.pdf
10. Anonymous Routing of Network Traffic Using Tor. https://witestlab.poly.edu/blog/anonymous-routing-of-network-traffic-using-tor/
11. How does VPN work. https://www.namecheap.com/vpn/how-does-vpn-virtual-private-network-work/
12. Multi-hop VPN. https://www.comparitech.com/blog/vpn-privacy/multi-hop-vpn/
13. Tools and Protocols for Anonymity on the Internet. https://www.cse.wustl.edu/~jain/cse571-11/ftp/anonym/index.html
14. How Much Anonymity does Network Latency Leak? https://www.freehaven.net/anonbib/cache/tissec-latency-leak.pdf
15. Crowds: anonymity for Web transactions—Reiter and Rubin 1998 https://dl.acm.org/doi/abs/https://doi.org/10.1145/290163.290168
16. Tor: The Second-Generation Onion Router—Dingledine et al (2004). https://apps.dtic.mil/dtic/tr/fulltext/u2/a465464.pdf
17. Project: AN.ON—Anonymity.Online—Federrath et al. (2006) https://anon.inf.tu-dresden.de/index_en.html
18. Invisible Internet Project: Jrandom et al. (2007). https://staas.home.xs4all.nl/t/swtr/documents/wt2015_i2p.pdf
19. Chaum DL (1981) Untraceable electronic mail, return addresses, and digital pseudonyms. Commun ACM 24(2):84–90. https://doi.org/10.1145/358549.358563
20. Universal Re-encryption: For Mix-Nets and Other Applications. https://www.slideserve.com/rollo/universal-re-encryption-for-mix-nets-and-other-applications-to-appear-ct-rsa-04
21. From a Trickle to a Flood: Active Attacks on Several Mix Types. https://www.freehaven.net/anonbib/cache/trickle02.pdf
22. High-latency Anonymity Systems. https://www.cse.wustl.edu/~jain/cse571-11/ftp/anonym/index.html#High
23. Mixmaster Protocol Version 2. https://tools.ietf.org/html/draft-sassaman-mixmaster-02
24. Challenges in deploying low-latency anonymity. https://pdfs.semanticscholar.org/b019/b5b97d8b85cff0281696918b411a2cadc161.pdf
25. Tor (anonymity network): https://en.wikipedia.org/wiki/Tor_(anonymity_network)
26. How does Tor Work: https://robertheaton.com/2019/04/06/how-does-tor-work/
27. TOR Nodes Explained! https://medium.com/coinmonks/tor-nodes-explained-580808c29e2d
28. How Much Anonymity does Network Latency Leak? https://www-users.cs.umn.edu/~hoppernj/tissec-latency-leak.pdf
29. Attacks-on-Tor: https://github.com/Attacks-on-Tor/Attacks-on-Tor

30. Tor Installation: https://tb-manual.torproject.org/installation/
31. Epic Browser: https://www.epicbrowser.com/
32. Brave Browser: https://brave.com/
33. SRWare Iron Browser: https://www.srware.net/iron/
34. Attribution accuracy when using anonymity in group support systems: https://www.sciencedirect.com/science/article/abs/pii/S1071581997901348
35. Online Aggression: The Influences of Anonymity and Social Modeling. https://digitalcommons.unf.edu/cgi/viewcontent.cgi?article=1472&context=etd
36. Who Is That? The Study of Anonymity and Behavior. https://www.psychologicalscience.org/observer/who-is-that-the-study-of-anonymity-and-behavior
37. Installing I2P: https://geti2p.net/en/download
38. I2P Router Console view. https://upload.wikimedia.org/wikipedia/commons/f/f6/I2P_router_console_0.9.31-0.png
39. PipeNet: https://www.weidai.com/pipenet.txt
40. https://www.sciencedirect.com/topics/computer-science/anonymous-network
41. Freenet https://freenetproject.org/index.html

Appendix
List of Standards

3DES, 3-DES or TDES	Triple Data Encryption Standard
ASPSP	Account Servicing Payment Service Provider
BEUC	The European Consumer Organization
DARPA	Defense Advanced Research Projects Agency
DES	Data Encryption Standard
GDPR	General Data Protection Regulation
NSA	National Bureau of Standards
PCI DSS	Payment Card Industry Data Security Standard

M. Chakraborty et al. (eds.), *The "Essence" of Network Security: An End-to-End Panorama*, Lecture Notes in Networks and Systems 163,
https://doi.org/10.1007/978-981-15-9317-8

Author Index

M. Chakraborty et al. (eds.), *The "Essence" of Network Security: An End-to-End Panorama*, Lecture Notes in Networks and Systems 163,
https://doi.org/10.1007/978-981-15-9317-8

The manufacturer's authorised representative in the EU is Springer Nature Customer Service Centre GmbH, Europaplatz 3, 69115 Heidelberg, Germany. If you have any concerns regarding our products, please contact ProductSafety@springernature.com

Printed and bound by CPI Group (UK) Ltd, Croydon, CR0 4YY
30/09/2026
02225251-0008